ELECTRICAL SYSTEMS FOR A&Ps

International Standard Book Number 0-89100-412-2
For sale by: IAP, Inc., A Hawks Industries Company
Mail To: P.O. Box 10000, Casper, WY 82602-1000
Ship To: 7383 6WN Road, Casper, WY 82604-1835
(800) 443-9250 ❖ (307) 266-3838 ❖ FAX: (307) 472-5106

IAP, Inc.
7383 6WN Road, Casper, WY 82604-1835

© 1992 by IAP, Inc.
All Rights Reserved

Except as permitted under the United States Copyright Act of 1976,
no part of this publication may be reproduced or distributed in any form
or by any means, or stored in a database or retrieval system,
without the prior written permission of the publisher.

Printed in the United States of America

Library of Congress Cataloging-in-Publication Data

Electrical systems for A&Ps.
    p. cm.
  ISBN 0-89100-412-2 (pbk.) : $18.95
    1. Airplanes--Electrical equipment. 2. Electronics.   I. IAP, Inc.
II. Title: Electrical systems for A and Ps.
TL690.E395 1992
629.135'4--dc20   92-54670
                                                                       CIP

Table of Contents

Chapter I	Theory and Principles of Basic Electricity	1

Discovery of Electricity; Electron Theory; Static Electricity; Magnetism; Electromagnetism; Sources of Electrical Energy; Current Electricity and Ohm's Law; Circuit Elements; Circuit Arrangements

Chapter II	Direct Current	27

Direct Current Terms and Values; Series DC Circuits; Parallel DC Circuits; Complex DC Circuits; Changing DC to AC

Chapter III	Alternating Current	37

Advantage of AC Over DC; Generation of Alternating Current Electricity; Alternating Current Terms and Values; Alternating Current in a Circuit Containing Only Resistance; Inductance; Capacitance; Series Alternating Current Circuits; Parallel AC Circuits; Resonance in an AC Circuit; Three-Phase Alternating Current; Converting AC into DC

Chapter IV	Electrical Measuring Instruments	71

The D'Arsonval Meter; Ammeters, Milliammeters and Microammeters; Voltmeters; Ohmmeter; Multimeters; Electrodynamometer; Repulsion-Type Moving Vane Meters; D'Arsonval Meters with Rectifiers; Thermocouple-Type Ammeters; Vibrating Reed Frequency Meters

Chapter V	Aircraft Batteries	85

Primary Cells; Secondary Cells

Chapter VI	Electrical Generators and Motors	97

DC Generators; DC Alternators and Their Controls; Alternators; DC Motors; AC Motors

Chapter VII	Electronic Control Devices	139

Basic Electronic Theory; Atoms, Crystals, and Energy States; Semiconductors; Diodes; Ideal Diodes; Diode Applications; Special Purpose Diodes; Transistors; Bipolar Transistors; Field Effect Transistor

Chapter VIII	Aircraft Electrical Systems	147

Wiring Installation; Wiring Diagrams; Wire Types; Wire Size Selection; Wire Identification; Wire Installation and Routing; Lacing and Tying Wire Bundles; Wire Termination; Electrical System Components; Switches; Relays and Solenoids; Current Limiting Devices; Aircraft Lighting Systems; Exterior Lights; Interior Lights; Maintenance and Inspection of Lighting Systems; Small Single-Engine Aircraft Electrical Systems; Battery Circuit; Generator Circuit; Alternator Circuit; External Power Circuit; Starter Circuit; Avionics Power Circuit; Landing Gear Circuit; Alternating Current Supply; Small Multi-Engine Aircraft; Paralleling with Vibrator-Type Voltage Regulators; Paralleling with Carbon-Pile Voltage Regulators; Paralleling Twin-Engine Alternator Systems; Large Multi-Engine Aircraft; AC Alternator Drive; Alternator Instrumentation and Controls; Automated AC Power Systems

Chapter IX	Types of Electrical Circuits	191

Analog Electronics; Rectifiers; Amplifiers; Oscillators; Digital Electronics; Digital Building Blocks; Microcomputers

Chapter X	Basic Semiconductor Circuits	201

The Rectifier; Half-Wave Rectifier; Full-Wave Rectifier; Bridge-Type, Full-Wave Rectifier; Full-Wave, Three-Phase Rectifier; Filters; Amplifiers; Transistor Amplifiers; Alphabet Classification of Amplifiers; Oscillators; Voltage Control Circuits; Full-Wave Voltage Doubler; Voltage Regulator

Chapter XI	Radio Transmitters and Receivers	213

Aviation Frequency Spectrum; Radio Waves; Composition; Propagation and Reception; Radio Receivers; Amplitude Modulated; Frequency Modulated; Radio Transmitters; Antenna; Principles of Operation; Length; Polarization and Field Pattern; Types; Transmission Lines

Chapter XII	Communication Systems	225

Short Range Communications; VHF Transmitters; VHF Receivers; Long Range Communications; SELCAL; Audio Integrating Systems; Emergency Systems

Chapter XIII	Navigation Systems	229

Short Range (VHF) Navigation; Omnirange Navigation; DME/TACAN; Area Navigation; Long Range Navigation Systems; Automatic Direction Finder; LORAN C; OMEGA/VLF Navigation; Inertial Navigation Systems; Global Positioning Satellite System; Multisensor Navigation Systems

Chapter XIV	Terminal Navigation and Collision Avoidance Systems	243

Instrument Landing System; Localizer; Glide Slope; Marker Beacons; Microwave Landing System; Radar Transponders; 4096 Code Transponders; Mode S Transponders; Traffic Alert and Collision Avoidance System; Datalink Communications

Chapter XV	Weather Warning Systems	251

Weather Radar; System Operation; Color Radar; Doppler Radar; Radar Components and Operation; Radar Frequency Ranges; Storm Scope Systems

Chapter XVI	Electronic Instrumentation	255

Computerized Fuel Systems; Radio Altimeter; Electronic Flight Instrument Systems; Electronic Attitude Director Indicator; Electronic Horizontal Situation Indicator; Electronic Systems Monitoring Displays

Chapter XVII	Electronic Flight Systems	263

Autopilot Basics; Autopilot Functions; Flight Management Systems; Autopilot System Maintenance

Chapter I

Theory and Principles of Basic Electricity

A. Discovery of Electricity

One of the first recorded mentions of electricity was by the Greek philosopher Thales in about 500 BC. He mentioned the fact that when substances such as amber and jet were rubbed, they would attract such light objects as feathers and bits of straw. Later in the eighteenth century, it was discovered that there were two kinds of forces, or charges, caused by rubbing certain kinds of materials. Charges of the same kind would repel each other while opposite charges would attract.

In about the middle of the eighteenth century, the practical mind of Benjamin Franklin found a way to prove that lightning was a form of electricity. In his famous kite experiment, he flew a kite into a thunderstorm and found that sparks would jump to the ground from a metal key attached to the wet string. Franklin made a logical assumption that whatever it was that came down the string was flowing from a high level of energy to a lower level. And so he assigned the term "positive" to the high energy, and "negative" to the lower level. It was not known what actually came down the string, but he used a term associated with the flow of water, and said that it was "current" that flowed down the string, from the positive charge to the negative.

This assumption of Franklin's was accepted until the discovery of the electron in 1897, and many textbooks in use today still speak of current as being from positive to negative. It was not until it was discovered that it is actually electrons, or negatively charged particles of electricity that move through a circuit, could we understand the true nature of electricity.

B. Electron Theory

1. Composition of Matter

When a component such as a light bulb is connected to a source of electrical energy by solid conductors, or wires, there *appears* to be no movement within the conductor, even though it is obvious that something is happening.

If we were able to see what is actually inside the wires that carry the electricity, we would see that they are not *really* solid at all, but actually contain far more empty space than matter. Matter can be defined as anything that has mass and occupies space. Matter may exist in any of four states: solid, liquid, gaseous, or plasma. The smallest particle of matter in any state that still possesses the characteristics of the matter is the atom.

a. The Atom

All of the material from which our universe is made is composed of atoms, the smallest particles of an element which can exist, either alone or in combination with other atoms. Each atom consists of a nucleus,

Figure 1-1. Benjamin Franklin's practical mind found a way to prove that lightning was a form of electricity with his famous kite experiment.

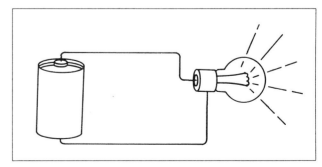

Figure 1-2. There appears to be no movement with the conductor, but it is obvious that something is happening.

made up, usually, of positively charged protons, and neutrons having no electrical charge but of the same weight as the protons. Surrounding this nucleus and traveling at incredible speed are electrons, each of which weighs only about $1/1845$ as much as a proton and which has a negative charge.

The most simple atom is that of hydrogen, which consists of only one proton and no neutrons in the nucleus. Spinning around the nucleus is a single electron.

All electrons are alike, as are all protons and all neutrons, but the number and arrangement of these elementary building blocks in the atom determine the material the atoms make up. For example, copper has a nucleus made up of 29 protons and 36 neutrons. surrounding the nucleus and spinning in four rings, or "shells", are 29 electrons. This combination gives us what is called an electrically balanced atom, as there are exactly the same number of positive charges (protons) as there are negative charges (electrons). The neutrons have no electrical charge, and so they do not enter into the flow of electricity.

We are familiar with the diagrams of atoms in our textbooks and training manuals, but these cannot possibly be drawn to scale. For example, to have the same *mass* as a proton, the electrons would have to be about 1,845 times as large, and the distance between the electrons and the nucleus would have to be about 100,000 times the diameter of the nucleus.

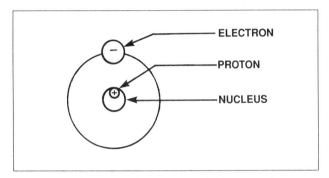

Figure 1-3. The simplest atom, hydrogen, has one electron spinning around the nucleus consisting of one proton. There are no neutrons in the nucleus.

All matter contains energy, and energy in the atom causes the electrons to spin around the nucleus. As they spin, centrifugal force tends to pull them away from the nucleus. But there is an electrostatic field within the atom that produces a force which exactly balances this centrifugal force and holds the electrons a specific distance away from the nucleus.

The electrons spin around the nucleus in rings or shells, and when energy is added to an atom, such as is done when the material is heated, the orbiting radius (the distance between the electrons and the proton) is increased, and the bond (force of attraction) between the proton and the electron is decreased.

b. Ions

A positive electrical force outside the atom can attract, and rob electrons from the outer ring and leave the atom in an unbalanced condition. This unbalanced condition leaves the atom with an electrical charge. Charged atoms are called ions. For example, copper has one electron in its outer ring, and if a positive force is applied to the atom, this outer ring electron, which is negative, will be drawn from the atom leaving it with more protons than electrons. It then becomes a positive ion and will attract an electron from a nearby balanced atom. If an atom possesses an excess of electrons, it is said to be negatively charged and is called a negative ion. Electrons constantly move about within a material from one atom to another, in a continuous but random fashion.

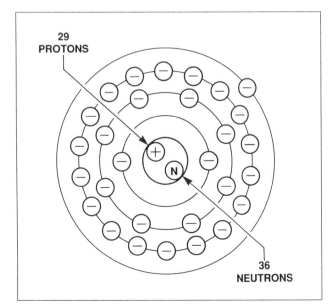

Figure 1-4. The nucleus of a copper atom consists of positively charge protons and electrically neutral neutrons. Spinning around the nucleus are negatively charged electrons.

c. Conductors and Insulators

Some materials which exist in nature have an atomic structure which enables then to easily permit the movement of electrons when conditions permit. Other materials have a structure that severely impede, or oppose, the movement of electrons.

Materials which permit the flow of electrons are referred to as conductors. The best four conductors are silver, copper, gold, and aluminum. Conductors are discussed in significant detail in a later section of this manual.

Materials which oppose the movement of electrons are called insulators. Insulators are often placed around conductors to prevent the inadvertent flow of electricity. Some common insulating materials are plastic, rubber, glass, ceramics, air (vacuum), and oil.

2. Electron Flow

Let's see what happens when a conductor made of copper is connected across a source of electrons. An electron is attracted from an atom by the positive terminal of the source, and it leaves the conductor. The atom which lost the electron has not become a positive ion and pulls an electron away from the next atom. This exchange continues until the electron that left the conductor is replaced by one from the negative terminal of the source.

It takes us a while to visualize what is happening inside the conductor, but electron movement takes place within the conductor at about the speed of light; that is, about 186,000 miles per second. This is not saying that a single electron moves through the conductor from one end to the other at this speed, but because of the domino effect, an electron entering one end of the conductor will almost immediately force an electron out of the other end.

a. Effects of Electron Flow

We cannot see the movement of electrons within the conductor, but as they move, they produce effects we can use. They cause a magnetic field to surround the conductor, and the greater the amount of flow, the stronger will be the field. Also, as they are forced to flow, the opposition to their flow produces heat within the conductor.

b. Direction of Flow

The effects of electricity were observed long before electrons were known to exist, and in explaining what was seen, an incorrect assumption was made. Electricity appeared to follow the rules of hydraulics in that pressure, flow, and opposition were present, and a definite relationship existed between the three. Since the flow of electricity could not actually be observed, it was only natural to assume that it flowed from a high level of energy to a lower level or, in electrical terms, from "positive" to "negative".

This theory worked well for years, and many texts have been written calling the flow of electrons "current flow", and assuming that whatever it was that flowed in the circuits moved from the positive terminal of the source to the negative terminal.

As we have gained knowledge of the atom, it has become apparent that it was the electron with its *negative* charge that actually moved through the circuit, and the texts have had to be revised to explain electron flow as being from the *negative* terminal of the source through the load, back into the source of the *positive* terminal.

There are two ways we can consider flow: electron flow, which is from negative to positive, and the flow of "conventional current", which, while technically incorrect, follows the arrows used on semiconductor symbols. You may use either method

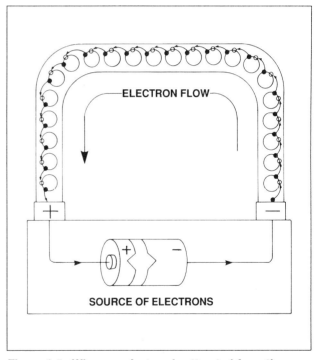

Figure 1-5. When an electron is attracted from the conductor by the positive charge of the source, it leaves a positive ion. This ion attracts an electron from an adjoining atom. This exchange continues through the conductor until an electron is furnished by the negative

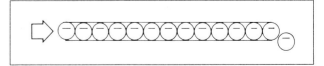

Figure 1-6. When one electron is forced into the conductor, it immediately forces an electron out of the opposite end of the conductor.

for tracing flow, but you must be consistent. *In this chapter, we will follow electron flow, from negative to positive, and we will use the terms electron flow and current interchangeably.* In chapters where we are dealing with semiconductor devices and using their symbols, we will follow the flow of "conventional current" from positive to negative, because the arrows used in semiconductor symbols point in the direction of conventional current flow.

3. Units of Electrical Measurement

a. Quantity

The electron is such an extremely small particle of electricity that an enormous number of them are required to have a measurable unit. The coulomb is the basic unit of electrical quantity and is equal to 6.28 billion, billion electrons. This is most generally written as 6.28×10^{18}, which means that the number of electrons is 6.28 followed by 16 zeros. The symbol for quantity is "Q".

b. Flow

When one coulomb of electrons flows past a point in one second, there is a flow of one ampere or one amp. It makes no difference whether we think of electron flow or conventional current, it is all generally called current and its symbol is "I".

c. Opposition

The ohm is the standard unit of resistance, or opposition to current flow, and is the resistance at a specific temperature of a column of mercury having a specified length and weight. More practically, it is the resistance through which a pressure of one volt can force a flow of one ampere. The symbol for resistance is "R".

d. Pressure

The volt is the unit of electrical pressure and is the amount of pressure required to force one amp of flow through one ohm of resistance. There are a number of terms used to express electrical pressure. They are voltage, voltage drop, potential, potential difference, EMF, and IR drop. These terms have slightly different shades of meaning, but are often used interchangeably. The symbol for electrical pressure is "E".

e. Power

The end result for practical electricity is power, and electrical power is expressed in watts. One watt is the amount of power dissipated when one amp of current flows under a pressure of one volt. The symbol for power is "P".

4. Metric Prefixes and Powers of Ten

a. Prefixes

So many terms used in the study of electricity deal with numbers that are either extremely large or extremely small. Because of this, metric prefixes have been adapted to them

These prefixes also help us in dealing with very small numbers. For example, the basic unit of capacitance, the farad, is much too large for practical use in aircraft electronics, and one of the commonly used capacitors has a capacity of 0.000,000,000,002 farad. A number such as this is awkward to work with, and its use encourages errors. A much more convenient way to express this same unit is to used the term two picofarads, or 2 pf. The term, by the way, has previously been called a micromicro farad, and you may still see it referred to in this way. It may be written 2 mmF or 2 μμF with the Greek letter μ (mu) used.

Another example is the emergency frequency used for aircraft communication which is 121,500,000

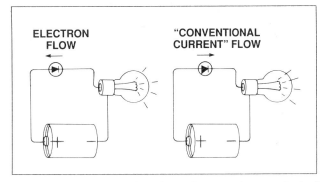

Figure 1-7. Electrons flow from the negative terminal of the source back to the positive terminal. Conventional current is said to travel from positive to negative. Electron flow is opposite the arrows used in semiconductor symbols. It is easier to think in terms of conventional current, which, even though it is technically incorrect, is easier to follow when semiconductor symbols are used.

VOLTAGE

VOLTAGE DROP

ELECTROMOTIVE FORCE (EMF)

POTENTIAL

POTENTIAL DIFFERENCE

IR DROP

Figure 1-8. All of these terms may be used to express electrical pressure.

hertz, or cycles per second. This number is large and unwieldy, so we can divide it by one million and use the metric prefix "mega". The number becomes 121.5 megahertz.

b. Powers of Ten

Multiplying and dividing very large and very small numbers is made easier by the use of powers of ten. In this method of handling numbers, convert every number into a number between one and ten by moving the decimal the proper number of places in the correction direction. For example, 0.000,000,002 can be converted into 2.0 by moving the decimal to the right nine places. Since the number is smaller than one, the number two will have to multiplied by a *negative* power of ten. When 0.000,000,002 is converted into a power of ten number it becomes 2×10^{-9}.

Numbers larger than one are converted in exactly the same way, except they are multiplied by a positive power of ten. One coulomb contains 6,280,000,000,000,000,000 electrons. This number is easier to work with when it is converted to 6.28×10^{18} by moving the decimal to the left 18 places and multiplying 6.28 by ten 18 times.

Numbers that have been converted into powers of ten may be multiplied or divided by performing the required work on the numbers, and then adding the powers of ten (the exponents) to multiply, or subtracting the powers to divide.

$0.0025 \times 5,000 = 2.5 \times 10^{-3} \times 5 \times 10^{3} = 12.5$

$0.125 \times 0.5 = 1.25 \times 10^{-1} \times 5 \times 10^{-1}$
$= 6.25 \times 10^{-2} = 0.0625$

$5,000,000 \div 250,000$
$= 5 \times 10^{6} \div 2.5 = 2 \times 10^{1}$
$= 20$

$0.125 \div 0.5 = 1.25 \times 10^{-1} \div 5 \times 10^{-1}$
$= 0.25 \times 10^{0} = 0.25$

C. Static Electricity

Electricity may be classified in two types: current and static. In current electricity, the electrons move through a circuit and perform work, either by the magnetic field created by their movement, or by the heat generated when they are forced through a resistance. Static electricity, on the other hand, normally serves little useful purpose, and is more often a nuisance rather than a useful form of electrical energy.

Static electricity is of real concern to us during the fueling operation of an aircraft. As the aircraft flies, friction between the air and the surface builds up a large static charge which cannot readily bleed off upon landing, because the rubber tires insulate the aircraft from the ground. Fuel trucks and fueling pits are grounded to the earth, so if the first contact with the aircraft is with the fuel nozzle in the open filler neck, a spark can jump in the explosive fumes and cause a serious fire. To prevent this, always connect the aircraft to the fuel truck and to the ground by the grounding cables which are provided on all fuel trucks for this purpose.

MULTIPLIER	PREFIX	SYMBOL
1,000,000,000,000	tera	t
1,000,000,000	giga	g
1,000,000	mega	M
1,000	kilo	k
100	hecto	h
10	deka	dk
0.1	deci	d
0.01	centi	c
0.001	milli	m
0.000,001	micro	μ (mu)
0.000,000,001	nano	n

Figure 1-9. Metric prefixes.

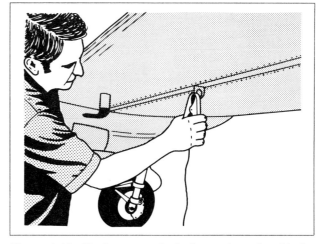

Figure 1-10. Airplanes and fuel trucks should be grounded together to neutralize the charge of static electricity before the fueling nozzle is put into the tank.

1. Positive and Negative Charges

We cannot see electricity, but the effects of both types are easy to observe. If we have a couple of balls of pith wood, we can suspend them by threads and observe the effects of static electricity as we charge them.

Rub a glass rod with a piece of wool or fur and the rod will pick up a lot of extra electrons and will become negatively charged. When we hold the rod close to a ball which has a neutral charge, it will be attracted to the rod. But as soon as the ball touches the rod, electrons will flow to the ball and give it a negative charge. Now, it will be repelled by the rod.

If we rub the glass rod with a piece of silk, the glass will give up electrons to the silk, and the rod will have a deficiency of electrons, or will become

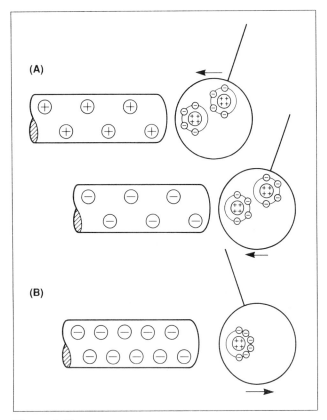

Figure 1-11.
- *(A) An uncharged pith ball will be attracted to a rod that has either a positive or a negative charge.*
- *(B) Once the charged rod has contacted the ball, the ball will assume the same charge as the rod and it will be repelled by the rod.*

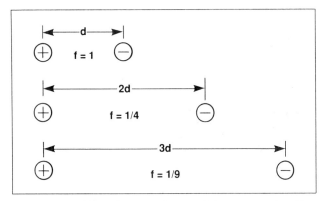

Figure 1-12. The force of repulsion or attraction decreases as the square of the distance between the charges.

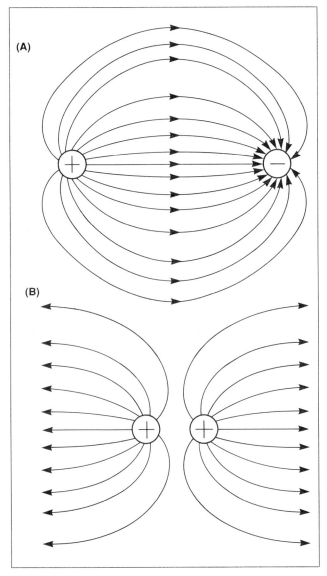

Figure 1-13.
- *(A) Lines of force leave the charged body at right angles to its surface. Since they are polarized alike, the lines spread apart. They enter the oppositely charged body at right angles to its surface.*
- *(B) Charged bodies reject lines of electrostatic force from other bodies having the same charge.*

positively charged. And when we hold it near one of the balls, it, too, will attract the ball, and once it has touched it, the ball will be repelled. This is the same thing that happens with the first ball, but there is a difference. One of the balls is charged *negatively* and the other is charged *positively*. When they are brought close together, they will be attracted to each other.

This demonstration proves that there are two types of charges and that like charges repel each other, while unlike charges attract. The strength of the repelling and attracting force varies as the square of the distance between the two charges. For example, if the distance between the two balls is doubled, the force of attraction will be reduced to one-fourth ($2^2 = 4$). If they are moved three times as far apart, the force will be only one-ninth of the original. On the other hand, if they are moved closer together, the force of attraction or of *repulsion* will increase as the square of the separation. If the distance is decreased to one-half, the force will be four times as great.

2. Electrostatic Fields

If we were able to see the lines of electrostatic force that exist between charges, we would see that the lines leave one charged object and, in our case, we will arbitrarily assume that they leave the one having the positive charge and enter the other, the one having the negative charge. If the charges are close together, all of the lines will link, and the two charges will form a neutral, or an uncharged, group. The lines of force from like charges repel each other and will tend to push the charges apart. Electrostatic fields are also known as dielectric fields.

3. Distribution of Electrical Charges

When a body having a smooth or uniform surface is electrically charged, the charge will distribute evenly over the entire surface. If the surface is irregular in shape, the charge will concentrate at the points or areas having the sharpest curvature.

This explains the action of static dischargers used on many aircraft control surfaces. As the airplane flies through the air, friction causes a large static charge to build up on the control surface. These surfaces are connected to the airframe structure by hinges which do not provide a particularly good conductive path, so the charge builds up in strength. Static discharges are attached to the control surface to dissipate this charge. They have sharp points on which the static charges concentrate and will discharge into the air before they can build up on the smooth surface sufficiently high to jump across the hinges and cause radio interference.

As a further aid in preventing radio interference, the control surface is bonded to the structure. This means that a flexible metal braid is attached to both the control surface and the structure to act as a good conductor, so the charge can be dissipated as it forms.

D. Magnetism

1. Magnetic Characteristics

One of the most useful devices in the production and use of electricity is the magnet. First discovered in the province of Magnesia in Asia Minor, a form of iron oxide demonstrated a strange property: when it was suspended in the air or floated on a chip, it would always align itself in a north-south direc-

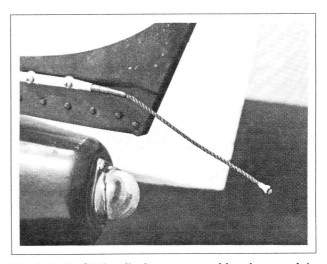

Figure 1-14. Static dischargers provide sharp points from which static charges are dissipated into the air before they can build up to a high potential on the control surface.

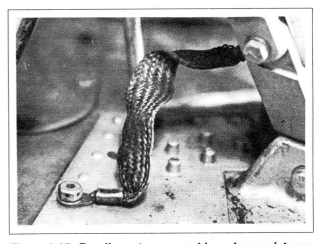

Figure 1-15. Bonding straps provide a low-resistance path between the control surface and the aircraft structure to prevent the build-up of a static charge on the control surface.

tion. This strange stone was used by early seafarers as a leading stone or "lodestone" to aid in navigation.

The strange characteristics of these natural magnets have been studied, and now we are able to produce artificial magnets. By definition, a magnet is a body that has the property of attracting iron and producing magnetic fields external to itself. And it is these magnetic fields that are of interest to us.

The north-seeking end of a magnet is labeled "N", and the opposite south-seeking end is labeled "S". It is important to remember that the labels refer to the direction sought by the pole of the magnet. If the "N" end of the magnet is referred to as the north-seeking rather than north, there is less chance for confusion.

Lines of magnetic force, or flux, are always complete loops that leave the magnet at right angles to its surface at the north pole, and since they are all polarized in the same direction, they repel each other and spread out. They draw closer together as they re-enter the magnet at the south pole, at right angle to its surface, and travel through the magnet to complete the loop.

There is no insulation against the lines of flux, as they will pass through any material; but if it is important that a device be protected from magnetic fields, it can be entirely surrounded by a soft iron shield. Iron, as we will see, has a very high permeability and provides a much easier path for the flux than the air, and since the flux lines travel the path of least resistance, they will flow through the iron and leave an area inside the shield that has no magnetic field.

The ends of the magnet where the lines of force leave and return are the poles, and are called the

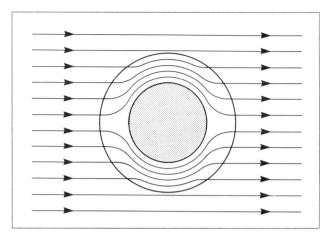

Figure 1-17. *The only way to shield an object from lines of magnetic force is to enclose it in a shield made of a highly permeable material. The lines of force will flow through the shield and bypass its center.*

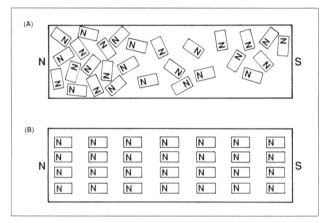

Figure 1-18.
(A) *In an unmagnetized material, all of the individual magnetic fields, or domains, are arranged in a random fashion and cancel each other.*
(B) *When the material is magnetized, all of the domains are aligned, and the material has a north and south pole, just like that of the individual domains.*

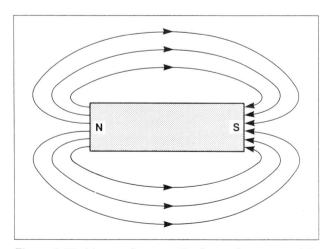

Figure 1-16. *Lines of magnetic force form complete loops, leaving the magnet at its north pole and returning at its south pole.*

north- and south-seeking poles or, more commonly known today as the north and south poles.

If we were able to see inside a piece of unmagnetized iron, we would see that it contains an almost infinite number of magnetic fields oriented in such a random fashion that they cancel each other. Now, if this piece of iron were placed in a strong magnetic field, all of these little fields, or domains as they are called, would align themselves with the strong field, and the iron would become a magnet, having a north and south pole, and lines of magnetic flux would encircle it.

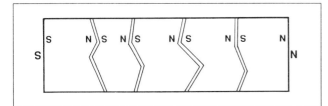

Figure 1-19. Regardless of how small the pieces of a magnet may be broken, each piece will have a north and a south pole.

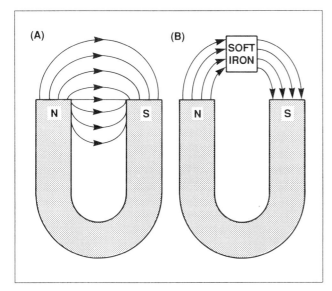

Figure 1-20.
(A) Lines of magnetic flux are assumed to leave the north pole of a magnet at right angles to its surface and travel to the south pole where they enter at right angles to its surface.
(B) The flux lines always seek the path of least resistance even traveling longer distances if they can travel through a material with a high permeability.

The domain theory of magnetism is supported by the fact that each magnet has both a north and south pole, regardless of the size of the magnet. If we break a bar magnet in two, each half will demonstrate the characteristics of the original magnet, and if we break each of these halves in two, all of the small pieces will still be magnets having north and south poles, with lines of magnetic flux surrounding them.

Soft iron has a very low retentivity, meaning that as soon as the magnetizing force is removed, the domains will lose their alignment, the fields will cancel each other, and the iron will no longer act as a magnet. Hard steel and some of the alloys of iron using aluminum, nickel, cobalt, and molybdenum have very high retentivities, and will retain the alignment of their domains long after the magnetizing force has been removed. Materials of this type are used for permanent magnets in aircraft magnetos, instruments, and radio speakers.

The number of lines of flux that loop through the magnet gives us an indication of its strength. One line of flux is called one maxwell. The flux density is the number of lines of flux for a unit area, and it is measured in gausses, with one gauss representing a density of one maxwell per square centimeter.

Lines of flux always follow the path of least resistance as they travel from the north pole to the south. They will even travel a longer distance if the traveling is easier. The measure of the ease with which the lines of flux can travel through a material or medium is measured in terms of permeability. Air is used as a reference and is given the permeability of one. Flux can travel through iron much more easily than through air, since it has a permeability of 7,000. Some of the extremely efficient permanent magnet alloys have permeability values as high as 1,000,000.

Lines of flux are invisible, but if we place a magnet under a piece of paper and sprinkle iron filings over it, they will form a definite pattern showing the lines they follow. They pass directly between the poles of a horseshoe magnet, but, if we place a piece of soft iron above the poles, the lines will enter the iron and flow through it to the south pole. This is because iron has so much greater permeability than air.

It is the characteristic of the lines of flux to pass through a permeable material that explains the attraction of a piece of iron to a magnet. Remember that the lines of flux always seek the path of least resistance between the poles, and since air has a very low permeability compared to iron, if a piece of iron gets within the field of a magnet, the lines of flux will travel through it rather than through the air around it. The lines of flux want to link the poles with the *shortest* possible loops, so they will exert a strong pull to get the piece of iron centered between the poles as the iron is pulled in closer, more lines of flux can pass through it and the pull will be stronger. When it is centered, it will resist any force that tries to lengthen the lines of flux by pulling the iron out of the field.

Almost any magnetic material regardless of its retentivity, will lose some of its magnetic strength if its lines of flux must pass through the air. Because of this, magnets whose strength is critical are stored with *keepers* of soft iron linking the poles to provide a highly permeable path for the flux.

Magnetism and lines of magnetic flux follow the same rules as charges of static electricity. Like poles repel each other, and the force of repulsion follows the inverse square law. This means that if the distance between the poles is doubled, the force of repulsion will be reduced to one-fourth. Unlike poles attract each other, and the force of attraction is squared as the distance is decreased. Halving the separation increases the force of attraction four times.

E. Electromagnetism

Though the effects of magnetism had been observed for centuries, it was not until 1819 that the relationship between electricity and magnetism was discovered. The Danish physicist Hans Christian Oersted discovered that the needle of a small compass would be deflected if it was held near a wire carrying electric current. This deflection was caused by an invisible magnetic field surrounding the wire.

We can see the effect of this field if we sprinkle iron filings on a plate that surrounds a current-carrying conductor. The filings will arrange themselves in a series of concentric circles around the conductor.

By observing the effect of the field and remembering what we have learned about electron flow, we can understand a great deal about electromagnets, and about the way they serve us in such useful ways.

Electrons are negative charges of electricity that can be forced to flow through a conductor. As they travel, they produce the lines we have just observed.

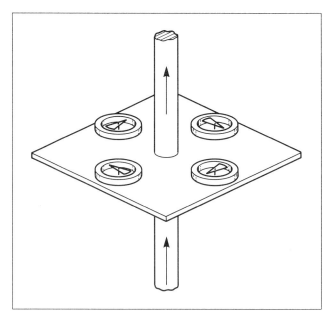

Figure 1-22. The relationship between magnetism and electricity was discovered when it was found that the needle of a magnetic compass was deflected when it was placed near a current-carrying conductor.

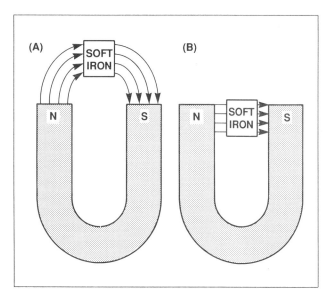

Figure 1-21.
(A) The lines of flux pass through a material having a high permeability. In their effort to keep all of the force loops as short as possible, a force is exerted on the soft iron to pull it into the center of the magnetic field.
(B) When the soft iron is centered between the poles, it will resist any attempt to lengthen the lines of force.

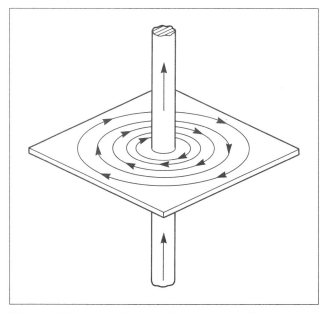

Figure 1-23. Lines of magnetic force encircle a current-carrying conductor. These lines are relatively weak and have no polarity.

The greater the amount of flow, the stronger will be the magnetic field.

The left-hand rule gives us the direction of these lines of flux, and by knowing their direction, we can determine the direction of rotation of motors and the polarity of electromagnets. If we grasp the conductor in our left hand with our thumb pointing in the direction of electron flow from negative to positive, our fingers will encircle the conductor in the direction of the lines of flux.

Now, this magnetic field around the conductor does not serve much of a practical purpose because, while it has direction, it does not have any poles, and it is relatively weak. To increase its strength, and make it useful, we can wind the conductor into the form of a coil. When we do this, we concentrate the lines of flux, and the coil attains the characteristics of a magnet.

In figure 1-25, the electrons flow into the coil from the right, and as the conductor passes over the top of the coil, the electrons are flowing away from us, as we see by the cross representing the tail of the arrow. Below the coil, the electrons flow toward us, as is seen from the dot representing the head of the arrow. When electron flow is away from us, the lines encircle the conductor in a counterclockwise direction, and when they come toward us, the field circles the conductor clockwise.

The lines of flux around each turn aid or reinforce the flux around every other turn, and there is a resultant field that enters the coil from the left and leaves it from the right. When we use the same terminology we used for permanent magnets, we find that in this electromagnet, the pole on the right is the north pole since the lines of flux are leaving it, and the pole of the left where the lines enter is the south pole.

From what we have just seen, we can apply another left-hand rule—the left-hand rule for coils. If we grasp a coil with our left hand in such a way that our fingers wrap around the coil in the direction of electron flow, our thumb will point to the north pole formed by the coil.

The strength of an electromagnet is determined by the number of turns in the coil and by the amount of current flowing through it. This strength, or the magnetomotive force produced by the coil can be compared to electromotive force or electrical pressure of an electrical circuit. Magnetomotive force is measured in gilberts. Gilberts are symbolized by the letters "Gb".

The unit of magnetic field intensity of the lines of flux is the gauss. An individual line of flux called a maxwell, in an area of 1 square centimeter produces a field intensity of 1 gauss.

The law for magnetic circuits tells us that, one gilbert is the amount of magnetomotive force produced by one maxwell flowing through a magnetic circuit having one unit of "reluctance", the opposition of the circuit of the flow of magnetic

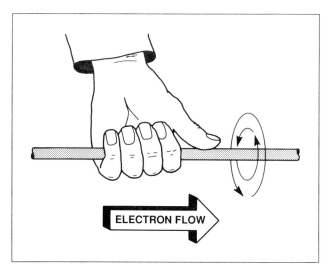

Figure 1-24. *If a current carrying conductor is grasped with the left hand with the thumb pointing in right direction of electron flow (from negative to positive), the lines of force will encircle the conductor in the same direction as the fingers are pointing.*

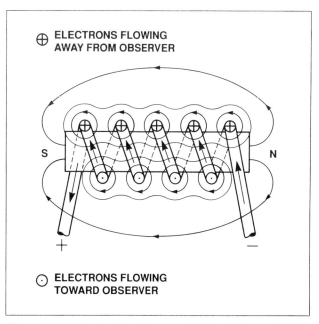

Figure 1-25. *By forming the conductor into a coil, the ines of flux can be concentrated, and the resulting coil will behave as a magnet and will have a north and a south pole.*

flux. Simply stated, an electromagnet having one ampere-turn will produce a magnetomotive force of 1.256 gilberts.

We can further increase the strength of an electromagnet by concentrating the lines of flux. This is accomplished by using some highly permeable material, such as soft iron, for the core.

There are two types of electromagnets used in practical applications: those having fixed cores, and those with movable cores. Fixed-core electromagnets are used in such devices as voltage regulators, where current proportional to the generator output voltage flows through the coil.

Movable-core electromagnets are called solenoids. A soft iron core is held out of the center of the coil by a spring, but as soon as current flows in the coil, the magnetic field will pass through the core and, in its efforts to keep the loops of force as small as possible, will pull the core into the center of the coil.

F. Sources of Electrical Energy

Man has not been given the prerogative of creating nor destroying energy but, within rather wide guidelines, energy may be converted from one form into another. This ability to exchange forms makes electrical energy all the more valuable to us. The conversion of magnetic, mechanical, thermal, chemical, and light energy into electricity, and the exchange of electricity back into these other forms of energy are commonplace and easily accomplished in the world today.

1. Magnetism

Lines of magnetic flux pass between the poles of a magnet, and if a conductor is moved through these lines of flux, they will transfer to the conductor and force electrons to flow through it. This is the principle used to generate most of our electricity today. Our aircraft carry generators or alternators to produce electricity by this method, and the atomic and hydro-electric powerplants produce power by the same procedure.

The amount of electricity that is generated depends on the *rate* at which the lines of flux are cut. This rate may be increased by increasing the number of

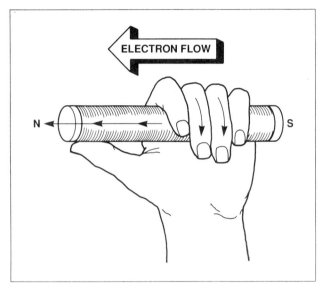

Figure 1-26. If a coil is grasped with the left hand in such a way that the fingers encircle it in the same direction as the electron flow (from negative to positive), the thumb will point to the north pole of the electromagnet formed by the coil.

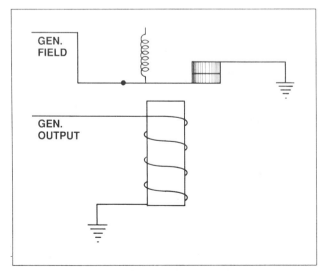

Figure 1-27. An electromagnetically operated switch with a fixed core is called a relay.

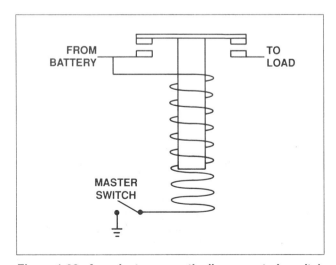

Figure 1-28. An electromagnetically operated switch with a movable core is called a solenoid.

lines of flux, by making the magnet stronger or by moving the conductor through the lines faster.

2. Chemical

In the chemical make-up of materials, there can exist an imbalance of electrons and protons. If a material having an excess of electrons is connected by a conductor to a material having a deficiency, electrons will be forced through the conductor.

For example, if a piece of aluminum and a piece of copper are immersed in a solution of hydrochloric acid and water, and the two pieces of metal are connected by a piece of wire, electrons will leave the aluminum and flow to the copper. The electrons which leave the aluminum are replaced by the negative chlorine ions from the acid. When the chlorine combines with the aluminum, it eats away part of the metal and forms a gray powdery material on the surface. Positive hydrogen ions will be attracted to the copper where they are neutralized by the electrons that came from the aluminum, and bubble to the surface as free hydrogen gas.

3. Thermal (Heat)

When certain combinations of wire, such as iron and constantan or chromel and alumel, are joined into a loop with two junctions, a thermocouple is formed. An electrical current will flow through the wires when there is a *difference* in the temperature of the two junctions. A cylinder head temperature measuring system has one junction held tight against the engine cylinder head by a spark plug (the hot junction), while the other junction is in the relatively constant temperature of the instrument panel (the cold junction).

4. Pressure

Crystalline material such as quartz has the characteristic that when it is bent or deformed by a mechanical force, an excess of electrons will accumulate on one surface, leaving the opposite surface with a deficiency. This is known as the piezoelectric effect and is made use of in crystal microphones and phonograph pickups. And the fact that this interchange between mechanical and electrical energy is reversible makes crystals useful for producing alternating current for radio transmitters. A piece of crystal has only one natural frequency at which it will vibrate, and if it is

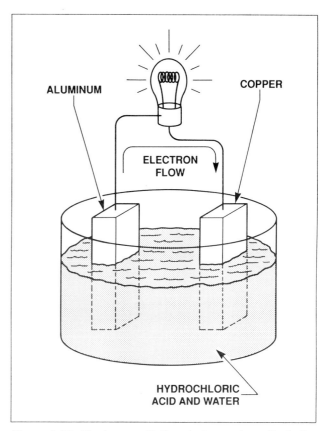

Figure 1-30. Electrons will flow between two dissimilar metals when they are connected by a conductor and are both immersed in an electrolyte.

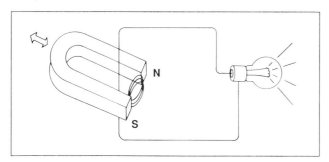

Figure 1-29. The amount of electricity generated by electromagnetic induction is determined by the rate at which the conductor cuts through lines of magnetic flux.

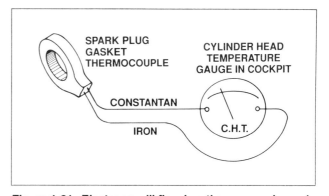

Figure 1-31. Electrons will flow in a thermocouple made of certain dissimilar metals when there is a temperature difference between the two junctions.

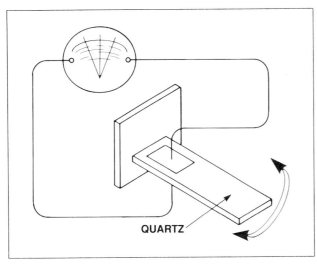

Figure 1-32. An electrical potential difference will be built up across the faces of certain crystalline materials when they are bent or otherwise subjected to mechanical pressure.

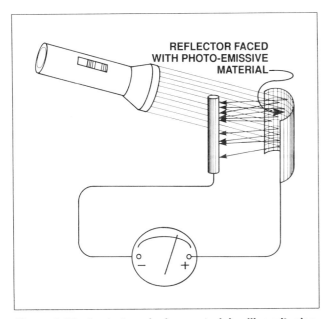

Figure 1-33. A photoemissive material will emit electrons when it is struck by light.

excited by pulses of electrical energy, it will vibrate at this frequency. As it vibrates, it produces between its faces an alternating voltage having an accurate frequency.

5. Light

Light is a form of energy. When it strikes certain photoemissive materials such as selenium, it imparts enough energy to the atoms that electrons are discharged from their bonds and are free to flow in a circuit and do work.

Switches may be controlled by light-sensitive devices to turn airport lights on at dusk and off at dawn, and sensitive light measuring meters are used in photography to determine the amount of light available, so that the proper exposure can be made.

6. Friction

Although not a practical method of producing electricity for power, friction between two materials can produce static electricity. Static electricity and electrostatic fields were discussed in an earlier section of this manual. The technician needs to be aware of the ease with which static electricity can be produced, and with methods for its control or elimination, as many electrostatic-discharge-sensitive (ESDS) devices are used in electronics equipment aboard many modern aircraft.

G. Current Electricity and Ohm's Law

1. Ohm's Law

In our study this far, we have seen that a concentration of electrons will produce an electrical pressure that will force electrons to flow through a circuit. And by assigning values to the pressure, flow, and opposition, we can understand the relationship that exists between them, and can accurately predict what will happen in a circuit under any given set of conditions.

It was the German scientist George Simon Ohm who proved the relationship between these values, and in 1826 he published his findings. Ohm's law is the basic statement which says in effect that the current that flows in a circuit is directly proportional to the voltage (pressure) that causes it, and inversely proportional to the resistance (opposition) in the circuit. The units we use make this relationship easy to see: one volt of pressure will cause one ampere of current to flow in a circuit whose resistance is one ohm.

For ease of handling these terms in formulas, voltage is represented by the letter E, current by the letter I, and resistance by the letter R. A statement of Ohm's law in the form of a formula is, therefore, $E = I \times R$. If we want to find the current, we use the formula $I = E/R$, and resistance may be found by the formula $R = E/I$.

Power in an electrical circuit is measured in watts, and one watt is the amount of power used in a circuit when one amp of current flows under a pressure of one volt.

The relationship between voltage, current, resistance, and power is such that any one value may be found when any two of the others are known. One easy way to find the correct formula is to use a series of divided circles representing the symbols in the formula. In figure 1-34, we see that the voltage in the circuit is equal to the product of the current and the resistance (E = I × R). The top half of the circle is equal to the bottom half. We also know that the current may be found by dividing the voltage by the resistance (I = E/R), and the resistance is equal to the voltage divided by the current (R = E/I).

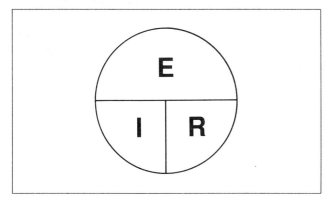

Figure 1-34. The relationship between voltage, current, and resistance.

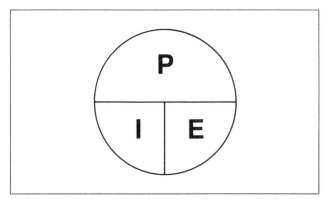

Figure 1-35. The relationship between power, current, and voltage.

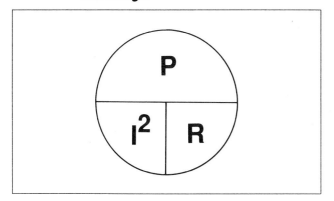

Figure 1-36. The relationship between power, current, and resistance.

The same relationship may be found between power, current, and voltage. We know that power is equal to voltage times current (P = I × E). Figure 1-35 shows us an easy way to find current by dividing power by voltage (I = P/E), and to find the voltage by dividing power by the current (E = P/I).

The other six relationships are just as easy to find when we use the circles of figures 1-36 and 1-37. In figure 1-36, we find that P = I² × R, and R = P/I². Now we have one very small problem, I² = P/R, but we want I, not I², so we must take the square root of both sides of the equation. When we do this, we end up with I = $\sqrt{P/R}$.

Figure 1-37 works in the same way with E² = P × R. To find E, we must take the square root of both sides. When we do, we get E = $\sqrt{PR}$. The other relationships give us P = E²/R and R = E²/P.

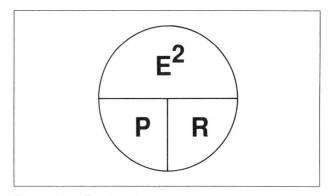

Figure 1-37. The relationship between voltage, power, and resistance.

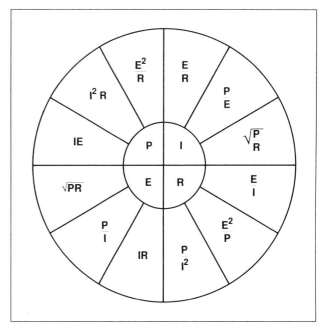

Figure 1-38. Summary of basic equations using the volt, ampere, ohm, and watt.

15

2. Mechanical Power in Electrical Circuits

Power, as we remember from basic physics, is the time-rate of doing work, and the practical unit of measurement is the horsepower which is the amount of power required to do 33,000 foot-pounds of work in one minute, or its equivalent of 550 foot-pounds of work in one second. A constant allows us to relate electrical power to mechanical power. The electrical equivalent of one horsepower is 746 watts.

If we have a 24-volt electric hoist, we can find the amount of current needed to raise a 1,000-lb. load 6′ in 30 seconds.

$$1{,}000 \times 6 = 6{,}000 \text{ foot-pounds}$$

$$\frac{6{,}000}{30} = 200 \text{ foot-pounds per second}$$

$$\frac{200}{550} = 0.364 \text{ HP}$$

$$746 \times 0.364 = 271.5 \text{ watts}$$

$$I = \frac{P}{E} = \frac{271.5}{24} = 11.31 \text{ amps}$$

3. Heat in Electrical Circuits

In circuits where mechanical work is not actually being done, power is still a very important consideration. For example, if you install a resistor in a light circuit to drop the voltage from 12 volts down to three, for a light bulb that requires 150 milliamps, you must find the resistance in ohms and the power in watts, this resistor must dissipate.

To solve this problem, we must first find the voltage to be dropped:

$$E = 12 - 3 = 9 \text{ volts}$$

Find the resistance required:

$$R = E/I = 9/0.15 = 60 \text{ ohms}$$

Find the power dissipated in the resistor:

$$P = I^2 \times R = 0.15^2 \times 60 = 1.35 \text{ watts}$$

The resistor will only have to 1.35 watts, but for practical purposes, you will most probably use a two-watt resistor.

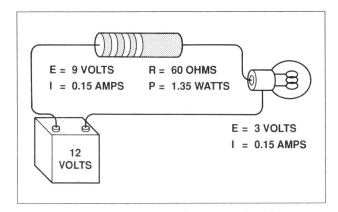

Figure 1-39. Determining the characteristics of a resistor needed to drop voltage in a circuit.

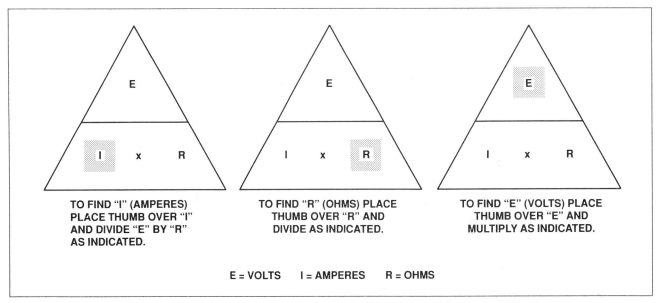

Figure 1-40. Ohm's law relationships.

H. Circuit Elements

All complete electrical circuits consist of at least a minimum of a source of electrical energy, a load device to use the electrical energy produced by the source, and conductors to connect the source to the load or loads in the circuit. These circuit elements do not comprise a practical electrical circuit, however. In order to make a circuit practical, a control device, such as a switch, must be placed in the circuit to allow the loads to be easily and safely energized and de-energized. Some type of protection must be provided for the circuit wiring, in the form of fuses or circuit breakers, to stop current in the event of an overload or other circuit malfunction.

The source may be a battery, a generator, or other apparatus that converts some form of energy into electricity, and these are discussed in the appropriate section of the text. The load may be anything that converts electrical energy into mechanical or chemical energy, or into heat or light. These devices will be discussed in the appropriate section. Here, we are primarily concerned with the conductors used to join the source and the load, and components used to control the flow of current and protect the circuit from damage due to shorts.

1. Conductors

The purpose of a conductor is to provide a path for the electrons to flow from the source, through the load, back to the source. It must do this with the minimum of resistance, but other factors must also be taken into consideration, so the choice of a conductor is often a compromise.

Most aircraft electrical systems are of the single-wire type, meaning that the aircraft structure itself provides the path through which the current returns from the load to the source. A great deal of weight is saved by using this type of system, but it is extremely important that a good connection capable of carrying all of the current is provided between the aircraft structure and the battery, generator, and all of the devices using the current.

The resistance of a conductor is affected by two things: its physical characteristics and its dimensions.

a. Physical Characteristics

1) Resistivity

For most practical aircraft circuits, we use two types of conductors, copper and aluminum. Copper wire has only about two-thirds of the resistance of the equivalent gage of aluminum wire, and is the one most generally used. But for applications requiring a great deal of current, aluminum wire is often used. Its resistivity is higher than that of copper, and a larger conductor is needed, but since aluminum weighs so much less than copper, a great deal of weight may be saved by its use.

2) Temperature

Most metals have what is known as a positive temperature coefficient of resistance. This means that the resistance of the material will increase as its temperature increases. This characteristic is used in some temperature measuring instruments where the resistance change in a piece of wire is used to measure temperature. For practical purposes, however, both copper and aluminum have such a small change in resistance with temperature over the temperature range encountered in flight that it is normally not considered to be a problem.

b. Dimensions

1) Length

For most common conductors, the resistance will vary directly with length. That is, as length increases for a given specific conductor, its resistance will increase.

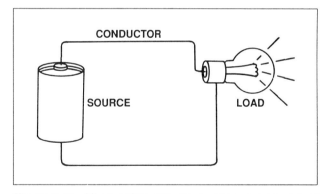

Figure 1-41. All complete circuits must have a source of electrical energy, a load to use the energy, and conductors to join the source and the load.

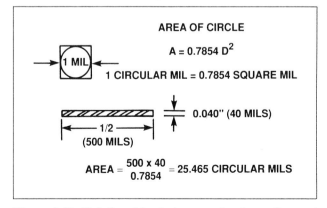

Figure 1-42. Relationship between circular mils and square mils.

2) Cross-Sectional Area

The resistance of a conductor varies *inversely* as its cross-sectional area. Aircraft wire is measured by the American Wire Gage system, with the larger numbers representing the smaller wires. The smallest size wire normally used in aircraft wiring is the 22-gage size which has a diameter of about 0.025". For carrying very large amounts of current, cables up to the 0000 gage size (spoken of as four aught size) are used. These gage cables have a diameter of about 0.52".

Since it is the area rather than the diameter of a conductor that determines its current-carrying capability, and since almost all conductors are round rather than square, the circular mil measurement is handy to use. A circular mil is the standard measurement of cross-sectional area, and is the area of a circle whose diameter is one mil, or $1/1,000$". To find the cross-sectional area of a conductor in circular mils, you need only square its diameter in mils. For example, if a round wire has a diameter of $1/8$" (0.125" or 125 mils) it would have an area of 15,625 circular mils.

COPPER ELECTRIC WIRE CURRENT-CARRYING CAPACITY				
Wire Size Specification MIL-W-5086	Single Wire in Free Air— Maximum Amperes	Wire in Conduit or Bundled— Maximum Amperes	Maximum Resistance— Ohms/1,000 Feet (20°C)	Nominal Conductor Area— Circular Mills
AN-20	11	7.5	10.25	1,119
AN-18	16	10	6.44	1,779
AN-16	22	13	4.76	2,409
AN-14	32	17	2.99	3,830
AN-12	41	23	1.88	6,088
AN-10	55	33	1.10	10,443
AN-8	73	46	.70	16,864
AN-6	101	60	.436	26,813
AN-4	135	80	.274	42,613
AN-2	181	100	.179	66,832
AN-1	211	125	.146	81,807
AN-0	245	150	.114	104,118
AN-00	283	175	.090	133,665
AN-000	328	200	.072	167,332
AN-0000	380	225	.057	211,954
ALUMINUM ELECTRIC WIRE CURRENT-CARRYING CAPACITY				
Wire Size Specification MIL-W-7072	Single Wire in Free Air— Maximum Amperes	Wire in Conduit or Bundled— Maximum Amperes	Maximum Resistance— Ohms/1,000 Feet (20°C)	Nominal Conductor Area— Circular Mills
AL-6	83	50	0.641	28,280
AL-4	108	66	.427	42,420
AL-2	152	90	.268	67,872
AL-0	202	123	.169	107,464
AL-00	235	145	.133	138,168
AL-000	266	162	.109	168,872
AL-0000	303	190	.085	214,928

Figure 1-43. Characteristics of aircraft copper and aluminum wire.

Since the circular mil is smaller than the square mil, to find the area of a square or rectangular conductor in circular mils, divide the square mil area by 0.7854 (the relationship between the area of a circle and the square of its diameter). For example, to find the cross-sectional area of a strap of copper 0.040" thick and ½" wide, you would first find its area in square mils, which is 500 × 40 = 20,000. Now divide this by 0.7854. The area is 25,465 circular mils, which is about the same as a 6-gage stranded copper wire.

2. Control Devices

a. Switches

1) Toggle or Rocker Switches

Switches used to control the flow of electrons in most aircraft circuits are of the enclosed toggle or rocker type. These switches are actuated by either moving the bat-shaped toggle or by pressing on one side of the rocker.

If the switch controls only one circuit and has only two connections and two positions, open and closed, it is called a single-pole, single-throw, or SPST switch. These may be used to control lights, either turning them ON or OFF.

Some switches are used to control more than one circuit and may be either the single- or double-throw type. The double-pole, single-throw (DPST) switch could be used to control both the battery and the generator circuit so they would both be turned ON and OFF at the same time. A double-pole, double-throw (DPDT) switch controls two circuits and has either two or three positions.

Both toggle and rocker switches may have one or both of their positions spring-loaded so they will return to the OFF position when your finger is removed.

2) Wafer Switches

When a switch is used to select one of a number of conditions, a wafer switch is often used. These switches may have several wafers stacked on the same shaft, and each wafer may have as many as twenty positions. Wafer switches are seldom used for carrying large amounts of current and are most

Figure 1-44. Most of the switches found in modern aircraft are either of the toggle or the rocker type.

Figure 1-45. Wafer switches are used when it is necessary to select any of a large number of circuit conditions.

Figure 1-46. Precision switches snap open or closed with an extremely small amount of movement of the operating control.

generally open with the wires soldered to the terminals on the wafers.

3) Precision Switches

Limit switches, uplock and downlock switches, and many other applications require the switch to be actuated by mechanical movement of some mechanism. In these applications it is usually important that the switch actuate when the mechanism reaches a *very definite* and *specific* location. These switches require an extremely small movement to actuate. Snap-acting switches of the type pioneered by Micro Switch division of Honeywell find a wide use in these applications. An extremely small movement of the plunger trips the spring so it will drive the contacts together. When the plunger is released, the spring snaps the contacts back open.

4) Relays and Solenoids

It is often necessary to open or close a circuit carrying a large amount of current from a remote location and with a small switch. An example is the starter circuit for an aircraft engine. The starter motor requires a great deal of current and the cable between the battery and the starter must be as short as possible. But the starter switch must be located inside the cockpit and is often incorporated in the ignition switch. A solenoid, or contactor, is used for this application. A small amount of current is used to energize an electromagnet which closes the contacts in the circuit carrying the large amount of current.

If the electromagnet has a fixed core that attracts a movable armature, the switch is called a relay, but if the core is movable, and is pulled into the hollow coil, it is called a solenoid.

3. Protective Devices

Protective devices are installed in electrical circuits to prevent damage caused by excessive current flow. Excessive current may be the result of overloading the circuit, or a short circuit.

Overloading the circuit is the result of connecting loads that are too large for the wiring, or of connecting too many smaller loads.

A direct short is when some part of the circuit in which full system voltage is present comes into direct contact with the return side of the circuit, or ground. This establishes a path for current flow with little or no resistance, other than that of the conductors themselves. This will result in very large current flow, heating, and likely damage.

a. Fuses

A fusible link made of a low melting point alloy enclosed in a glass tube is used as a simple and reliable circuit protection device. The load current flows through the fuse, and if it becomes excessive, will melt the link and open the circuit. Some fuses are designed to withstand a momentary surge of

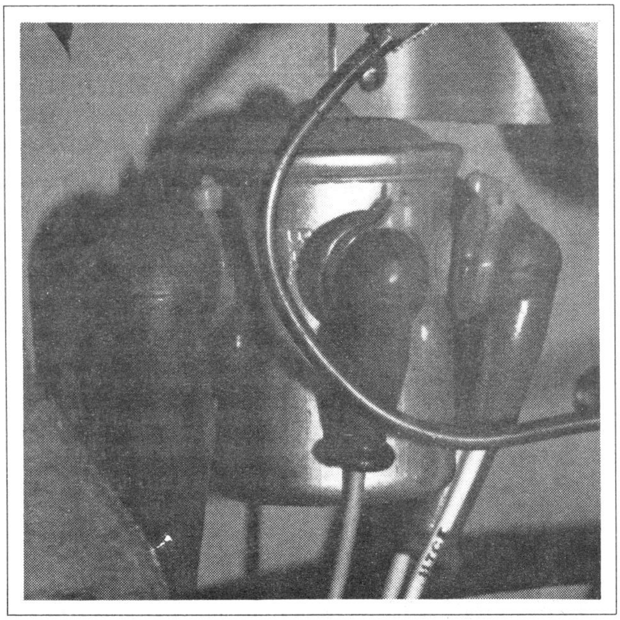

Figure 1-47. Starter solenoid switches control large amounts of current, but they are operated by a very small current that flows through the coil.

Figure 1-48. The heat caused by excess current melts the fusible link and opens the circuit. Slow-blow fuses will stand a momentary excess of current, but sustained overload will soften the link and the spring will pull apart.

Figure 1-49. Most circuit protection for modern aircraft is provided by circuit breakers that may be reset in flight.

current, but will melt if the current is sustained. These slow-blow fuses have a small spring attached to the link so when the sustained current softens the link, the spring will pull it in two and open the circuit.

b. Circuit Breakers

It is often inconvenient to replace a fuse in flight, so most aircraft circuits are protected by circuit breakers that will automatically open the circuit if the current becomes excessive, but may be reset by moving the operating control, which may be a toggle, a push-button, or a rocker. If the excess current was caused by a surge of voltage, or by some isolated and non-recurring problem, the circuit breaker will remain in and the circuit will operate normally. But if an actual fault such as a short circuit does exist, the breaker will trip again, and it should be left open.

Aircraft circuit breakers are of the trip-free type which means that they will open the circuit irrespective of the position of the operating control.

With this type of breaker, it is impossible to hold the circuit closed, if an actual fault exists.

There are two operating principles of circuit breakers. Thermal breakers open the circuit when the excess current heats an element in the switch and it snaps the contacts open. The other type is a magnetic breaker which uses the strength of the magnetic field caused by the current to open the contacts.

Automatic re-set circuit breakers that open a circuit when excess current flows but automatically close it again after a cooling-off period are usually not used in aircraft circuits.

4. Resistors

It is often necessary in electrical circuits to control current flow by varying voltage by converting some of the electrical energy into heat. And these resistors may be classified as fixed or variable.

a. Fixed Resistors

1) Composition

The great majority of resistors used to control small amounts of current are made of a mixture of carbon and an insulating material. The relative percentage of the two materials in the mix determines the amount of resistance a given amount of the material will have. Small amounts of material are used to dissipate small amounts of power and, for more power, more material is used. Composition resistors are normally available in sizes from $\frac{1}{8}$ watt up to two watts. The larger the physical size of the resistor, the more power it will dissipate.

Most modern resistors are of the axial-lead type; that is, the leads come directly out of the ends of the resistor. The ohmic value of this type of resistor is indicated by three or four bands of color around one end.

The band nearest the end of the resistor is the first significant figure of the resistance, the second band represents the second figure, and the third band tells the number of zeros to add to the two numbers. For example, if the first band is green (5), the second is brown (1), and the third is yellow (0000), the resistor would have a resistance of 510,000 ohms.

If there is not a fourth color band, the resistor has a tolerance of plus or minus 20%. But if there is a fourth band of silver, the resistor is within plus or minus 10% of the indicated value. If the fourth band is gold, it is within plus or minus 5% of the value shown by the color code.

Resistors having gold or silver in the fourth band are rated according to the Electronic Industries Association (EIA) preferred values, and are those most commonly used. Certain precision resistors, however, have much closer tolerance than the 5%, 10%, and 20% of the preferred values, and these have a fourth band of one of the same colors as used in the first three bands. These resistors are marked according to the EIA alternative values and their tolerances are shown in the table of figure 1-50(C).

Low-resistance composition resistors may have silver or gold as the third band. If the third band is silver, multiply the two significant figures by 0.01, and if it is gold, multiply them by 0.1. For example, a resistor having the first band of yellow, the second band of violet, and a third band of gold would have a resistance of 4.7 ohms. If the third band had been silver, the resistance would have been 0.47 ohms.

Some composition resistors have their leads coming off of the body radially instead of parallel to the resistor axis. These radial-lead resistors are color coded with the same colors, but the color of the body is the first significant figure of the resistance, the color of the end is the second significant figure, and a dot or band of paint around

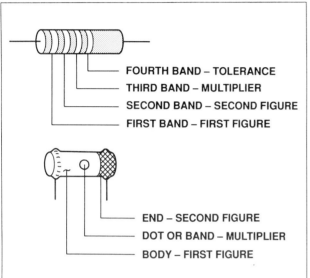

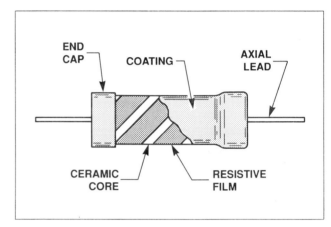

Figure 1-51. Film resistor construction.

Color	Number	Multiplier	Tolerance EIA Preferred	Tolerance EIA Alternate
Black	0			
Brown	1	10		1%
Red	2	100		2%
Orange	3	1,000		3%
Yellow	4	10,000		4%
Green	5	100,000		5%
Blue	6	1,000,000		6%
Violet	7	10,000,000		7%
Gray	8	100,000,000		8%
White	9	1,000,000,000		9%
Gold		0.1	5%	
Silver		0.01	10%	
No Color			20%	

Figure 1-50.
  (A) Color code marking for axial-lead resistors.
  (B) Color code markings for radial-lead resistors.
  (C) Resistor color code values.

Figure 1-52. Wire-wound resistors are used when there is a great deal of power that must be dissipated.

the middle of the resistor is painted silver, the resistance is within plus or minus 10%. If it is gold, it is within plus or minus 5%, and if there is no gold or silver paint on the end, the tolerance is plus or minus 20% of the indicated resistance. As an example, if the body of the resistor is brown, one end is black, and the dot is green, and there is no silver or gold paint on the resistor, the resistance is 1,000,000 ohms or, as this is more commonly called, one megohm, plus or minus 20%.

2) Film Resistors

Carbon resistors have been replaced in some modern electronics equipment by film resistors. Film resistors are manufactured by placing a thin layer, or film, of resistive material around a nonconductive core material such as ceramics. The resistor leads, usually of the axial type, are inserted into a cap and placed onto the ends of the ceramic core as shown in figure 1-51.

Film resistors usually are of the low power variety, ranging from one-tenth to two watts, and are generally available in the same range of resistance values as carbon resistors.

3) Wire-Wound Resistors

When more power needs to be dissipated than can be handled by a composition resistor, special resistors made of resistance wire wound over hollow ceramic tubes are used. Some of these resistors are tapped along the length of the wire to provide different values of resistance, and others have a portion of the wire left bare, so a metal band can be slid over the resistor, allowing it to be set to any desired resistance. When the screw is tightened, the band will not move from the selected resistance.

b. Variable Resistors

When it is necessary to change the amount of resistance in a circuit, variable resistors may be used.

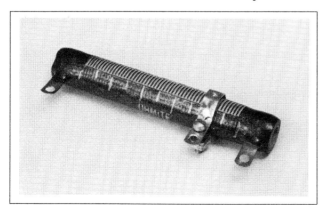

Figure 1-53. Wire wound resistors may have a portion of the wire exposed and incorporate a movable tap.

These may be of either the composition or the wire-wound type. In the composition resistor, the mix is bonded to an insulating disk, and a wiper, or sliding contact, is rotated by the shaft to vary the amount of material between the two terminals. We saw earlier in the discussion of resistance that the resistance varies with the length of the conductor,

Figure 1-54. Variable resistors allow the amount of resistance in a circuit to be changed by rotating the shaft.

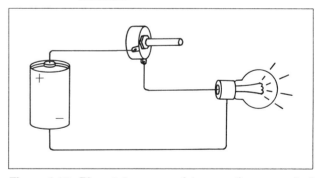

Figure 1-55. Rheostats are used to vary the amount of resistance in a circuit.

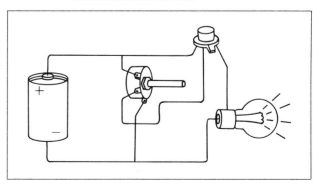

Figure 1-56 . Potentiometers are used as voltage dividers in a circuit.

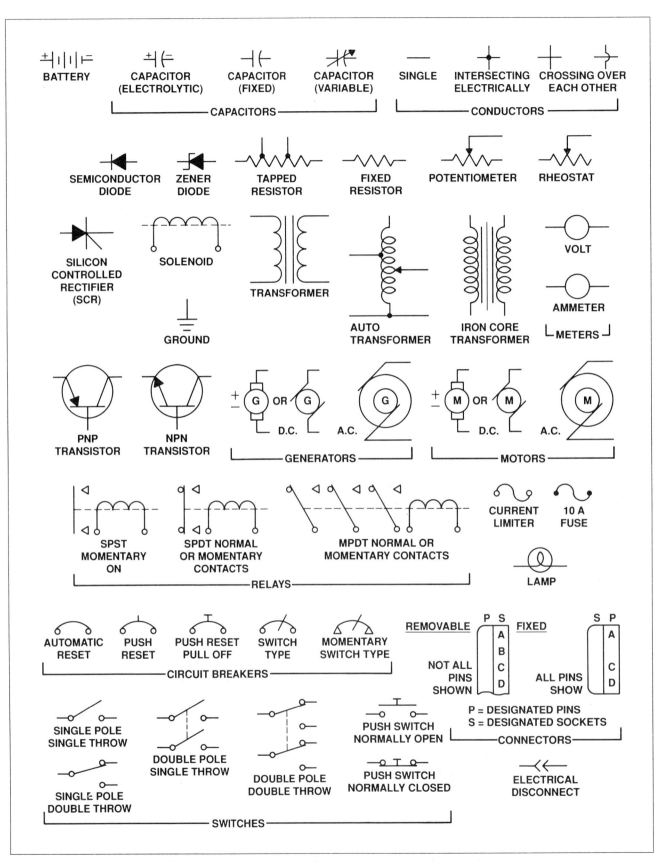

Figure 1-57. Electrical symbols.

and the farther the sliding contact is from the fixed contact, the greater will be the resistance.

Resistance wire may be wound around a form which is shaped so that the sliding contact will touch the wire at the edge of the form. As the contact is rotated, the length of wire between the terminals varies and the resistance changes.

Variable resistors having only two terminals, one at the end of the resistance material, and the other, the sliding contact, are called rheostats and are used to vary the amount of resistance in a circuit.

If the resistor has three terminals, one for either end of the resistance material and one for the slider, it is called a potentiometer and may be used as a voltage divider, of which we will say more in the section on circuits.

I. Circuit Arrangements

All electrical circuits must contain three components: the source of electrical energy, a load device to use the energy, and conductors to connect the source and the load. Control devices such as switches and resistors and such protective devices as fuses or circuit breakers tailor the circuit to its purpose.

For a circuit to be complete, there must be at least one continuous path from one of the terminals of the source, through the load and back to the other terminal. If there is any interruption or break in this path, the circuit is said to open, and there can be no flow of electrons. If, on the other hand, there is a path from one terminal of the source to the other without passing through the load the circuit is *shorted*; and not only is there no work being done, but the lack of resistance in the circuit will allow excessive current to flow. In this case, unless a protective device such as a fuse opens the circuit, the wiring and even the source are likely to be damaged.

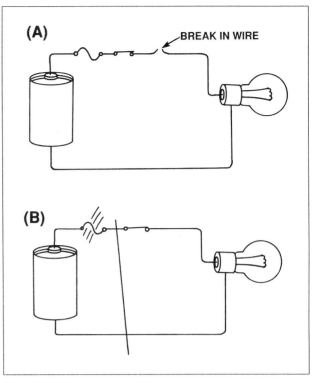

Figure 1-58.
(A) An open circuit allows no current flow and the circuit cannot function.
(B) A short circuit causes excessive current flow and can damage the circuit components or wiring unless a protecting device such as a fuse opens the circuit.

Chapter II

Direct Current

A. Direct Current Terms and Values

1. Direct Current

Direct current (DC) is current which is non-varying in nature, such as that obtained from a battery or filtered power supply. This type of current may be referred to as "pure DC", meaning that no alternating current, noise, etc., is present.

2. Pulsating DC

Pulsating DC is a current (or voltage) which varies from a zero reference level to a maximum or peak value, never dropping below the zero reference. Pulsating DC is produced by a rectifier in a power supply and is filtered to remove the pulses or variations, thereby producing pure DC.

3. Average Value

The average value of DC is the average of the current or voltage excursion made by a pulsating DC waveform as it moves from zero to its maximum value. The average value is computed by multiplying the maximum value of the pulsating waveform by 0.637. Example: assume the maximum value for the waveform below is 20 V. Multiplying 20 × 0.637, we obtain the average value of 12.74 V.

4. Polarity

The polarity of DC is expressed as being either positive or negative. Polarity is determined by establishing a reference point (usually ground) and measuring a voltage in reference to that point. For example, if we measure battery voltage, we measure from battery negative, which is connected to ground, to battery positive and determine that battery voltage is +12 V. That is, the battery voltage has a value of 12 volts and has positive polarity with respect to ground.

B. Series DC Circuits

1. Basic Series Circuit Analysis

If there is only one path for electrons to flow, the circuit is called a series circuit. In figure 2-3 we have a typical circuit in which the battery is the source of power, and the lamp, which is our load, is in series with the rheostat, the switch, and the fuse. Since there is only one path for electron flow, if either the switch is open or the fuse is blown, the lamp cannot burn.

The rheostat is in series with the lamp, and when it is set for a minimum of resistance, the maximum amount of current will flow through the lamp, and it will burn with full brilliance. But when its resistance is increased, part of the power from the battery is dissipated in the resistor in the form of heat. Less power is available for the lamp, and it will burn with less than full brilliance.

There are two laws propounded by the German physicist Gustav Robert Kirchhoff that go a long way in explaining the behavior of voltage and current in electrical circuits. His voltage law helps us understand the action in series circuits. Kirchhoff's voltage law states that *the algebraic sum of the applied voltage and the voltage drop around any*

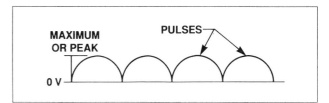

Figure 2-1. Pulsating DC waveform.

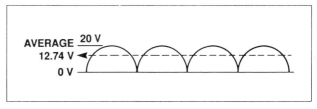

Figure 2-2. Average value of pulsating DC.

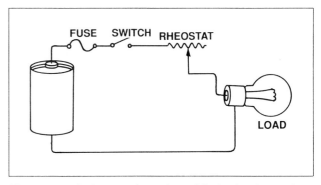

Figure 2-3. A rheostat in series with the load can drop part of the voltage and control the brightness of the light.

closed circuit is equal to zero. What this means simply is that the voltage that is dropped across every part of the load must be exactly the same as the voltage supplied by the source. In figure 2-4 we have a good example of this. A six-volt battery in series with a switch, three lamps, and a milliammeter. Each lamp is rated at 2 volts and requires 300 milliamps of current to burn at full brilliance. When the switch is closed, all of the lamps light up fully bright.

We can analyze our circuit and prove Kirchhoff's voltage law with a voltmeter that is capable of measuring up to 6 volts. Place the positive lead at C. With the switch closed, it offers no resistance, and according to Ohm's law (E = I × R) if the resistance is zero the voltage will also be zero, and so there is no voltage reading across the switch. But when the switch is opened, there will be no current flowing in the circuit, and so there will be no voltage dropped by the lamps, and the meter will read the full source voltage of 6 volts.

Now, with the switch closed and the lamps burning, measure between D and E. The meter should read two volts, indicating that the resistance of the lamp has dissipated enough power to drop this much pressure (voltage).

Measure the voltage between points E and F. It also should be 2 volts, as should the voltage between points F and G. And the voltage between points D and G should be exactly the same as between A and H, proving that Kirchhoff's voltage law is true.

Ohm's law can also be applied to this circuit. Using a chart similar to that in figure 2-5, we can fill in the blanks. Since this is a series circuit with only one path for the electrons to flow, the current will be the same through each lamp as it is through the source, in this case it is 300 milliamps.

The source voltage is 6 volts, and we have just found the voltage across each lamp to be 2 volts. The sum of the voltage dropped across the lamps does, in fact, equal the source or applied voltage.

Now to find the resistance of the lamps. The operating resistance cannot be measured by an ohmmeter, because the resistance of a lamp is quite different when it is hot than when it is cold, so we find the resistance by Ohm's law.

$$R = E/I = 2/0.3 = 6.67 \text{ ohms}$$

There are three of these lamps in series, so we have a total circuit resistance of 20 ohms. To prove that this is correct, apply Ohm's law to the source:

$$E = I \times R = 0.3 \times 20 = 6 \text{ volts}$$

This is our source of voltage.

The power dissipated by each lamp may be found by the formula:

$$P = I \times E = 0.3 \times 2 = 0.6 \text{ watt}$$

The three lamps will dissipate 1.8 watts, which is exactly the same as we find when we multiply by source voltage by the total current.

To briefly summarize the characteristics of a series circuit:

1. There is only one path for the electrons to follow from the source, through the load back to the source.
2. The current is the same wherever it is measured in the circuit.
3. The sum of all the voltage drops equals the source voltage.
4. The total resistance of the circuit is the sum of the individual load resistances.
5. The total power dissipated in the circuit is the sum of the power dissipated in each of the individual load resistances.

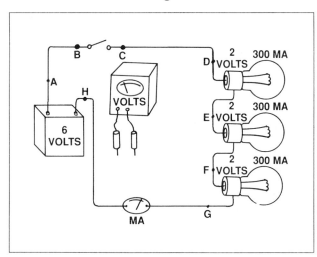

Figure 2-4. This series circuit operates according to Kirchhoff's voltage law.

Component	Voltage Volts	Current MA	Resistance Ohms	Power Watts
A	2	300	6.67	0.6
B	2	300	6.67	0.6
C	2	300	6.67	0.6
Source	6	300	20	1.8

Figure 2-5. Circuit conditions in a series circuit.

2. Voltage Dividers

In electronic work, it is often necessary to have a series of different voltages, using a voltage divider. A voltage divider consists of a series of resistors across the power source. In figure 2-6, we have three 1,000-ohm resistors across a 24-volt battery. When there is no load attached to any of the terminals, there will be a current of 8 milliamps (0.008 amp) flowing through the divider. This current will produce an 8-volt drop across each of the resistors.

When a load is placed across any of the terminals of the divider, it will act in parallel with that portion of the resistance and lower the total resistance. This will increase the current through the circuit. If, for example, a load made up of a 1,000-ohm resistor is placed between the ground terminal and terminal A, the resistance between A and G will drop to 500 ohms, and the current will increase from 8 milliamps to 9.6 milliamps. The voltage across R_3 and the load will be 4.8 volts instead of the 8 volts that was between these same terminals without the load.

It is sometimes necessary in electronic circuits to have voltages that operate on either side of a reference value. For example, if we want voltages of –8 to +16 volts without a load, we can use a voltage divider made up of three 1,000-ohm resistors across our 24-volt source. Rather than using the negative terminal of the battery as our ground (reference point) as we did previously, we can use the junction between R_2 and R_3.

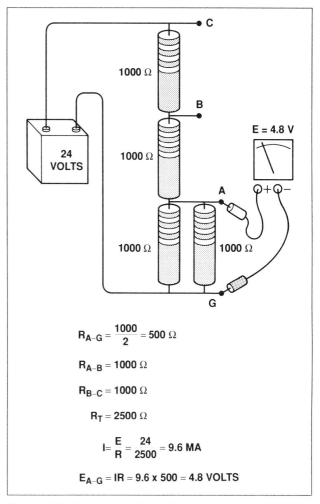

$R_{A-G} = \dfrac{1000}{2} = 500\ \Omega$

$R_{A-B} = 1000\ \Omega$

$R_{B-C} = 1000\ \Omega$

$R_T = 2500\ \Omega$

$I = \dfrac{E}{R} = \dfrac{24}{2500} = 9.6\ \text{MA}$

$E_{A-G} = IR = 9.6 \times 500 = 4.8\ \text{VOLTS}$

Figure 2-7. A loaded voltage divider.

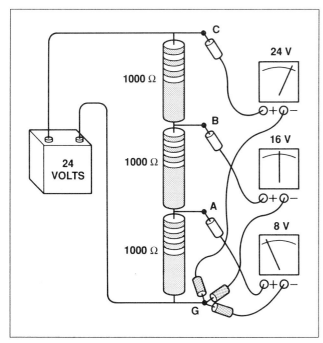

Figure 2-6. An unloaded voltage divider.

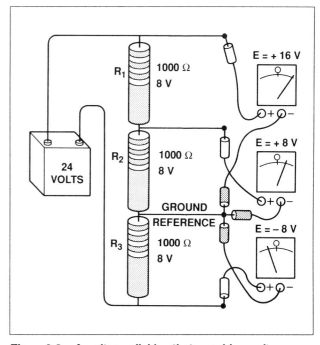

Figure 2-8. A voltage divider that provides voltages on either side of ground (reference voltage).

The lower end of R₃ is 8 volts negative with respect to the ground, the junction between R_2 and R_1 is 8 volts positive with respect to ground, and the top of R_1 is 16 volts positive with respect to ground. These values will, of course, change when a load is applied, but the polarity with respect to ground will not change.

C. Parallel DC Circuits

The most widely used circuit arrangement is the parallel circuit. In this type of circuit, all of the load components are directly across the source, and if one should fail, it will have no effect on the others. You will remember, in a series circuit, if one of the lamps burned out, none of them could burn, but in a parallel circuit, if a lamp burns out, it will have no effect on the others.

The behavior of a parallel circuit is explained in part by Kirchhoff's current law which states that *the algebraic sum of the currents at any junction of conductors is zero*. This means that all of the current that arrives at any junction within the circuit must leave that junction.

In figure 2-9, we have a simple three-lamp parallel circuit using the same components we had in the previous series circuit, except that the source is only 2 volts instead of 6. All of the current flows through the fuse and the switch, and then it splits up, with some passing through each of the lamps. The amount that passes through each lamp is determined by its resistance, and in our example, we are assuming that all of the lamps are the same.

Notice that we have three separate paths from the source through part of the load and back to the source. Each route must obey Kirchhoff's voltage law. Since each path has only one load device, the voltage dropped must equal the source voltage, so each lamp has the full 2 volts across it. Measurement may be made with the voltmeter between A and H, B and C, D and E, F and G, and all will read 2 volts. From this we see that in a parallel circuit, the voltage across each path is the same as the source voltage.

The milliammeters in each of the paths indicate that 300 milliamps flows through each lamp, and the total current is 900 milliamps, verifying Kirchhoff's current law, and showing that the total current is the sum of the current flowing in each of the paths.

We can find the resistance of each lamp by Ohm's law:

$$R = E/I = 2/0.3 = 6.67 \text{ ohms}$$

Now, since the resistance of each of the lamps is the same, the total resistance will be just one-third of the resistance of one lamp. This is because the voltage is the same across all of the lamps, but only one-third of the current flows through each of them. The total resistance is then 2.22 ohms.

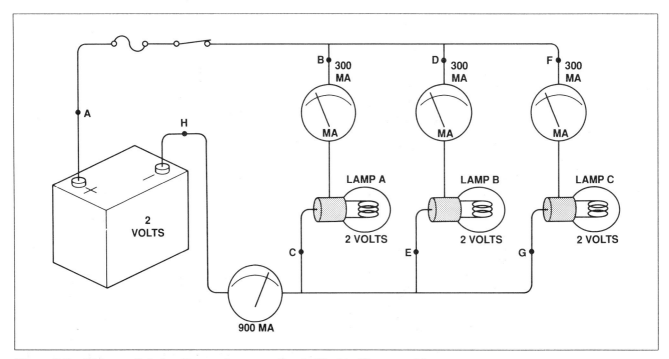

Figure 2-9. This parallel circuit operates according to Kirchhoff's current law.

We can check this answer by using Ohm's law on the source and find that:

$$P = I \times E = 0.3 \times 2 = 0.6 \text{ watt}$$

The total power found by adding the power dissipated in each lamp is 1.8 watts, and this may also be checked by using $P = I \times E$ for the source.

$$P = 2 \times 0.9 = 1.8 \text{ watts}$$

We can summarize the characteristics of a parallel circuit:
1. There is more than one path for the electrons to follow from the source, through part of the load, back to the source.
2. The voltage is the same across any of the paths.
3. The current through each path is inversely proportional to the resistance of the path.

Component	Voltage Volts	Current MA	Resistance Ohms	Power Watts
A	2	300	6.67	0.6
B	2	300	6.67	0.6
C	2	300	6.67	0.6
Source	2	900	2.22	1.8

Figure 2-10. Circuit relationships in a parallel circuit.

4. The total current is the sum of the current flowing through each of the individual paths.
5. The total resistance of the circuit is less than the resistance of any of the paths.
6. The total power dissipated in the circuit is the sum of the power dissipated in each of the individual load resistances.

1. Resistance in a Parallel Circuit

a. All Resistances Have the Same Value

We saw in the previous problem that if the resistances are all the same in a parallel circuit, all you need to do to find the total resistance is to divide the value of a single resistor by the number of resistors. The formula looks like:

$$R_T = r/n$$

r = resistance of one resistor
n = number of resistances

b. Two Unlike Resistances

If there are two unlike resistors in a parallel circuit, you may find the total resistance by dividing the product of the individual resistance by their sum:

$$R_T = \frac{R_1 \times R_2}{R_1 + R_2}$$

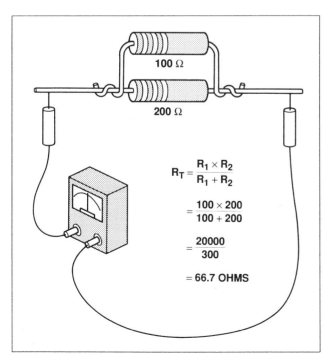

Figure 2-11. Finding the equivalent resistance of two unlike resistors in parallel.

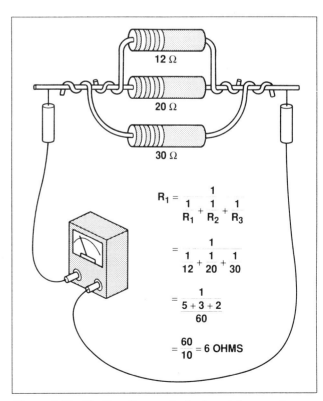

Figure 2-12. Finding the equivalent resistance of more than two unlike resistors in parallel.

If we have a 100-ohm resistor in parallel with one having a resistance of 200 ohms, their total resistance found by this formula is 66.7 ohms. This, you will notice, is less than the resistance of the smaller resistor.

c. More Than Two Unlike Resistances

When we have more than two unlike resistors in parallel, we can find the total resistance, by using a formula that gives us the reciprocal of the sum of the reciprocals of the individual resistances. What this means is that we use a formula such as the one in figure 2-12. We have a 12-ohm, a 20-ohm, and a 30-ohm resistor in parallel. When we use this formula, we find that these three resistors have a total resistance of 6 ohms. Again, this is less than the smallest of the three resistors.

D. Complex DC Circuits

Many of our circuits are more complex than the simple series and parallel circuits we have just discussed, but by carefully analyzing them, we can usually combine the series and parallel components and find equivalent values that allow us to rearrange the entire circuit into a single equivalent resistance. Then, by applying Ohm's law, we can find all of the values we need to know. The arrangement in figure 2-14 is a good exercise in circuit analysis. We want to complete the chart in figure 2-13 with all of the values that are missing.

Our first step is to draw the circuit using the conventional symbols (figure 2-15). And, now, since we know all of the resistance values, our next step is to find the equivalent resistance of each of the parallel combinations. Starting with R_2 and R_3, we find their equivalent resistance to be 12 ohms.

Item	Voltage Volts	Resistance Ohms	Current Amps	Power Watts
R_1		4		
R_2		20		
R_3		30		
R_4		12		
R_5		24		
R_6		40		
R_7		60		
Source	24			

Figure 2-13. Find all of the values that are missing in this chart.

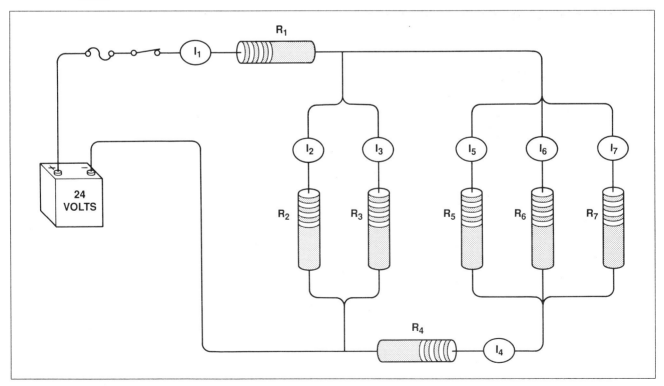

Figure 2-14. A complex circuit having both series and parallel combinations of resistance.

Now, our next step is to find the equivalent resistance of R_5, R_6, and R_7. This also happens to be 12 ohms.

To make the problem easy to follow, re-draw the circuit, using the two equivalent resistances instead of the original combinations. When we do this, it becomes obvious to us that the next step is to combine R_{5-6-7} with R_4. This gives us an equivalent resistance of 24 ohms.

We now have a parallel combination of $R_{4-5-6-7}$ and R_{2-3}. The equivalent of these two combinations is 8 ohms.

All we have to do now is to find the equivalent of the series combination of R_1 and $R_{2-1-4-5-6-7}$, which is 12 ohms. This is equivalent resistance of the entire circuit.

Now that we know the equivalent resistance for the entire circuit, we can work the problem back into its original form and find all of the missing values.

We start by finding the total current.

$$I_T = E/R = 24/12 = 2 \text{ amps}$$

All of the current flows through resistor R_1, and since it has a resistance of 4 ohms, the voltage dropped across it is 8 volts.

When 8 volts is dropped across R_1, it leaves 16 volts across both the combinations R_{2-3} and R_{5-6-7}.

The current through the first combination is 1.33 amps, and that through the second is 0.67 amp.

We can now find the current through resistors R_{5-6-7}.

$$I_5 = E/R = 8/24 = 0.33 \text{ amp}$$

$$I_6 = E/R = 8/40 = 0.2 \text{ amp}$$

$$I_7 = E/R = 8/60 = 0.14 \text{ amp}$$

The total current through these three resistors is the same 0.67 amp that flowed through R_4.

To find the current through R_2 and R_3, we use the formula:

$$I_2 = E/R = 16/20 = 0.80 \text{ amp}$$

$$I_3 = E/R = 16/30 = 0.54 \text{ amp}$$

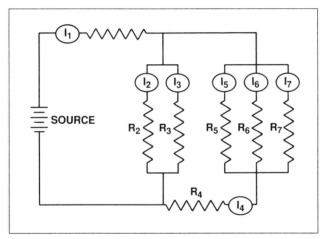

Figure 2-15. The complex circuit of figure 2-14, using conventional symbols.

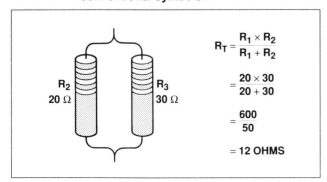

Figure 2-16. Find the equivalent resistance of R_2 and R_3.

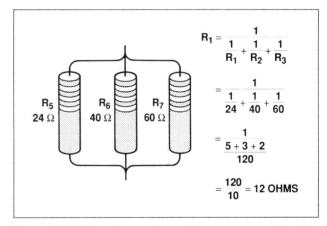

Figure 2-17. Find the equivalent resistance of R_5, R_6, and R_7.

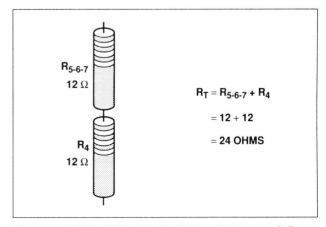

Figure 2-18. Find the equivalent resistance of R_{5-6-7} and R_4.

We can find the power dissipated in each resistor by simply multiplying the current through each of the resistors by the voltage drop across each resistor. This is done in figure 2-26.

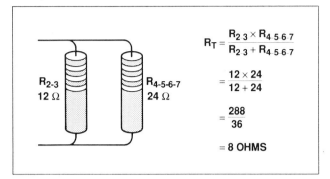

Figure 2-19. Find the equivalent resistance of $R_{2\text{-}3}$ and $R_{4\text{-}5\text{-}6\text{-}7}$.

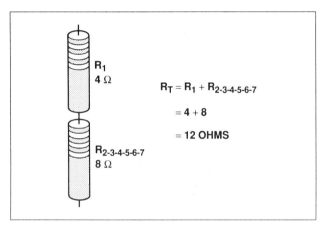

Figure 2-20. Find the total equivalent resistance by adding R_1 to $R_{2\text{-}1\text{-}4\text{-}5\text{-}6\text{-}7}$.

$$I_T = \frac{E}{R} = \frac{24}{12} = 2$$

Figure 2-21. Find the total current flowing in the circuit.

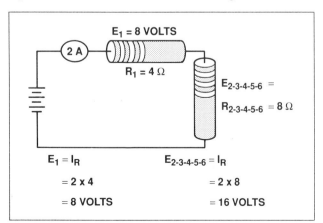

Figure 2-22. Find the voltage dropped across R_1 and across $R_{2\text{-}1\text{-}4\text{-}5\text{-}6}$.

All of the values we found by analyzing the complex circuit are in figure 2-28.

We can check our problem by determining that the following statements are true:

1. The total power is equal to the sum of the power dissipated in each of the resistors.

   48 watts = 48 watts

2. The voltage drops across R_1, R_4, and $R_{5\text{-}6\text{-}7}$ must equal the source voltage.

   24 volts = 24 volts

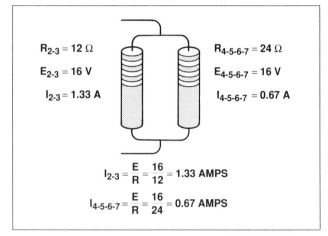

Figure 2-23. Find the current flowing through $R_{2\text{-}3}$ and $R_{4\text{-}5\text{-}6\text{-}7}$. The entire 0.67 amp flows through R_4, and the voltage drop across it is 8 volts. This same 0.67 amp flows through the combination $R_{5\text{-}6\text{-}7}$, and since its equivalent resistance is 12 ohms, there will also be 8 volts dropped across it.

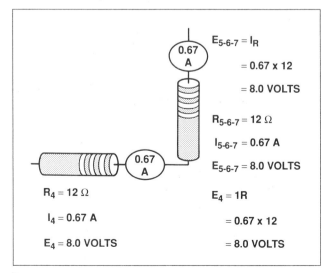

Figure 2-24. Find the voltage drop across R_4 and across $R_{5\text{-}6\text{-}7}$.

3. The voltage drops across R_1, and either R_2 or R_3 must equal the source voltage.

$$24 \text{ volts} = 24 \text{ volts}$$

4. Current through R_4 is the same as the sum of the current through R_2, R_3, R_5, R_6, and R_7.

$$0.67 \text{ amp} = 0.67 \text{ amp}$$

5. The current through R_1 is equal to the sum of the current through R_2, R_3, R_5, R_6, and R_7.

$$2 \text{ amps} = 2 \text{ amps}$$

E. Changing DC to AC

Often, it is necessary to change direct current to alternating current. For example, many aircraft require alternating current to power equipment such as flight instruments, navigation receivers, etc. During an emergency, when normal aircraft power is not available, power is taken from the battery to operate all electrical loads. Since batteries are capable of only storing direct current, a means must be provided to change DC to AC.

The device used to change DC to AC is the inverter. Inverters may be either of two distinct types: the rotary inverter or the static inverter. Rotary inverters are essentially DC motors with an AC generator built in. They are powered by a DC source and have AC as an output.

Static inverters are electronic devices containing a specialized circuit known as an oscillator. An oscillator is capable of changing DC to AC through electronics, which will be discussed in greater detail in a later section of this manual. Oscillators are used in conjunction with amplifiers to produce the correct value of AC from the DC input provided to it. The static inverter has replaced the rotary inverter in most applications, as it is much quieter and more efficient.

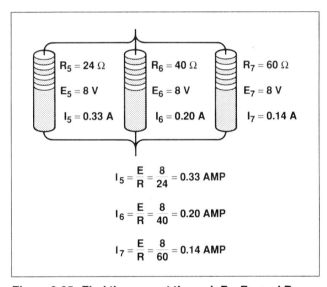

Figure 2-25. Find the current through R_5, R_6, and R_7.

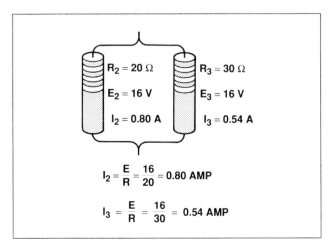

Figure 2-26. Find the current through R_2 and R_3.

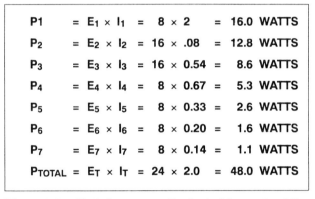

P_1	= $E_1 \times I_1$ =	8 × 2	=	16.0 WATTS
P_2	= $E_2 \times I_2$ =	16 × .08	=	12.8 WATTS
P_3	= $E_3 \times I_3$ =	16 × 0.54	=	8.6 WATTS
P_4	= $E_4 \times I_4$ =	8 × 0.67	=	5.3 WATTS
P_5	= $E_5 \times I_5$ =	8 × 0.33	=	2.6 WATTS
P_6	= $E_6 \times I_6$ =	8 × 0.20	=	1.6 WATTS
P_7	= $E_7 \times I_7$ =	8 × 0.14	=	1.1 WATTS
P_{TOTAL}	= $E_T \times I_T$ =	24 × 2.0	=	48.0 WATTS

Figure 2-27. Find the power dissipated in each of the resistors.

Item	Voltage Volts	Resistance Ohms	Current Amps	Power Watts
R_1	8	4	2.00	16.0
R_2	16	20	0.80	12.8
R_3	16	30	0.54	8.6
R_4	8	12	0.67	5.3
R_5	8	24	0.33	2.6
R_6	8	40	0.20	1.6
R_7	8	60	0.14	1.1
Source	24	12	2.00	48.0

Figure 2-28. Circuit relationships in a complex circuit.

Chapter III

Alternating Current

The advantages offered by alternating current have resulted in it being adopted almost exclusively for commercial power systems. Even on a small scale, AC is used in medium and large aircraft systems for a number of applications.

A. Advantage of AC Over DC

Alternating current, which continually changes its value and periodically reverses its direction of flow, has many advantages over direct current, in which the electrons flow in one direction only. AC is much easier to generate in the large quantities needed for our homes and industries and for our large transport aircraft.

More important, though, is the ease with which we can change the values of current and voltage to get the most effective use of our electrical energy. For example, in our homes and shops, most of our AC electricity has a pressure of about 115 volts, and if we need a kilowatt of power, almost 9 amps of current must flow. The current flowing in a conductor determines the amount of heat generated, and therefore the size of conductor needed. So, if we can get the same amount of power with less current, we can use smaller conductors which will save both money and weight.

Figure 3-1. Transformers are used to change the values of alternating current and voltage.

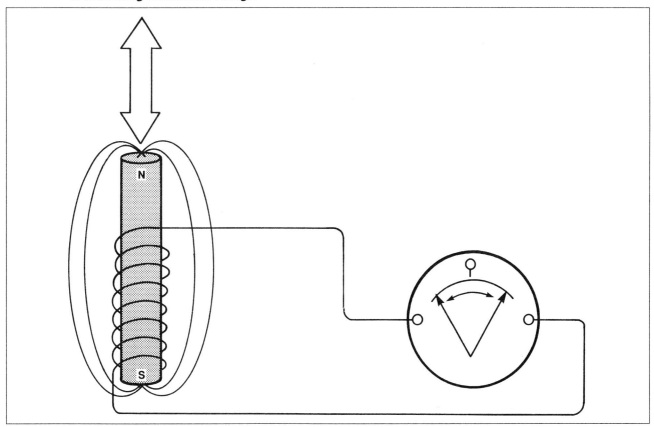

Figure 3-2. Current will flow in a conductor when lines of magnetic flux from a permanent magnet cut across it.

In order to use the smallest conductor possible for cross-country transportation of electrical power, the voltage of the electricity carried in our transmission lines is boosted up to several thousand volts. For example, at 15,000 volts, we will need to transmit only 0.067 amp for one kilowatt of electrical power. Before the electricity is brought into our homes or shops, it is transformed down to a usable value of around 115 volts, so it will be safer and more convenient for us to handle. Between the generation of alternating current and its final use, the voltage and current will be changed many times, and the transformers that do this changing are quite efficient and very little energy is lost.

B. Generation of Alternating Current Electricity

We have seen in our study of direct current electricity that there is a close relationship between magnetism and electricity. Any time electrons flow in a conductor, a magnetic field surrounds the conductor, and the strength of this field is determined by the amount of electron flow. We have also found that when a magnetic field is moved across a conductor, electrons are forced to flow in it, and the amount of this flow is determined by the rate at which the lines of magnetic flux are cut by the conductor. Increasing the number of lines of flux by making the magnet stronger or increasing the speed of movement between the conductor and the magnet will increase the amount of electron flow.

If a conductor, wound in the form of a coil, is attached to an electrical measuring instrument, and a permanent magnet is moved back and forth through the coil, the meter will deflect from side to side. This indicates that the electrons flow in one direction when the magnet is moved into the coil then reverse and flow in the opposite direction when the magnet is withdrawn. This is alternating current (AC).

The AC electricity with which we are most familiar has been generated by a rotary generator in which the conductor in the form of a coil is rotated inside a magnetic field. If we were to watch on an oscilloscope the changing values of the voltage as the coils are rotated, we would see that they start at zero, rise to a peak, and then drop back off to zero. As the coils continue to rotate, the voltage builds up in

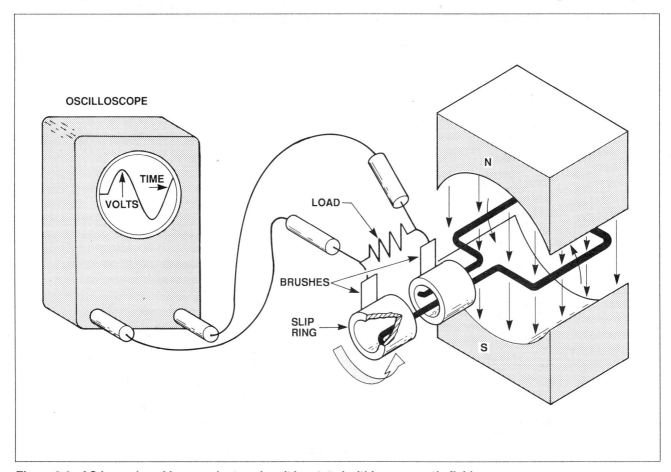

Figure 3-3. AC is produced in a conductor when it is rotated within a magnetic field.

the opposite direction to a peak and then back to zero. There is one complete cycle of voltage changes for each complete revolution of the coil.

The wave form of alternating current, or voltage produced by a rotary generator, is called a sine wave. The voltage or current changes according to the sine of the angle through which the generator has rotated.

The sine of an angle is the relationship between the length of the hypotenuse of a right triangle to the length of the side opposite the angle we are considering. When the coil is in position A of figure 3-4, it does not cut any lines of flux from the magnet as it moves, so no voltage is generated. This is our starting point or the 0° angle.

By the time the coil rotates 45° (point B), it cuts across some of the lines of flux, and the voltage generated at this point is 0.707 times the peak amount. As the coil continues to rotate, it reaches point C where it cuts across a maximum number of lines of flux for each degree it rotates, and here it produces the peak voltage. Further rotation decreases the number of lines of flux cut for each degree of rotation until, at point D, the coil is cutting no flux, and the output is again zero.

Rotation beyond this point brings the opposite side of the coil down through the flux near the south pole of the magnet, and the voltage builds up in the opposite direction. The change in voltage is continuous and smooth, and its instantaneous

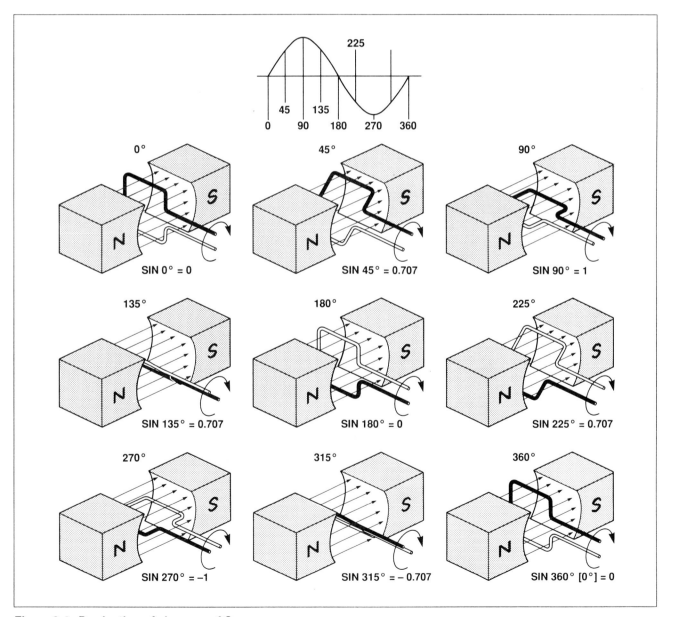

Figure 3-4. Production of sine wave AC.

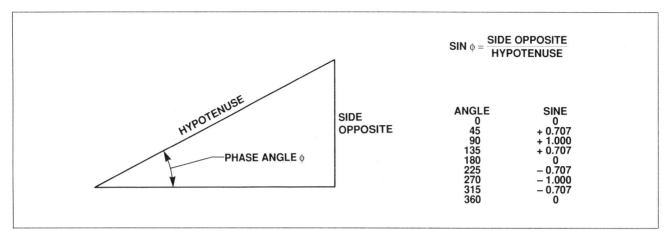

Figure 3-5. The sine of the phase angle is the ratio of the length of the side opposite the angle to the length of the hypotenuse.

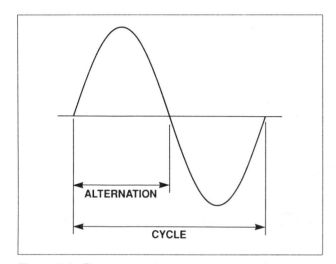

Figure 3-6. Sine wave values.

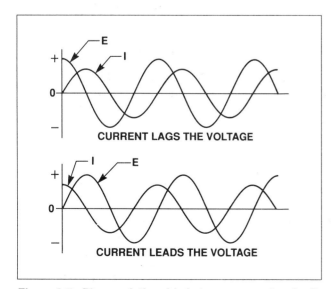

Figure 3-7. Phase relationship between current and voltage in an AC circuit.

value may be found by multiplying its peak value by the sine of the angle through which the generator has rotated since its zero-voltage position.

C. Alternating Current Terms and Values

1. Cycle

A cycle is one complete sequence of voltage or current changes from zero through a positive peak to zero, then through a negative peak, back to zero where it can start over and repeat the sequence.

2. Alternation

An alternation is one-half of an AC cycle in which the voltage or current rises from zero to a peak and back to zero.

3. Period

The time required for one cycle of events to occur is known as the period of the alternating current or voltage.

4. Frequency

The number of complete cycles per second is the frequency of the AC, and it is express in hertz. One hertz is one cycle per second. The frequency of the alternating current produced by a generator is determined by the number of pairs of magnetic poles in the generator and the speed in revolutions per minute of the rotating coils. Frequency may be found by the formula:

$$\text{Frequency (Hz)} = \frac{\text{Poles}}{2} \times \frac{\text{RPM}}{60}$$

The frequency of commercial alternating current in the US is 60 Hz, while in some foreign countries it is 50 Hz. The AC power used in most aircraft is 400 Hz, for reasons we will discover later.

5. Phase

In a DC circuit, a change in the voltage will cause an immediate change in the current, and so the voltage and current are said to be in phase.

In alternating current where the values are constantly changing, certain circuit components cause a phase shift between the voltage and the current. Some components cause the current to change before the voltage changes, and the current is said to lead the voltage, while other components cause the voltage to change before the current, and the current is then said to lag the voltage.

6. Power

a. Apparent Power

You will remember in our study of direct current electricity, we saw that electrical power is the product of the voltage and the current and it is measured in watts. In alternating current, this is not true because the current is not necessarily in phase with the voltage, and the product of the voltage and the current is the apparent power and is expressed in volt-amps rather than in watts.

If the current is in phase with the voltage, we have the condition we see in figure 3-9. The power developed at any instant is the product of the voltage and the current, and as long as the voltage and the current are in phase, the power will be positive; that is, the generator is supplying power to the load. We will remember from our study of mul-

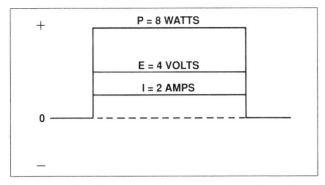

Figure 3-8. In a DC circuit, the current and the voltage are in phase, and the power is the product of the current and the voltage.

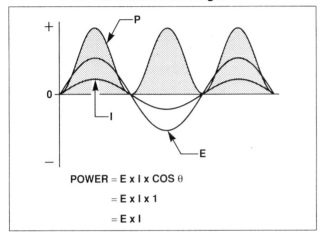

Figure 3-9. In a purely resistive circuit, the current and the voltage are in phase. The power is the product of the current and the voltage.

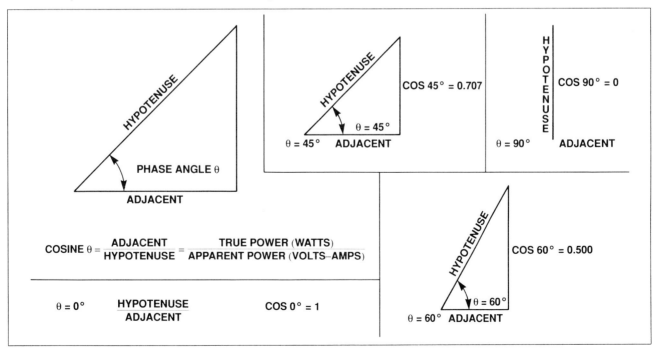

Figure 3-10. The power factor of an AC circuit is the cosine of the phase angle, which is the ratio of the length of the side adjacent the angle to the length of the hypotenuse.

DEG.	SIN.	COS.	TAN.	COT.	
0	.0	1.000	.0	+ ∞	90
1	.0175	.9999	.0175	57.29	89
2	.0349	.9994	.0349	28.64	88
3	.0523	.9986	.0524	19.08	87
4	.0698	.9976	.0699	14.30	86
5	.0872	.9962	.0875	11.43	85
6	.1045	.9945	.1051	9.514	84
7	.1219	.9926	.1228	8.144	83
8	.1392	.9903	.1405	7.115	82
9	.1564	.9877	.1584	6.314	81
10	.1737	.9848	.1763	5.671	80
11	.1908	.9816	.1944	5.145	79
12	.2079	.9782	.2126	4.705	78
13	.2250	.9744	.2309	4.331	77
14	.2419	.9703	.2493	4.011	76
15	.2588	.9659	.2680	3.732	75
16	.2756	.9613	.2868	3.487	74
17	.2924	.9563	.3057	3.271	73
18	.3090	.9511	.3249	3.078	72
19	.3256	.9455	.3443	2.904	71
20	.3420	.9397	.3640	2.747	70
21	.3584	.9336	.3839	2.605	69
22	.3746	.9272	.4040	2.475	68
23	.3907	.9205	.4245	2.356	67
24	.4067	.9136	.4452	2.246	66
25	.4226	.9063	.4663	2.145	65
26	.4384	.8988	.4877	2.050	64
27	.4540	.8910	.5095	1.963	63
28	.4695	.8830	.5317	1.881	62
29	.4848	.8746	.5543	1.804	61
30	.5000	.8660	.5774	1.732	60
31	.5150	.8572	.6009	1.664	59
32	.5299	.8481	.6249	1.600	58
33	.5446	.8387	.6494	1.540	57
34	.5592	.8290	.6745	1.483	56
35	.5736	.8192	.7002	1.428	55
36	.5878	.8090	.7265	1.376	54
37	.6018	.7986	.7536	1.327	53
38	.6157	.7880	.7813	1.280	52
39	.6293	.7772	.8098	1.235	51
40	.6428	.7660	.8391	1.192	50
41	.6561	.7547	.8693	1.150	49
42	.6691	.7431	.9004	1.111	48
43	.6820	.7314	.9325	1.072	47
44	.6947	.7193	.9657	1.036	46
45	.7071	.7071	1.0000	1.000	45
	COS.	SIN.	COT.	TAN.	DEG.

Figure 3-11. Trigonometric function table.

tiplication of signed numbers that when we multiply signed numbers having like signs, whether they are positive or negative, the product will be positive.

b. True Power

If the current and the voltage are not continually in phase—that is, if the current either leads or lags the voltage—there will be at least part of the cycle in which the voltage is positive while the current is negative. The product of unlike signed numbers is always negative and the power that is produced during this portion of the cycle is negative power; that is, the load will be forcing power back into the generator.

True power is expressed in watts and is the product of the voltage and only that portion of the current that is in phase with the voltage.

c. Power Factor

"Power factor" is a term used to indicate the amount of the current that is in phase with the voltage, and it may be found as the ratio between the true

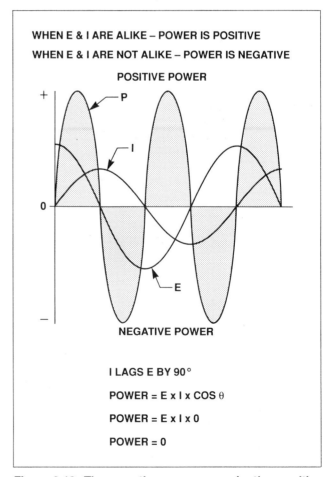

Figure 3-12. The negative power equals the positive power when the current and the voltage area 90° out of phase.

power and the apparent power. If the power factor is 0.5, only 50% of the current is in phase with the voltage. If all of the current is in phase with the voltage, as it is in a circuit having no opposition other than resistance, the power factor will be 1.0, meaning that all of the current is in phase with the voltage.

If the amount of phase shift between the voltage and the current is known, we can find the power factor, as it is the cosine of the phase angle. In figure 3-10, we see that in a right triangle, the ratio of the length of the side adjacent the phase angle to the length of the hypotenuse is the cosine of the angle. When the current and the voltage are in phase, or the phase angle is zero, the side adjacent the 0° angle and the hypotenuse are the same length, and so the ratio is 1.00. The real power is equal to the apparent power.

For a 45° phase angle, the length of the side adjacent (the true power) will be only 0.707 as long as the hypotenuse (the apparent power). When the current and the voltage are 60° out of phase, the true power is only 0.50 as much as the apparent power. And if the current and voltage are 90° out of phase, the adjacent side will have a length of zero, and the power factor will also be zero. There can be no real power produced in the circuit, even though voltage is present and current is flowing.

The true power in a circuit is found by multiplying the product of the voltage and the current by the power factor.

Power (watts) = E × I × power factor

As we will see very soon, circuits that have inductive or capacitive reactance as the chief opposition to the current flow have very low power factors and while a great deal of current flows, very little true power is produced.

7. Sine Wave Values

a. Peak Value

This is the maximum value of voltage or current in either the positive or the negative direction.

b. Peak-to-Peak Value

This is the maximum difference between the positive and the negative peak values of alternating voltage or current, and is equal to two times the peak value.

c. Average Value

If all of the instantaneous values of current or voltage in one alternation (or half-cycle of sine wave AC) are averaged together, they will have a value of 0.637 times the peak value. The average value has very little practical use and we seldom find it in practical computations.

d. Effective Value

If we square all of the instantaneous values of voltage or current in one alternation and find the average of these squared values, then take the square root of this average, we will have the root mean square, or rms, value of the sine wave. This value is 0.707 of the peak value and is equal to the amount of direct current required to produce the same amount of heat the peak value of AC will produce. We can see that 70.7 volts of DC will force enough current through a circuit to produce the same amount of heat as sine wave AC having a peak value of 100 volts. For this reason, the rms value of AC is called its effective value.

An oscilloscope shows the full waveform of AC. When it is used to measure voltage, it gives us peak-to-peak voltage, but this same voltage measured on an AC voltmeter will indicate the rms voltage. Two hundred peak-to-peak volts is the same as 70.7 volts rms.

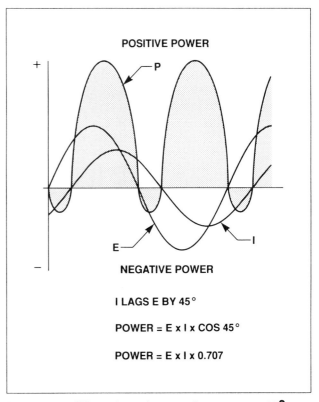

Figure 3-13. When the voltage and current are 45° out of phase, the true power is 0.707 times the apparent power.

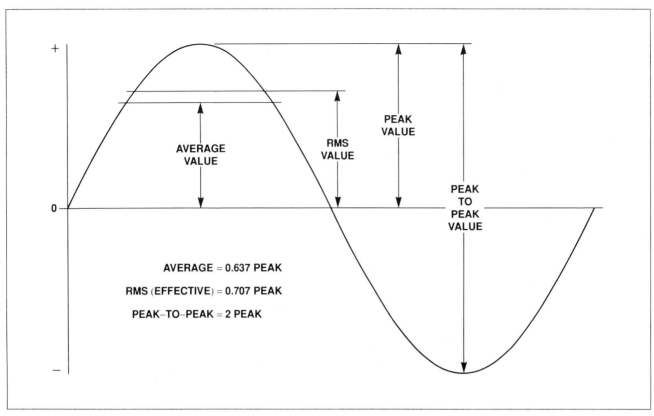

Figure 3-14. Values of sine wave AC.

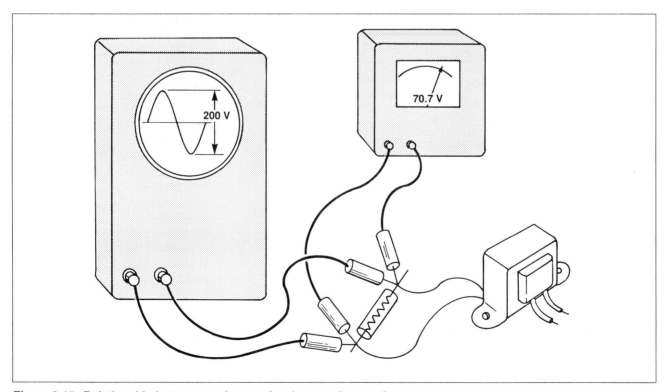

Figure 3-15. Relationship between peak-to-peak voltage and rms voltage.

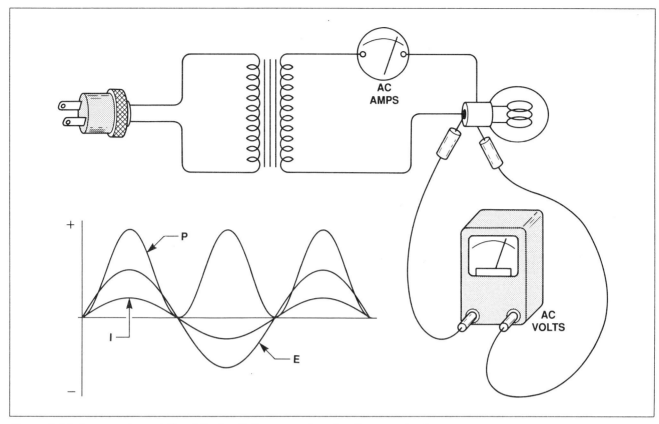

Figure 3-16. In a purely resistive AC circuit, the current and voltage are in phase.

D. Alternating Current in a Circuit Containing Only Resistance

Circuit components such as light bulbs, heaters, and composition resistors do not normally cause a phase shift in an AC circuit. Circuits containing only these types of devices are called resistive circuits, and the current and voltage will be in phase; that is, they will both pass through zero at the same time, going in the same direction.

In a purely resistive circuit, the power factor is one, and so the apparent power and the true power are the same. The power in watts is the product of the effective value of the voltage times the effective value of the current.

E. Inductance

Anytime current flows in a conductor, a magnetic field surrounds it and the strength of this field is determined by the amount of current flow. The direction of the lines of flux around the conductor may be found by the left-hand rule for generators, which states that if the conductor is held in the left hand so that the thumb points in the direction of electron flow (from the negative to the positive terminal of the source), the fingers will encircle the conductor in the direction of the lines of flux.

As the amount of current flow changes, the magnetic field expands or contracts. As it does, the flux cuts across the conductor and induces a voltage into it. According to Lenz's law, the voltage that is induced into the conductor is of such a polarity that it opposes the change that caused it. For example, as the voltage begins to rise and the current increases, the expanding lines of flux cut across the conductor and induce a voltage into it that opposes, or slows down, the rise. When the current flow in the conductor is steady, lines of flux surround it, and since there is no change in the amount of current, these lines do not cut across the conductor, and so there is no voltage induced into it. When the current decreases, the lines of flux cut across the conductor as they collapse, and they induce into it a voltage that opposes the decrease.

When a conductor carries alternating current, both the amount and the direction of the current continually change, and so an opposing voltage is constantly induced into the conductor. This induced voltage acts as an opposition to the flow of current and we will discuss it in detail under its proper name, inductive reactance.

1. Factors Affecting Inductance

Inductance opposes a change in current by the generation of a back voltage, and all conductors have the characteristic of inductance, since they all generate back voltage any time the current flowing in them changes. The amount of inductance is increased by anything that concentrates the lines of flux, or causes more of the flux to cut across the conductor. If the conductor is formed into a coil, the lines of flux surrounding any one of the turns cut not only across the conductor itself, but also across each of its turns, and so it generates a much greater induced current to oppose the source current.

If a soft iron core is inserted into the coil, it will further concentrate the lines of flux and cause a still higher induced current, which allows less source current to flow.

Inductance is measured in henries, and one henry will generate one volt of induced voltage when the current changes at the rate of one ampere per second.

The inductance of a coil is determined by the number of turns in the coil; by the spacing between the turns; by the number of layers of winding; and by the wire size. The ratio of the diameter of the coil to its length and the type of material used in the core also affect the amount of inductance in the coil. Since all of these factors are variable, there is no simple formula we can use to find the inductance of a coil.

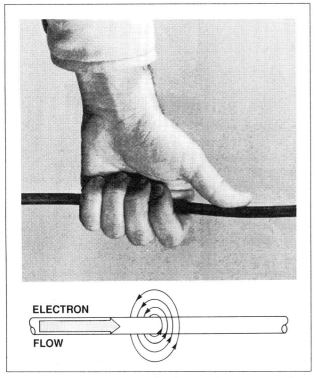

Figure 3-17.
(A) When a conductor is grasped by the left hand with the thumb pointing in the direction of electron flow, from negative to positive, the fingers will encircle the conductor in the same direction as the lines of magnetic flux.
(B) The relationship between the direction of electron flow and the lines of magnetic flux.

2. Series and Parallel Inductors

It is sometimes necessary to use more than one inductor, or coil, in a circuit. If they are connected in series in such a way that the changing magnetic field of one does not affect the others, the total inductance will be equal to the sum of their individual inductances.

$$L_T = L_1 + L_2 + L_3 \ldots$$

When inductors are connected in parallel, the total inductance will be less than that of any of the individual inductors. The formulas used for finding the

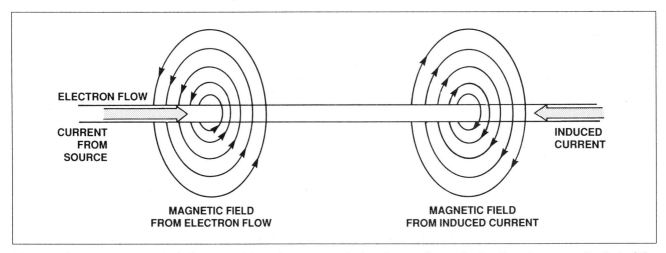

Figure 3-18. The magnetic field from the induced current encircles the conductor in the direction opposite that of the magnetic field produced by the source current.

total inductance of parallel connected inductors are the same as those used for finding the total resistance of parallel connected resistors. If the inductors are all the same, the total inductance may be found by dividing the inductance, in henries, of one inductor by the number of inductors:

$$L_T = \frac{L}{n}$$

If there are only two inductors, the total inductance may be found by dividing the product of the two inductors by their sum:

$$L_T = \frac{L_1 \times L_2}{L_1 + L_2}$$

When more than two inductors having different amounts of inductance are connected in parallel, the formula used is the same as that for finding

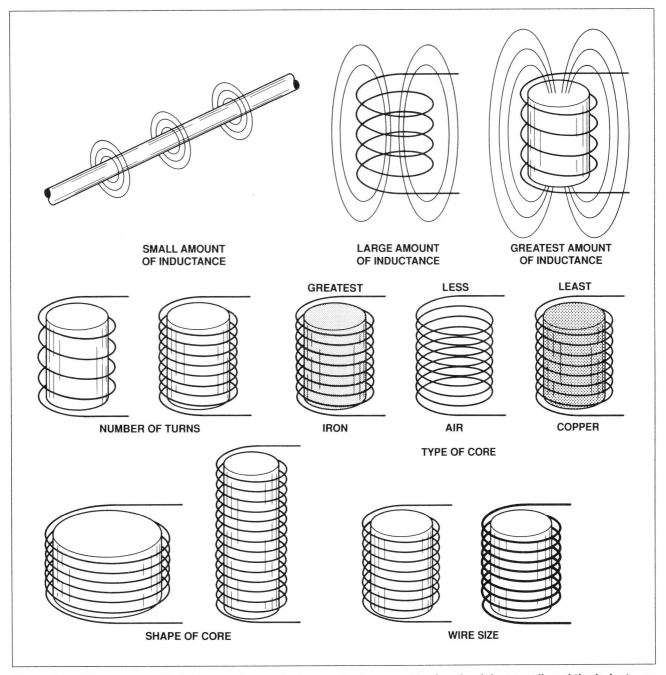

Figure 3-19. *The amount of inductance of a conductor may be increased by forming it into a coil, and the inductance of the coil is determined by the number of turns, the shape of the coil, the wire size, and the type of core.*

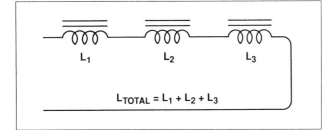

Figure 3-20. Inductances in series.

the total resistance of several unlike resistors connected in parallel; the reciprocal of the sum of the reciprocal of the inductances:

$$L_T = \cfrac{1}{\cfrac{1}{L_1} + \cfrac{1}{L_2} + \cfrac{1}{L_3} \cdots}$$

3. Time Constant of Inductors

If a source of DC is placed across an inductor and controlled by a switch (S_1) (figure 3-22(A)), we see that the current does not rise instantly. At the instant the switch is closed, the current finds a minimum of opposition and starts to flow at its greatest rate. But this great change in the rate of current flow from zero to maximum induces a maximum back voltage that opposes the current and causes the current rise to decrease. The time required for the current to rise 63.2% of its peak value is known as the time constant of the circuit and is determined by the value of inductance and resistance in the circuit.

$$\text{Time Constant} = \frac{L \text{ (henrys)}}{R \text{ (ohms)}}$$

For example, a circuit containing 2.0 henries of inductance and 50 ohms of resistance would have a time constant of 0.04 second, which would normally be spoken of as 40 milliseconds.

The current will rise to 63.2% of its peak value in a period of time equal to one time constant (0.04 second). In two time constants (0.08 second) it will rise to 86.5% of its peak value. In three time constants (0.12 seconds), it will rise to 95%, in four (0.16 seconds) to 98%, and it will take five time constants or 0.2 second for the current to rise to the peak value.

When the switch is opened, the current starts to drop off immediately, but the changing current induces a voltage that opposes the change and so drop-off is slowed down. In a period of time equal to one time constant, it will drop off to 36.8% of its peak value. In two time constants, it will drop to 13.5%; in three, to

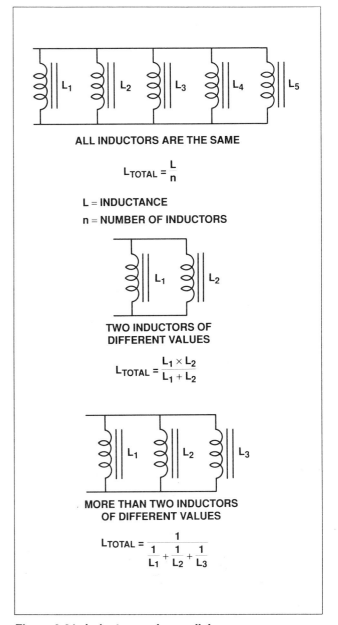

Figure 3-21. Inductances in parallel.

5%; in four time constants, to 2%; and it will take five time constants for it to drop off completely to zero.

4. Inductive Reactance

The opposition to the flow of alternating current caused by the generation of a back voltage as the magnetic flux cuts across the conductor is called inductive reactance, and its symbol is X_L. It is measured in ohms, and it varies directly as the frequency of the AC that produces it. The formula for inductive reactance is:

$$X_L = 2\pi f L$$

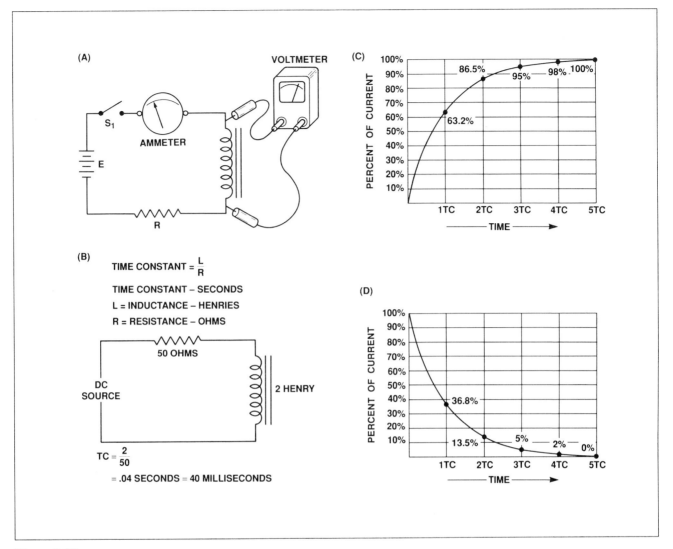

Figure 3-22.
(A) *Circuit for measuring the time constant of an R-L circuit.*
(B) *Computing the time constant of an R-L circuit.*
(C) *Time constant curve of an R-L circuit as the voltage builds up.*
(D) *Time constant curve of an R-L circuit as the voltage drops off.*

Two pi is the constant 6.28, the frequency is measured in hertz, and the inductance, in henries. From this formula, we see that for zero frequency, as we have in DC, there is no inductive reactance. This is because there is no changing magnetic field. As the frequency increases, the amount of change in the magnetic field increases, and the inductive reactance increases. An infinite frequency would produce an infinite amount of inductive reactance.

Remember that the opposition is caused by the generation of a counter, or back, electromotive force. Unlike the opposition caused by resistance, no heat is generated in the circuit by this opposition, and therefore no power is dissipated.

If the circuit is purely inductive—that is, there is no resistance present—the current will not begin to flow until the voltage has risen to its peak value. The amount of current is determined by the rate of change of the voltage, and we can see in figure 3-24 that when the voltage is at its peak, there is no change, and so the current is zero. As the voltage begins to drop off, the current rises until at the point the voltage passes through zero its rate of change is the greatest and the current is the maximum. From this, we can see that in a purely inductive circuit, the change in the current lags the change in voltage by 90°, and with a 90° phase angle, the power factor is zero.

49

There can be no true power in a purely inductive circuit, because the negative power equals the positive power. The load returns as much power as it receives from the source.

If an inductor of the proper size is placed in series with a light bulb, the inductive reactance will cause most of the source voltage to be dropped across the inductor, and the bulb will burn very dimly, if at all.

Most aircraft use 400-Hz AC, because the inductive reactance at this frequency is high enough to allow smaller transformers and motors to be used. If a transformer designed for 400-Hz AC is used in a 60-Hz circuit, the lower inductive reactance caused by the lower frequency will allow enough current to flow that the transformer will be burned out. But if a 60-Hz transformer is used in a 400-Hz circuit, there will be so much inductive reactance that the efficiency of the transformer will be too low for practical use.

5. Mutual Inductance

When alternating current flows in a conductor, the changing lines of flux radiate out and cut across any other conductor that is nearby, and anytime they cut across a conductor, they generate a voltage in it even though there is no electrical connection between the two. This voltage is said to be generated by mutual inductance, and is the basis for transformer action that is so important to us in our

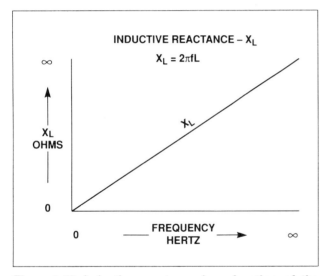

Figure 3-23. Inductive reactance is a function of the amount of inductance and the frequency of the circuit. X_L increases with an increase in both frequency and inductance.

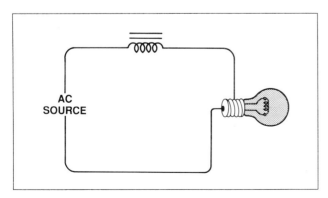

Figure 3-24. An inductor in series with the load in an AC circuit can prevent current reaching the load.

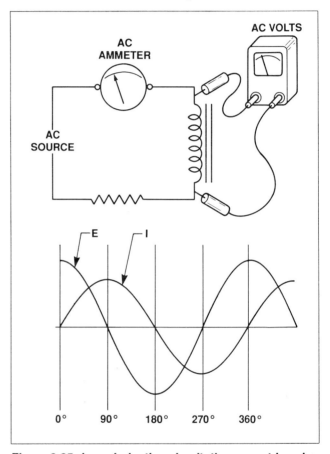

Figure 3-25. In an inductive circuit, the current lags behind the voltage.

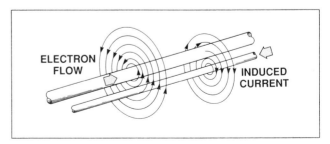

Figure 3-26. Mutual induction causes a voltage to be induced into a conductor not electrically connected to the conductor through which the source current flows.

use of alternating current, as it allows us to change the values of AC voltage and current in our circuits.

Consider in figure 3-27, two coils of wire (a primary and a secondary) wound around a common core, but not connected electrically.

When an alternating current flows in the primary, a voltage will be induced into the secondary, and current will flow in it.

Since we can consider these windings to be purely inductive, we see that the current in the primary winding lags the source voltage by 90°. The voltage induced into the secondary winding will be greatest when the change in current is the greatest, and it is therefore 90° out of phase with the current in the primary. So when the two phase shifts are added, the voltage in the secondary winding will be 180° out of phase with the voltage in the primary.

The amount of voltage generated in the secondary winding of a transformer is equal to the voltage in the primary, times the turns ratio between the primary and the secondary windings. For example, if there are 100 turns in the primary winding and 1,000 turns in the secondary, we have a turns ratio of 1:10, and if there are 115 volts across the primary, there will be 1,150 volts across the secondary. A transformer does not generate any power, so the product of the voltage and the current in the secondary must be the same as that in the primary. Because of this, there must be a flow of one ampere in the primary winding to produce a flow of 100 milliamps in the secondary.

A transformer may have its primary winding connected directly across the AC power line, and as long as there is an open circuit in the secondary winding, the back voltage produced in the coil will block the source voltage enough that there will be almost no current flowing through the primary winding. But, as we see in figure 3-29, when the push button in the secondary

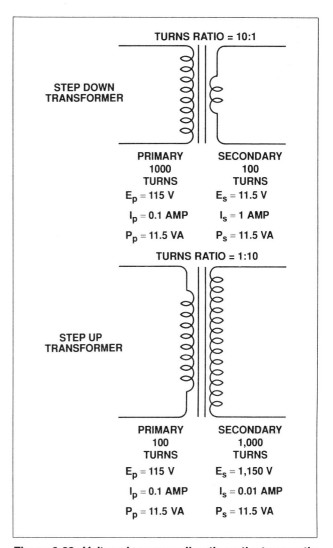

Figure 3-28. Voltage increases directly as the turns ratio between the primary and secondary windings of a transformer, while the current decreases in proportion to the turns ratio.

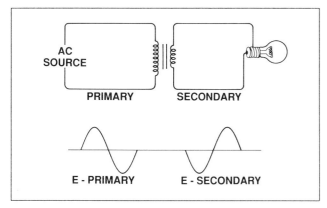

Figure 3-27. The voltage induced into the secondary of a transformer by mutual induction is 180° out of phase with the source voltage.

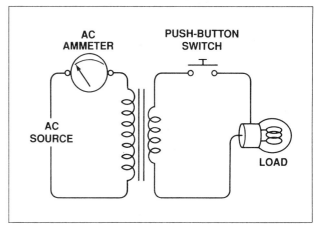

Figure 3-29. Primary current flows only when the secondary circuit is completed.

circuit is pressed to complete the circuit for the light, secondary current will flow and its flux will oppose that which created the back voltage so source current will flow in the primary. Only when the switch in the secondary circuit is pressed, will there be any current flow indicated on the AC ammeter in the primary circuit.

6. Types of Practical Inductors

a. Chokes

It is often necessary to install an inductor, or a choke, in a circuit to impede the flow of alternating current of a particular frequency, while not affecting the flow of AC below that frequency.

When AC in the power frequencies of 50 or 60 Hz is rectified (changed into DC) the output is in the form of pulsating DC, and if an inductor is placed in series with the load, the changing current will induce a back voltage that will tend to smooth out the pulsations (or ripples). Chokes of this type have laminated iron cores and often have an inductance of more than one henry.

b. Transformers

We can get almost any voltage of alternating current by using a transformer. The primary winding is designed to accept the voltage and frequency of the power source, and there may be one or more secondary windings needed for the particular application.

1) Step-Up or Step-Down Transformers

When there are more turns in the secondary than in the primary, the transformer is called a step-up transformer, but if the secondary has fewer turns, it is called a step-down transformer. Step-down transformers are often used to get the high current necessary for operating some motors.

2) Autotransformers

An autotransformer is a form of variable transformer such as we have in figure 3-31. There is only one winding, the primary, which is connected across

Figure 3-30. Forms of commonly used inductors.

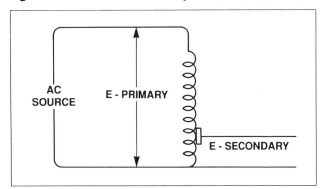

Figure 3-31. An autotransformer uses all of the winding for the primary and a portion of it for the secondary.

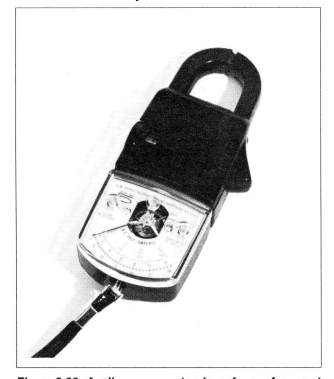

Figure 3-32. A clip-on ammeter is a form of current transformer.

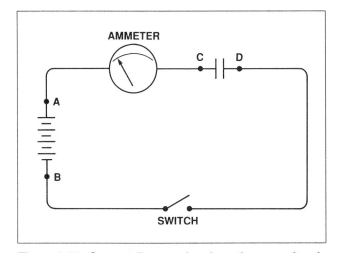

Figure 3-33. Current flows only when the capacitor is charging or discharging.

the source. One lead of the secondary is common with the primary, and the other secondary lead is connected to a brush which is movable and makes contact with a bare spot in the primary winding. The amount of secondary voltage is determined by the position of the brush. The greater the number of turns between the two secondary leads, the greater will be the secondary voltage.

3) Current Transformers

It is possible to measure the amount of AC flowing in a line by the use of a clamp-on type ammeter which uses a current transformer. The probe consists of a coil that can be opened up, and it is clamped over one of the wires carrying the current to be measured. The flow of alternating current induces a much smaller current into the coil that is proportional to the amount of load current flow. The meter that reads this induced current is calibrated in terms of the load current.

F. Capacitance

1. Energy Stored in Electrostatic Fields

Electrical energy, as we have seen, may be stored in the magnetic field which surrounds a conductor through which electrons are moving. It may also be stored in electrostatic fields caused by an accumulation of electrical charges that are not moving, but are static. The electromagnetic field strength is determined by the amount of current flowing in the conductor, but the strength of the electrostatic field is determined by the amount of pressure (voltage) on the static charges.

A capacitor, sometimes called a condenser, is a device that stores electrical energy in the electrostatic fields that exist between two conductors that are separated by an insulator, or a dielectric. Let's consider the circuit in figure 3-33 where we have two flat metal plates arranged so they face each other but are separated by an insulator. One of the plates is attached to the positive terminal of the power source and the other to the negative terminal. When the switch is closed, electrons will be drawn from the plate attached to the positive terminal and will flow to the plate attached to the negative. There can be no flow across the insulator, but the plates will become charged. If the voltmeter reading were taken across points C and D, it would be found to be exactly the same as that taken across points A and B. Current flow would be indicated by the ammeter during the time the plates are being charged, but when they become fully charged, no more current will flow.

In the circuit of figure 3-34, we have a power source connected to a capacitor through a resistor, which limits the amount of current that initially flows into the capacitor, but does not prevent the voltage across the capacitor rising to that of the source.

When the voltage across the capacitor rises to the voltage of the source, no more current will flow. Now the switch can be moved to position B to complete the circuit across the neon light, and the capacitor will immediately discharge through the light and cause it to flash. If the switch is placed in its neutral position, when the capacitor is charged, the capacitor will remain charged until the electrons eventually leak off through the dielectric.

2. Factors Affecting Capacitance

A capacitor is a device to store an electrical charge, and its capacity is measured in farads, with one farad the amount of capacity that will hold one coulomb of electricity (6.28×10^{18} electrons) under a pressure of one volt.

$$C = \frac{Q}{E}$$

C = capacity in farads
Q = charge in coulombs
E = voltage in volts

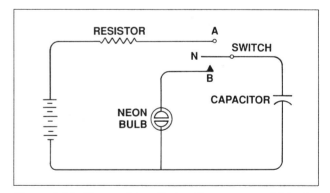

Figure 3-34. The resistor limits the rate at which the capacitor will charge when the switch is in position A. When it is in position B, the capacitor will discharge through the neon bulb, causing it to flash.

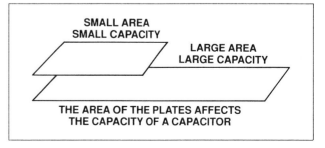

Figure 3-35. The greater the area of the plates, the greater will be the capacity of the capacitor.

The farad is such a large unit that it is seldom used in practical circuits. Instead, most capacitors are measured in microfarads which are millionths of a farad, or in picofarads which are millionths of millionths of a farad. Picofarads have formerly been called micro-micro farads and may still be referred to in this way in some texts. The Greek letter mu (μ) is used to represent the prefix micro.

1 microfarad (μF) = 1×10^{-6} Farad

1 picofarad (pF or μμf) = 1×10^{-12} Farad

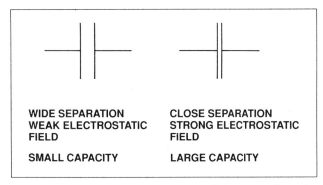

Figure 3-36. *The closer the plates of the capacitor, the greater will be the capacity.*

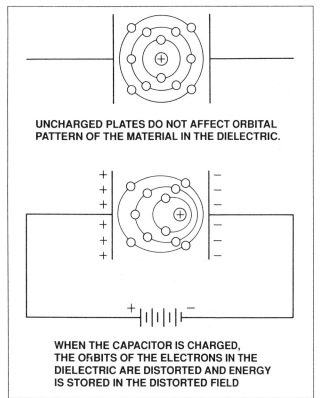

Figure 3-37. *The dielectric constant of the insulation between the plates of a capacitor determines its capacity. Energy is stored in the stress of the dielectric.*

The capacity is affected by three variables: the area of the plates, the separation between the plates, and the dielectric constant of the material between the plates.

It only stands to reason that the larger the plates, the more electrons can be stored. One very common type of capacitor has plates made of two long strips of metal foil separated by waxed paper and rolled into a tight cylinder. This construction provides the maximum plate area for its small physical size.

The distance the plates are apart determines the strength of the electrostatic field between them, and this affects the capacity. If the plates are widely separated, the field will be weak, and it will not pull very many electrons onto the negative plate. If, on the other hand, they are very close together, the attraction caused by the unlike charges will produce a very strong field in the dielectric, and many electrons will be held on the negative plate. The strength of the electrostatic field increases inversely as the separation between the plates. When the space between the plates is cut in half, the strength of the electrostatic field will double.

One problem with getting the plates too close together is the possibility of the field becoming so

KIND OF DIELECTRIC	APPROXIMATE* K VALUE
Air (at atmospheric pressure)	1.0
Bakelite	5.0
Beeswax	3.0
Cambric (varnished)	4.0
Fibre (red)	5.0
Glass (window or flint)	8.0
Gutta Percha	4.0
Mica	6.0
Paraffin (solid)	2.5
Paraffin Coated Paper	3.5
Porcelain	6.0
Pyrex	4.5
Quartz	5.0
Rubber	3.0
Slate	7.0
Wood (very dry)	5.0

* These values are approximate, since true values depend upon quality or grade of material used, as well as moisture content, temperature, and frequency characteristics of each.

Figure 3-38. *Table of dielectric constants.*

strong that electrons will be pulled across the insulator and actually flow to the positive plate. When this happens, in most capacitors, the dielectric will be damaged and a conductive path set up, shorting the capacitor and making it useless. For this reason, all capacitors are rated with regard to their working voltage, which is a DC measurement indicating the strength of the dielectric.

The third factor which affects the capacity of a capacitor is the material of the dielectric. More specifically, it is the dielectric constant of the insulating material. Energy is stored not only in the stress across the dielectric, but by the distortion of the orbits of the electrons in the material of which the dielectric is made. Air is used as the reference for measuring the dielectric constant and it is given a value of one. If glass, which has a dielectric constant of eight, is substituted for air as an insulator, the capacity will increase eight times because of the energy stored in the distortion of the electron orbits within the glass.

3. Series and Parallel Connection of Capacitors

a. Capacitors in Series

It is often necessary to connect multiple capacitors into a circuit and, when they are installed in series, the effect is comparable to increasing the separation between the capacitor plates, thus the total capacity will be less than that of any of the series capacitors. The formulas for finding the total series capacitance are the same as those used for finding the total resistance of resistors connected in parallel. When multiple capacitors of equal value are connected in series, the total capacitance is found by dividing the value of one capacitor by the number of capacitors in series:

$$C_T = \frac{C}{n}$$

If there are two unlike capacitors, the total may be found by dividing the product of the two by their sum.

$$C_T = \frac{C_1 \times C_2}{C_1 + C_2}$$

When there are more than two unlike capacitors, their total capacity is the reciprocal of the sum of the reciprocals of the individual capacitors.

$$C_T = \frac{1}{\frac{1}{C_2} + \frac{1}{C_2} + \frac{1}{C_3} \ldots}$$

b. Capacitors in Parallel

When capacitors are connected in parallel, the effect is the same as adding the areas of their plates. So, the total capacity is the sum of that of the individual capacitors.

$$C_T = C_1 + C_2 + C_3 \ldots$$

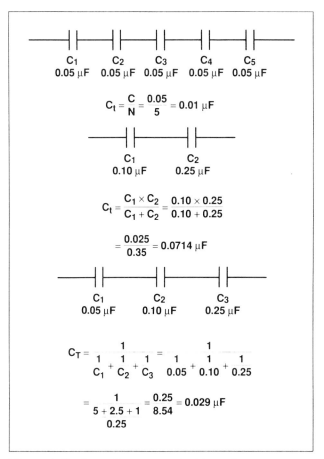

Figure 3-39. *Capacitors connected in series.*

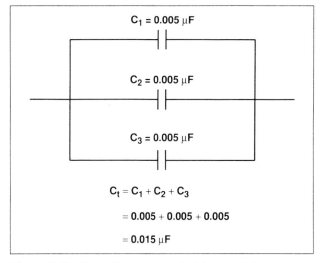

Figure 3-40. *Capacitors connected in parallel.*

4. Time Constant of Capacitors

We saw in our study of inductors that when a voltage source was placed across the inductor, the inductance slowed the rise of current in the circuit and, as a result, the changes in current lagged behind the changes in the voltage.

In a capacitive circuit, such as we have in figure 3-41, the results are the opposite. Changes in the current lead the changes in the voltage. When the switch is closed, current immediately begins to flow as electrons are attracted away from the positive plate and forced onto the negative plate. The voltage across the plate does not jump up immediately, but it rises as the plates become charged.

The time constant of a capacitive circuit is the time, in seconds, required for the voltage across the capacitor to reach 63.2% of the source voltage and is determined by both the capacitance and the resistance of the circuit.

$$TC = R \times C$$

TC = Time constant in seconds
R = Resistance in ohms
C = Capacitance in farads

Timing circuits are often made using a capacitor and a resistor in series. In figure 3-44, we have a 100,000-ohm resistor in series with a 100-μf capacitor, and both of these across a 100-volt power source. When the switch is closed, current will begin to flow, but it will be limited in its rate by the opposition caused by the resistor. In 10 seconds, the voltage will rise to 63.2 volts; in 20 seconds, it will be up to 86.5 volts; in 30 seconds, to 95 volts; in 40 seconds to 98 volts; and in 50 seconds the voltage will be equal to the source of 100 volts, and no more current will flow.

When the switch in figure 3-44 is placed in the discharge position, the voltage will begin to drop as the capacitor starts to discharge. It will follow the curve of figure 3-43, and in 10 seconds it will be down to 36.8 volts, and then its drop will slow down. In 20 seconds, it will be down to 13.5 volts; in 30 seconds, to 5 volts; in 40 seconds to 2 volts; and it will not be down to zero volts until the current has flowed through the resistor for 50 seconds or five time constants.

5. Capacitive Reactance

a. Factors Affecting Capacitive Reactance

Capacitive Reactance (X_C) is the opposition to the flow of AC caused by the capacitance in a circuit. For DC (frequency = zero) the opposition is infinite because no current can flow *through* a capacitor.

As the frequency increases from zero, current flows into and out of the capacitor, and, as you have noticed in the time constant curve, the greatest rate of flow occurs when the current flows into a discharged capacitor or out of a fully charged capacitor.

The amount of current flow in a capacitive circuit increases with either an increase in the size of the capacitor or with an increase in the frequency of the AC. And since capacitive reactance is the *opposition* to the flow of current, the larger the capacitor, or the higher the frequency, the lower will be the capacitive reactance. The formula for capacitive reactance shows us this relationship:

$$X_C = \frac{1}{2\pi f C}$$

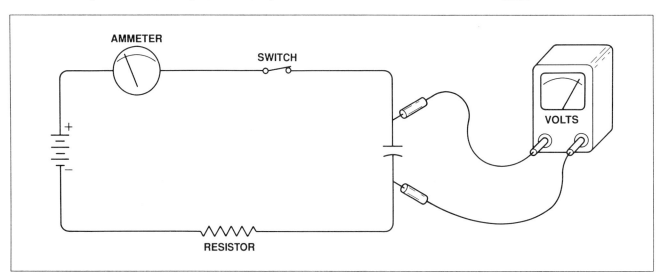

Figure 3-41. Circuit for measuring the time constant of a R-C circuit.

b. Phase Shift in a Capacitive Circuit

From figure 3-46, we can see that in a purely capacitive circuit, the changes in current *lead* the changes in voltage by 90°. The current must flow into the capacitor before the voltage across it can rise. When the capacitor is fully charged, the voltage will be maximum and the current will be zero. Now, as the capacitor begins to discharge, the current begins to flow out of it and the voltage across it starts to drop. The current flow will be the greatest as the voltage passes through zero, and as the voltage begins to build up in the opposite direction, the current flow starts to drop off until the capacitor is fully charged and the current will again be zero.

c. Power in a Capacitive Circuit

Since the current leads the voltage by 90 in a purely capacitive circuit, the power factor will be zero and there will be no real power produced, even though there is apparent power or volt-amps in the circuit. In figure 3-47, we see that twice in each cycle, the current and voltage are in phase with each other, and the source supplies power to the load. But also twice in each cycle, the current and voltage are out of phase with each other, and during this time the load will return power to the source. Since the power supplied and that returned are the same, no real power is produced in a purely capacitive circuit.

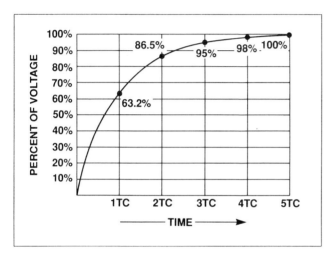

Figure 3-42. Time constant curve of a capacitor during charge.

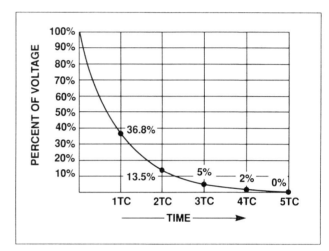

Figure 3-44. Measurement of the time constant of a R-C charge.

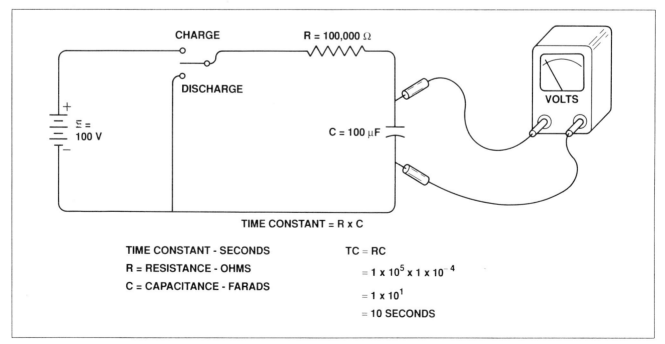

Figure 3-43. Time constant curve of a capacitor being discharged.

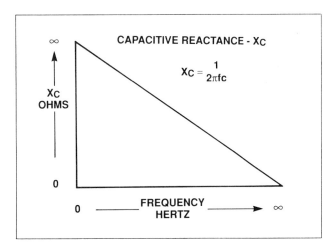

Figure 3-45. Capacitive reactance is a function of frequency and the capacity of the circuit. Capacitive reactance decreases as either the frequency or the capacity increases.

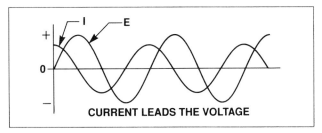

Figure 3-46. The current leads the voltage in a capacitive circuit.

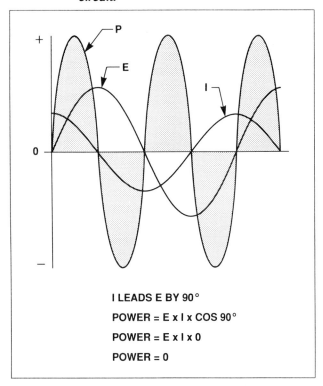

Figure 3-47. In a purely capacitive circuit, the negative power equals the positive power, and no true power is produced.

6. Types of Practical Capacitors

a. Fixed Capacitors

Capacitors may be divided basically into two types: fixed and variable. The fixed capacitors may be further divided into electrolytic and non-electrolytic types.

1) Nonelectrolytic Capacitors

When relatively low values of capacitance are needed, nonelectrolytic capacitors are used, and one of the most common types of nonelectrolytic capacitors is the paper capacitor. The plates are made of two strips of very thin metal foil and are

Figure 3-48. A magneto capacitor is a paper capacitor sealed in a metal container.

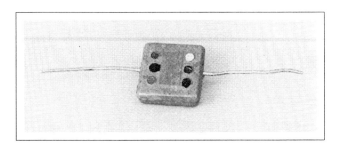

Figure 3-49. Block-type mica capacitor.

Figure 3-50. High voltage, oil-filled paper capacitor.

separated by a strip of waxed paper. The sandwich is coiled into a tight roll, and wire leads are attached to the plates. The assembly may be encapsulated in plastic, or, in the case of an aircraft magneto capacitor, sealed in a metal can.

Capacitors requiring a smaller capacity, but a higher working voltage, are made using stacks of thin metal foil sandwiched between thin sheets of mica. This stack-up is encapsulated in plastic and is familiar to us as the rectangular block-like capacitor.

For high voltage applications, paper capacitors may be enclosed in a metal container filled with an insulating oil. If a voltage surge should break through the insulator, the oil will flow in and restore its insulating characteristics. These are called self-healing capacitors.

High voltage, low capacitance capacitors may be made of either a disc or a tube of ceramic material plated with silver on each side to form the plates. The leads are attached to the silver, and the entire unit is covered with a protective insulation.

2) Electrolytic Capacitors

Electrolytic capacitors are used when it is necessary to have a large amount of capacity with a relatively low working voltage. These capacitors are polarized, meaning that they can be used only in DC circuits because they act as capacitors only when they are properly connected into the circuit. If they are installed with the wrong polarity, current will flow through them, causing them to overheat and actually explode.

The reason electrolytic capacitors have such a high capacity for their small physical size is their extremely thin dielectric. The positive plate is made of aluminum foil and has electrolytically deposited on its surface an extremely thin oxide film which serves as the dielectric. A liquid or paste electrolyte, which will conduct current, saturates a gage which is held in contact with both the positive plate and the negative can in which the capacitor is sealed. The moist electrolyte forms the second plate of the capacitor. This combination of metal plate, oxide film, and conductive liquid or paste makes a capacitor which has the maximum capacity for its size.

b. Variable Capacitors

The capacity of a capacitor is determined by three things: the area of the plates, the separation of the plates, and the type of dielectric. If we are able to change any of these three factors, we can change the capacity.

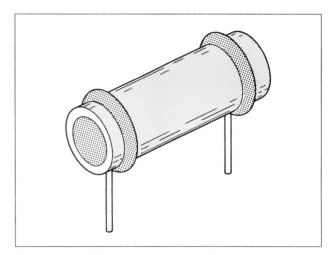

Figure 3-51. Tubular ceramic capacitor.

Figure 3-52. Electrolytic capacitor.

Figure 3-53. Variable plate area tuning capacitor using air as the dielectric.

1) Changing the Area of the Plates

Most radios have variable capacitors in which we change the area of the plates. The plates are made of thin sheets of aluminum and are meshed together with one group of plates fixed and serving as the stators. The other plates (the rotors) are mounted on a shaft and mesh with the fixed plates, but do not touch them. The air between the plates serves as the dielectric. When the plates are fully meshed, the capacity is maximum. As the shaft is rotated, the area of the plates that are meshed decreases and the capacity becomes less.

2) Changing the Spacing Between the Plates

While variable area capacitors are used for the main tuning capacitor for a radio, small trimmer and padder capacitors are used for fine tuning. These small capacitors are made up of a stack of metal foil plates separated by thin sheets of mica. A screw adjustment allows the plates to be squeezed tightly together to increase the capacitance or relaxed to decrease the capacitance.

3) Changing the Dielectric Constant

The most popular fuel quantity measuring system used in our modern aircraft is the capacitance system in which the measuring units are capacitors in the form of probes in the fuel tanks. Each of these probes is made up of two concentric tubes which fit across the tank from top to bottom (figure 3-55). Each tube acts as one of the plates of the capacitor, and both the area and the separation between the plates are fixed. When the tank is empty, the dielectric is air which has a dielectric constant of one. When the tank is full, the dielectric is the fuel which has a dielectric constant of approximately two. The indicator for this system measures the capacity of the probes and converts it into terms of gallons or pounds of fuel in the tanks.

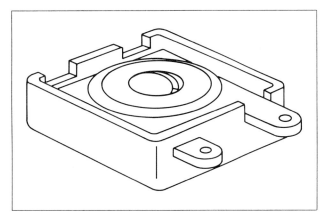

Figure 3-54. Variable capacitor in which it is possible to vary the separation of the plates.

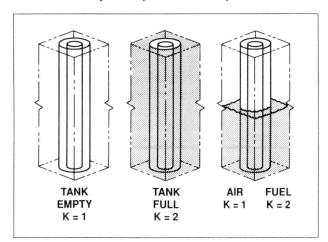

Figure 3-55. The capacitance-type fuel gaging system uses capacitors as probes in the fuel tank. When the tank is empty, the dielectric is air. When the tank is full, fuel is the dielectric.

G. Series Alternating Current Circuits

1. Impedance

The flow of current in an AC circuit is opposed by three things: by resistance which converts the electrical energy into heat, and by inductive and capacitive reactance which oppose the flow but do not produce heat. All circuits have some resistance, some inductance, and some capacitance, but the total opposition is not just the arithmetic sum of the three individual oppositions. Since both the inductance and the capacitance cause phase shifts between the voltage and the current, the total opposition will be the vector sum of the three oppositions.

Let's begin our study of Ohm's law in an AC circuit by examining a 400-Hz circuit in which the capacitance is negligible, but one which has a total resistance of 100 ohms and an inductance of 20 millihenries.

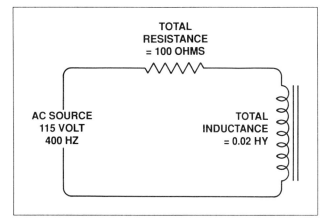

Figure 3-56. A series R-L circuit.

A vector is a quantity that has both magnitude and direction, and for solving AC problems, we will plot the opposition caused by the resistance of the circuit on a horizontal line extending to the right of the zero point on our graph (figure 3-57). Draw this to scale with a length representing 100 ohms of resistance. This is the total resistance of the circuit and includes not only any resistors in the circuit, but the resistance of the inductor as well.

We must find the inductive reactance of the 20-millihenry inductor, and to do this, we use the formula:

$$X_L = 2\pi fL$$

When we multiply the constant 2π (6.28) by the frequency of 400 Hz, and the inductance of two millihenries, we get an inductive reactance of 50.24 ohms (figure 3-58). We plot the vector for the inductive reactance vertically upward from the zero point.

We now have the vector quantities for resistance and inductive reactance at right angles to each other, and their vector sum is the circuit impedance (Z) which is the square root of the sum of their squares.

$$Z = \sqrt{X^2 + R^2}$$

By using the formula shown in figure 3-59, we find that the total circuit opposition (impedance) is 112 ohms.

2. Power Factor

Not all of the current and voltage in this circuit are in phase. Power is produced only by that portion of the voltage and current that are in phase, so we must find this percentage which is the power factor. We saw earlier in this study that the power factor was the ratio of the true power to the apparent power. It is also the ratio of the resistance of the circuit to the impedance (figure 3-60). In this problem, the power factor is 0.89 or 89% of the current is in phase with the voltage.

3. Phase Angle

The phase angle is the angle whose cosine is the power factor, and by looking at the trigonometric function chart of figure 3-11, we see that 0.89 is the cosine of 27°. In this circuit, the current lags behind the voltage by 27°.

4. Current

Rather than using $E = I \times R$ as an expression of Ohm's law in an AC circuit, this law must be stated as $E = I \times Z$ because we must consider the total opposition, the impedance, rather than just the resistance.

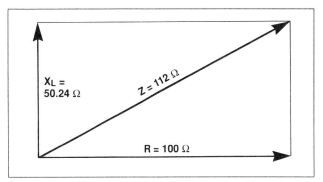

Figure 3-57. The total opposition (impedance) of an R-L circuit is the vector sum of the inductive reactance and the resistance.

$$X_L = 2\pi fL$$
$$= 6.28 \times 4 \times 10^2 \times 2 \times 10^{-2}$$
$$= 50.24 \text{ OHMS}$$

Figure 3-58. The formula for finding the inductive reactance.

$$Z = \sqrt{X_L^2 + R^2}$$
$$= \sqrt{50.25^2 + 100^2}$$
$$= \sqrt{2,525 + 10,000}$$
$$= \sqrt{12,525}$$
$$= 112 \text{ OHMS}$$

Figure 3-59. The formula for finding impedance.

tance. In our circuit, we found the impedance to be 112 ohms, and we know the source voltage is 115 volts, so we can find the total current through the series circuit to be $I = E/Z = 115/112 = 1.03$ amps.

5. True Power

The true power developed in this circuit is found by the formula $P = E \times I \times$ power factor, and is equal to $115 \times 1.03 \times 0.89 = 105.4$ watts.

6. Apparent Power

This is the product of the source voltage times the total current and is 115×1.03, or 118.5 voltamps.

7. Voltage

The voltage dropped across the inductor is not in phase with the current through the inductor, but the voltage and current are in phase through the resistor. Because of this, the sum of the voltage across the two components will be greater than the source voltage.

$$E_R = I_T \times R = 1.03 \times 100 = 103.0 \text{ volts}$$

$$E_L = I_T \times X_L = 1.03 \times 50.24 = 51.7 \text{ volts}$$

8. Circuits Having Capacitance and Resistance

Anytime we analyze a series circuit, we should make a chart similar to the one in figure 3-61 to record all of the information about the circuit. Now, let's look at a circuit in which there is negligible inductance, but there is a measurable capacitance and resistance, and we will complete the chart of figure 3-63.

In the circuit of figure 3-62, we have a 115-volt, 400-Hz power source and a total circuit resistance of 100 ohms and a capacitance of 5mf. Using the formula:

$$X_C = \frac{1}{2\pi f C}$$

we find the capacitive reactance is 79.6 ohms. The impedance is found by the formula:

$$Z = \sqrt{X^2 + R^2}$$

to be 128 ohms.

The total circuit current is found by the formula:

$$I = \frac{E}{Z}$$

to be 0.9 amp, and since this is a series circuit, the same current will flow through the resistor and the capacitor.

The voltage across the resistor is found by:

$$E = I \times R$$

to be 90 volts, and since this is considered to be a pure resistance, the voltage and the current are in phase.

$$\text{POWER FACTOR} = \frac{\text{TRUE POWER}}{\text{APPARENT POWER}} = \frac{R}{Z}$$
$$= \frac{100}{112}$$
$$= 0.893$$

Figure 3-60. The formula for finding the power factor.

E	= 115 VOLTS	I_T	= 1.03 AMPS
F	= 400 Hz	I_R	= 1.03 AMPS
R	= 100 OHMS	I_L	= 1.03 AMPS
L	= 0.02 HENRY	E_R	= 103 VOLTS
X_L	= 50.24 OHMS	E_L	= 51.7 VOLTS
Z	= 112 OHMS		

P_{TRUE} = 105.4 WATTS
$P_{APPARENT}$ = 118.5 VOLT-AMPS
POWER FACTOR = 0.89 [89%]
PHASE ANGLE = 27° LAGGING

Figure 3-61. Values for a series R-L circuit.

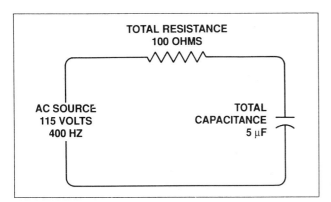

Figure 3-62. A series R-C circuit.

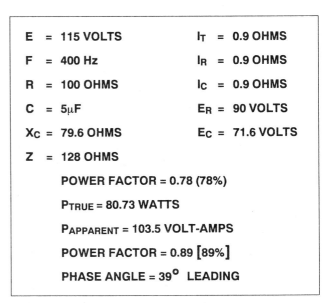

E	= 115 VOLTS	I_T	= 0.9 OHMS
F	= 400 Hz	I_R	= 0.9 OHMS
R	= 100 OHMS	I_C	= 0.9 OHMS
C	= 5μF	E_R	= 90 VOLTS
X_C	= 79.6 OHMS	E_C	= 71.6 VOLTS
Z	= 128 OHMS		

POWER FACTOR = 0.78 (78%)
P_{TRUE} = 80.73 WATTS
$P_{APPARENT}$ = 103.5 VOLT-AMPS
POWER FACTOR = 0.89 [89%]
PHASE ANGLE = 39° LEADING

Figure 3-63. Values for a series R-C circuit.

The voltage dropped across the capacitor, found by the formula $E = I \times X_C$, is 71.6 volts. And since the current changes must occur before there can be any voltage change, the current leads the voltage. The power factor is found by dividing the circuit resistance by the Impedance and is 0.78 or 78%. By referring to the table of trigonometric functions of figure 3-11 on page 42, we see that the angle whose cosine is nearest to 0.78 is 39°, and in this circuit, the current is leading the voltage.

The apparent power is the product of the source voltage and the total current and is 103.5 volt-amps, while the true power, which is the product of the apparent power and the power factor, is 80.73 watts.

9. Circuits Having Inductance, Capacitance, and Resistance

If all three of the variables are measurable in an AC circuit, we can analyze it in the same way we have just done. The only difference is that we must find the total reactance and use it to find the impedance.

$$Z = \sqrt{X_C^2 + R^2}$$
$$= \sqrt{79.6^2 + 100^2}$$
$$= \sqrt{6{,}336 + 10{,}000}$$
$$= \sqrt{16{,}336}$$
$$= 128 \text{ OHMS}$$

$$I = \frac{E}{Z} = \frac{115}{128} = 0.9 \text{ AMPS}$$

$$X_C = \frac{1}{2\pi fC}$$
$$= \frac{1}{6.28 \times 4 \times 10^2 \times 5 \times 10^{-6}}$$
$$= \frac{1}{6.28 \times 4 \times 5 \times 10^{-4}}$$
$$= \frac{10{,}000}{125.6}$$
$$= 79.6 \; \Omega$$

$$E_R = IR = 0.9 \times 100 = 90 \text{ VOLTS}$$
$$E_{X_C} = IX_C = 0.9 \times 79.6 = 71.6 \text{ VOLTS}$$
$$PF = \frac{R}{Z} = \frac{100}{128} = 0.78 = 78\%$$

PHASE ANGLE =
ANGLE WHOSE COSINE IS 0.78 = 39°
CURRENT IS LEADING

$$P_{APPARENT} = E \times I$$
$$= 115 \times 0.9 = 103.5 \text{ VOLT-AMPS}$$

$$P_{TRUE} = E \times I$$
$$PF = 115 \times 0.9 \times 0.76$$
$$= 80.73 \text{ WATTS}$$

Figure 3-64. Computation of the values for the series R-C circuit.

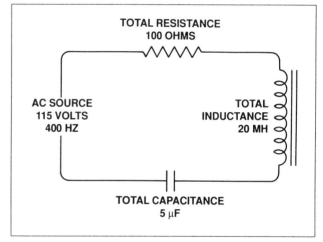

Figure 3-65. A series R-L-C circuit.

E	= 115 VOLTS	I_R	= 1.1 AMP
F	= 400 Hz	I_L	= 1.1 AMP
R	= 100 OHMS	I_C	= 1.1 AMP
L	= 0.02 HENRY	E_R	= 110 VOLTS
C	= 5μf	E_C	= 87.6 VOLTS
X_L	= 50.2 OHMS	E_L	= 55.2 VOLTS
X_C	= 79.6 OHMS	P_F	= 0.96 (96%)
X	= 29.4 OHMS-CAPACITIVE		
Z	= 104 OHMS		
I_T	= 1.1 AMP		

$P_{APPARENT}$ = 126.5 VOLT-AMPS
P_{TRUE} = 121.4 WATTS
PHASE ANGLE = 16° LEADING

Figure 3-66. Values for the series R-L-C circuit.

In a capacitive circuit, the current leads the voltage by 90°. In an inductive circuit, it lags by 90°. The current in the two components is 180° out of phase, and as far as the circuit is concerned, the currents cancel each other, and so do the reactances. The total reactance will then be the difference between the capacitive and the inductive reactance. If we consider a circuit having the same components we have just been working with, but having all three of them in the circuit at the same time, we can find all of the values on the chart of figure 3-67.

The total reactance will be the difference between the inductive and the capacitive reactance and is 29.4 ohms. Since the capacitive reactance is the greater, this total reactance is capacitive.

The impedance is found by using the total reactance, and it is found to be 104 ohms.

The total current is 1.1 amps, and since this is a series circuit, the current through all of the components is the same. The voltage dropped across each of the components is found by multiplying the total current by the resistance and the inductive and the capacitive reactance.

The power factor, phase angle, and the true and apparent powers are all found in the same way they were found in the previous problems.

H. Parallel AC Circuits

Almost all of the AC circuits in our shops and homes have the components connected in parallel rather than in series, and parallel AC circuits are handled in much the same way as parallel DC circuits, with the exception that we must take into consideration the phase shifts that occur between the current flowing in each of the three components.

In the circuit of figure 3-68, we have the same three components we have used in our series circuit, only this time they are connected in parallel.

The circuit has a 115-volt, 400-Hz power source and consists of a 100-ohm resistor, a 20-millihenry inductor, and a 5 µf capacitor, all in parallel.

We can follow the work in figure 3-70 as we analyze the circuit, starting with the resistor. The voltage across the resistor is the same as the source voltage (115 volts). The current through the resistor is found by the formula:

$$I = \frac{E}{R}$$

to be 1.15 amps.

The reactance of the inductor for the 400-Hz voltage is 50.24 ohms, and since the inductor has the full source voltage applied across it, the current through it found by the formula:

$$I = \frac{E}{X_L}$$

is 2.29 amps.

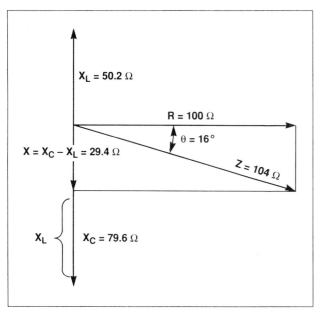

Figure 3-67. Graphic determination of impedance in a series R-L-C circuit.

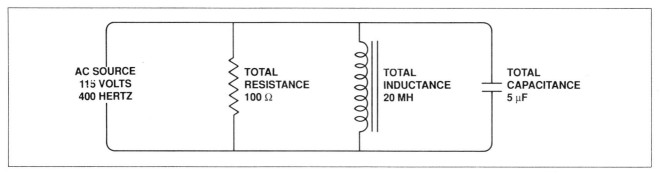

Figure 3-68. A parallel R-L-C circuit.

The reactance of the capacitor is found to be 79.6 ohms, and the current through it is 1.44 amps.

Before we can find the circuit impedance, we must find the total current which is the vector sum of the current flowing through the three components. The current through the inductor is 180° out of phase with that flowing through the capacitor, so the total reactive current is the difference between the two, which is 0.85 amps, and since the inductive reactance is greater than the capacitive reactance, the total reactance is inductive.

The total current is found as the square root of the square of the reactive current plus the square of the resistive current. This is 1.43 amps.

The impedance of the circuit is now found by dividing the total voltage by the total current, and this is 80.42 ohms.

The power factor is the ratio of the current through the resistor to the total circuit current, and it is 0.80. Eighty percent of the voltage and current are in phase with each other.

The phase angle is found by referring to the table of trigonometric functions (figure 3-11). The angle whose cosine is nearest 0.80 is 37°, and since the circuit is more inductive than capacitive, the current lags behind the voltage.

The apparent power is the product of the total voltage and the total current, and is 164.45 volt-amps. And the true power, which is the product of the apparent power and the power factor, is 131.56 watts.

E	= 115 VOLTS	E_L	= 115 VOLTS
f	= 400 Hz	E_C	= 115 VOLTS
R	= 100 OHMS	I_R	= 1.15 AMPS
L	= 0.02 HENRY	I_L	= 2.29 AMPS
C	= 5μF	I_C	= 1.44 AMPS
X_L	= 50.2 OHMS	I_X	= 0.85 AMPS
X_C	= 79.6 OHMS	I_T	= 1.43 AMPS
E_R	= 115 VOLTS	Z	= 80.42 OHMS
P_F	= 0.80 (80%)		

PHASE ANGLE = 37° LAGGING
$P_{APPARENT}$ = 164.45 VOLT-AMPS
P_{TRUE} = 131.56 WATTS

Figure 3-69. Values for a parallel R-L-C circuit.

I. Resonance in an AC Circuit

Inductive reactance in a coil is zero when the frequency is zero, and it increases smoothly until at an infinite frequency, it is infinity; that is, the back voltage generated in the inductor is equal to the source voltage, and no current can flow.

The reactance in a capacitor changes in exactly the opposite way as that in an inductor. At zero frequen-

$$X_L = 2\pi fL = 6.28 \times 4 \times 10^2 \times 2 \times 10^{-2}$$
$$= 50.2 \text{ OHMS}$$

$$X_C = \frac{1}{2\pi fC}$$
$$= \frac{1}{6.28 \times 4 \times 10^2 \times 5 \times 10^{-6}}$$
$$= \frac{10,000}{125.6} = 79.6 \text{ OHMS}$$

$$I_T = \sqrt{I_R^2 + (I_L - I_C)^2} = \sqrt{1.15^2 + 0.85^2}$$
$$= \sqrt{1.32 + 0.72} = \sqrt{2.04}$$
$$= 1.43 \text{ AMP}$$

$$I_R = \frac{E_T}{R} = \frac{115}{100} = 1.15 \text{ AMPS}$$

$$I_L = \frac{E_T}{X_L} = \frac{115}{50.2} = 2.29 \text{ AMPS}$$

$$I_C = \frac{E_T}{X_C} = \frac{115}{79.6} = 1.44 \text{ AMPS}$$

$$I_X = I_{XL} - I_{XC} = 2.29 - 1.44 = 0.85 \text{ AMPS}$$

$$Z = \frac{E_T}{I_T} = \frac{115}{1.43} = 80.42 \text{ OHMS}$$

$$PF = \frac{I_R}{I_T} = \frac{1.15}{1.43} = 0.80 = 80\%$$

PHASE ANGLE =
ANGLE WHOSE COSINE IS 0.80 = 37°
THE CURRENT LAGS THE VOLTAGE

$$P_{APPARENT} = E_T \times I_T$$
$$= 115 \times 1.43 = 164.45 \text{ VOLT-AMPS}$$

$$P_{TRUE} = E_T \times I_T \times PF$$
$$= 115 \times 1.43 \times 0.80$$
$$= 131.56 \text{ WATTS}$$

Figure 3-70. Computation of the values for a parallel R-L-C circuit.

cy, no current can flow through the capacitor, so the reactance is infinite. But as the frequency increases, the capacitive reactance decreases, until at an infinite frequency, there is no capacitive reactance. Figure 3-71 shows the way the two types of reactance vary with the frequency.

The lines representing the two reactances cross at the resonant frequency of the circuit. Here the inductive and the capacitive reactances are equal.

It makes no difference whether the capacitor and inductor are connected in series or in parallel, the resonant frequency is the same, and it may be found by the formula:

$$F_R = \frac{1}{2\pi \sqrt{LC}}$$

The resonant frequency in hertz is found by dividing 1 by the constant 2π times the square root of the product of the inductance in henries and the capacitance in farads.

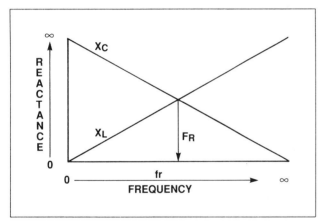

Figure 3-71. The resonant frequency of a circuit is that frequency at which the inductive and capacitive reactances are equal.

1. Series Resonant Circuit

At the resonant frequency of a series R-L-C circuit, the current flowing in the coil and the capacitor are equal, but they are 180° out of phase with each other. The inductive and capacitive reactances are also exactly the same, but, because of the phase difference, they cancel each other, and the total reactance is zero. The total opposition the circuit offers to the flow of AC is that of the resistance, and so the circuit impedance is minimum at resonance and is equal to the circuit resistance.

The voltage drop across the entire circuit is very low, or is zero, and the current flow is maximum. But the voltage across either of the reactances is quite high and can even be higher than the source voltage. Remember, the voltage across the inductor and the voltage across the capacitor are out of phase with each other so the sum of the individual voltages does not equal the source voltage, as it does in a DC circuit.

A series resonant circuit acts as a pure resistance, and the source voltage and current are in phase, so the power factor of the circuit is one.

2. Parallel Resonant Circuit

In a parallel R-L-C circuit at its resonant frequency, there is a large amount of current flowing between the capacitor and the inductor, storing energy first in an electrostatic field in the capacitor and then in an electromagnetic field around the inductor. If there were no resistance in the circuit, once the exchange of energy between the two types of fields had started, the circulating current, as this current is called, would continue to flow back and forth indefinitely. But in practice, all circuits have some resistance which causes this current to die down unless extra energy is added from the source.

At the resonant frequency, the circulating current in the inductor and capacitor is high, but there is almost no current supplied from the source, so

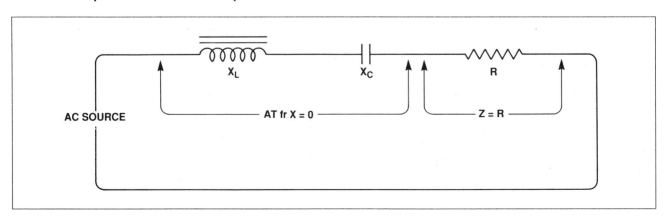

Figure 3-72. In a series R-L-C circuit at its resonant frequency, the inductive and capacitive reactances cancel each other.

the source sees the parallel circuit as having a high impedance. The reactances cancel each other, and so the opposition is purely resistive, and the power factor of the circuit is one.

J. Three-Phase Alternating Current

When it is necessary to get the maximum amount of power from an alternating current circuit, three-phase generators, or alternators, as they are more properly called, are used. These alternators have three sets of output windings excited by a single rotating field, and the voltage in each winding is 120° out of phase with that in the other windings. There are several advantages of using three-phase AC, and one of the main advantages lies in the fact that when the AC is rectified, or changed into direct current, there are three times as many pulses of rectified current as there are in single-phase AC, and these pulses overlap so that the current

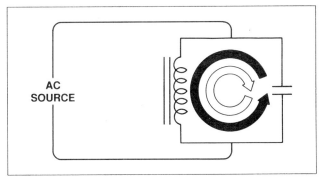

Figure 3-73. In a parallel L-C circuit at its resonant frequency, the circulating current is maximum, but the source current is minimum.

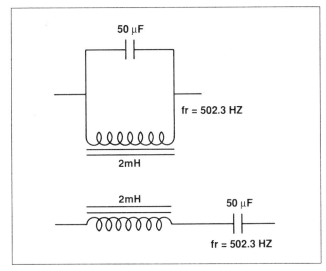

Figure 3-74. The resonant frequency is the same for a series or a parallel L-C circuit.

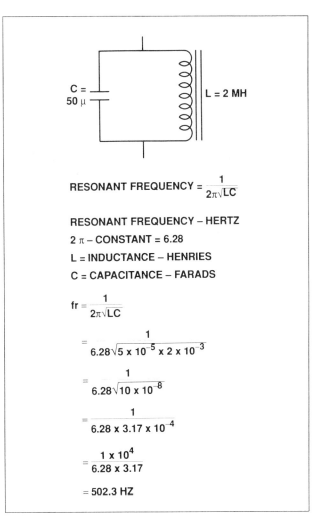

Figure 3-75. Computation of the resonant frequency of an L-C circuit.

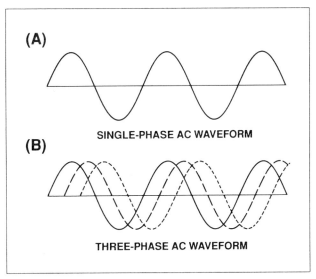

Figure 3-76.
(A) Single-phase sine wave AC waveform.
(B) Three-phase sine wave AC waveform.

never drops to an instantaneous value of zero. The higher frequency of these pulses also makes it easier to filter, or smooth out the ripples.

1. Y-Connected Alternators

There are two ways of connecting the three output windings of an alternator. We will first discuss the Y-connected windings in which one end of each of the three windings is connected together at a common, or neutral point. The other ends of the three windings are brought out of the alternator as the output leads, each of which is across two of the windings in series. Since the voltage across each winding is 120 electrical degrees out of phase with that in the other windings, the output voltage will never be twice that of one of the phase windings. It will instead be 1.73 times that of a single-phase winding. Many three-phase alternators produce 120 volts across each phase, and the voltage between any two of the output leads is 208 volts.

Since the windings are in series between the output leads, the output current is the same as the phase current.

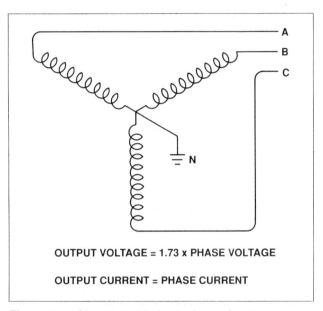

Figure 3-77. Y-connected, three-phase alternator.

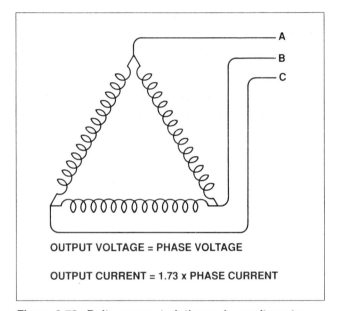

Figure 3-78. Delta-connected, three-phase alternator.

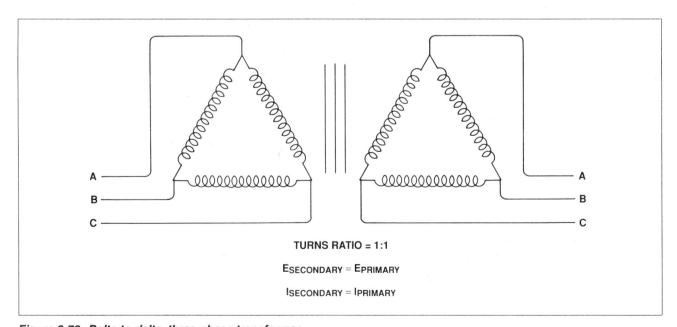

Figure 3-79. Delta-to-delta, three-phase transformer.

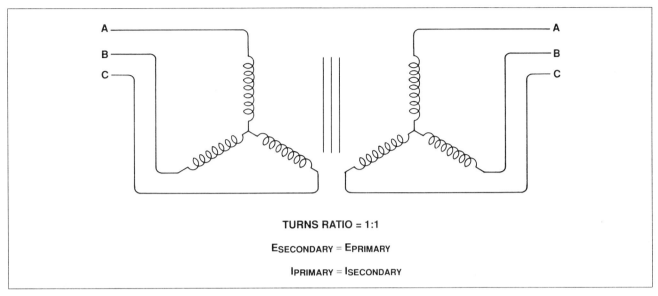

Figure 3-80. Y-to-Y, three-phase transformer.

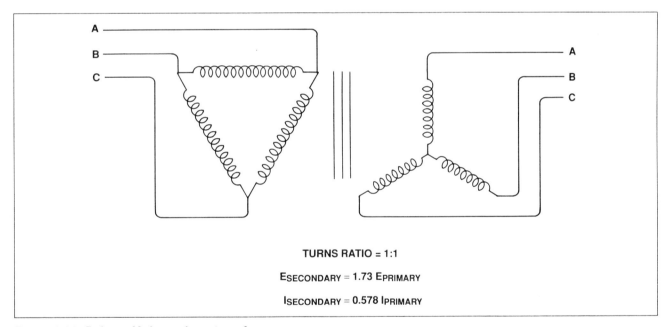

Figure 3-81. Delta-to-Y, three-phase transformer.

2. Delta-Connected Alternators

Both ends of each phase winding may be connected to the ends of the other windings to form a loop or a delta connection (so-called because of the triangular symbol formed by the coils resembling the Greek letter delta, Δ). An output lead is brought from each junction so that the output voltage will always be the same as the phase voltage. But, as you see in figure 3-78, there are two coils in series across, or in parallel, with each of the phase windings, and since the current in each of these windings is 120° out of phase with that in the other windings, the output current is 1.73 times that of the current in the phase winding.

3. Three-Phase Transformers

Three-phase transformers may have their primary and secondary windings connected in either a Y or a delta to provide the needed output.

a. Delta-to-Delta

In this type of connection, both the primary and secondary windings have their leads connected in the delta form, and if there are the same number of turns in the secondary as there are in the primary, both the secondary voltage and current will be the same as the primary voltage and current.

b. Y-to-Y

If both the primary and secondary windings are connected Y-fashion, the secondary voltage and current will be the same as the voltage and current in the primary.

c. Delta-to-Y

It is possible, by connecting the primary as a delta and the secondary as a Y, to have a secondary voltage that is 1.73 times as high as the primary voltage. But since a transformer is not capable of producing power, when the secondary voltage is higher than the primary, the secondary current must be lower. The current in the secondary will, with this form of connection, be only 0.578 times the primary current.

d. Y-to-Delta

If the primary winding is connected as a Y and the secondary as a delta, the secondary voltage will be only 0.578 of the primary voltage, but the secondary current will be 1.73 times the primary current.

K. Converting AC into DC

It is often necessary to convert AC into DC to power various circuits in the aircraft, and within electronics equipment. The conversion of AC to DC is accomplished by a circuit referred to as a rectifier. Rectifier circuits employ vacuum tube or solid-state diodes, which allow current flow in only one direction. Rectifier circuits are discussed at considerable length in the section of this manual which presents basic circuits.

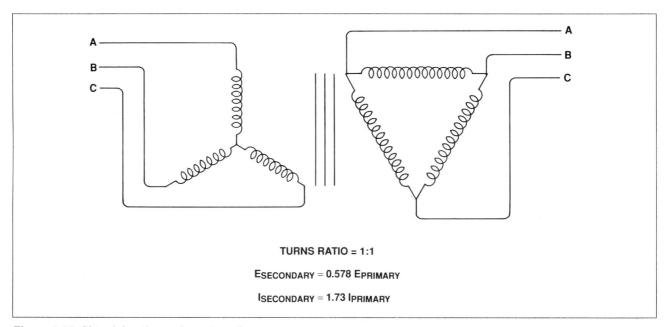

TURNS RATIO = 1:1
$E_{SECONDARY} = 0.578\ E_{PRIMARY}$
$I_{SECONDARY} = 1.73\ I_{PRIMARY}$

Figure 3-82. Y-to-delta, three-phase transformer.

Chapter IV

Electrical Measuring Instruments

A. The D'Arsonval Meter

In order to get the results we need from electrical circuits and components, we must be able to measure each of the four variables: current, voltage, resistance, and power.

There have been a number of principles used for these measurements, but by far the one most commonly found is that which uses electromagnetism, and is based on two fundamental assumptions:

1. The strength of an electromagnetic field is proportional to the amount of current that flows in the coil.
2. Voltage, resistance, and power all relate to a flow of current, and if the amount of current is known, the other values may be found.

The most widely used meter movement is the D'Arsonval movement whose pointer deflects an amount proportional to the current flowing through its moving coil. A reference magnetic field is created by a horseshoe-shaped permanent magnet, and its field is concentrated by a cylindrical keeper in the center of the open end.

Surrounding the keeper and supported by hardened steel pivots riding in smooth glass jewels, is a coil through which the current to be measured flows. The current enters and leaves the coil through calibrated hairsprings, one surrounding each of the pivots. Current flowing through this coil creates a magnetic field whose polarity is the same as that of the permanent magnet, and thus the two fields oppose each other. The opposing force rotates the coil on its pivots until the force of the hairspring exactly balances the force caused by the magnetic fields.

Oscillation of the pointer is minimized by electromagnetic damping. The moving coil is wound around a thin aluminum bobbin, or frame, and as this frame moves back and forth in the concentrated magnetic field, eddy currents are generated within the bobbin that produce their own fields which oppose the movement.

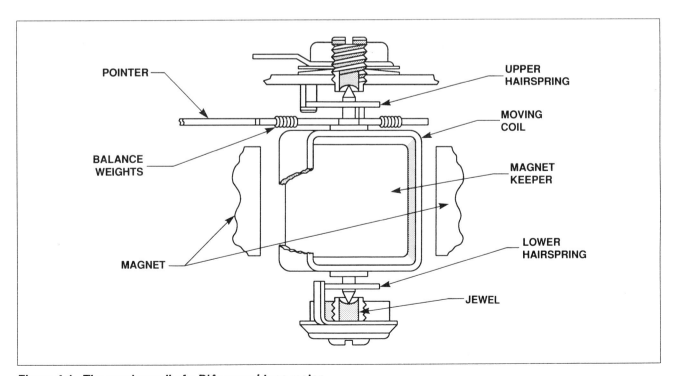

Figure 4-1. The moving coil of a D'Arsonval-type meter.

1. Meter Ratings and Terms

In order to use the basic D'Arsonval meter movement to measure the different variables, we must know some of its characteristics.

a. Full-Scale Current

This is the amount of current that must flow through the meter coil to deflect the pointer over the full calibrated scale.

b. Ohms Per Volt Sensitivity

This measurement of meter sensitivity is the reciprocal of the full-scale current and is the total amount of resistance for each volt of pressure needed to produce full-scale current. A meter that requires one milliamp ($1/1,000$ amp) of current for full-scale deflection would require one thousand ohms in the meter circuit to limit the current through the meter to one milliamp. This meter is said to have a sensitivity of one thousand ohms per volt.

Many multimeters have a sensitivity of 20,000 ohms per volt and require $1/20,000$ amp (50 microamps) of current to move the pointer full scale.

c. Meter Resistance

The total resistance of the meter must be considered when making any computations regarding the current through the meter. Both the moving coil and the hairsprings have resistance, and in some meters there is a temperature compensating resistor in series with the coil. This resistor is made of a material whose resistance decreases with an increase in temperature which is opposite to the change in resistance of the coil. As a result, the meter resistance remains constant as the temperature changes.

B. Ammeters, Milliammeters and Microammeters

The D'Arsonval meter is a current measuring instrument, and one with the proper sensitivity may be used in a circuit without any additional components.

But, if the range of current to be measured is greater than the full-scale current of the meter, a shunt must be installed in parallel with the meter. The load current flowing through the shunt will produce a voltage drop that is proportional to the amount of current, and the meter will read the voltage drop across the shunt and display this voltage in terms of amps, milliamps, or microamps on its scale. The standard aircraft shunt produces a voltage drop of fifty millivolts when its rated current flows through it.

It is sometimes necessary to extend the range of an ammeter by using a precision resistor as a shunt, and we can determine the resistance needed by the simple use of Ohm's law. Let's assume that we want our meter to deflect full scale when ten milliamps flows through the meter and shunt combination. If the meter requires one milliamp for full-scale deflection and has an internal resistance of 50 ohms, it will deflect full scale when it is connected across a voltage of 50 millivolts. The shunt, therefore, must produce a voltage drop of 50 millivolts when 9 milliamps flows through it. Remember that one

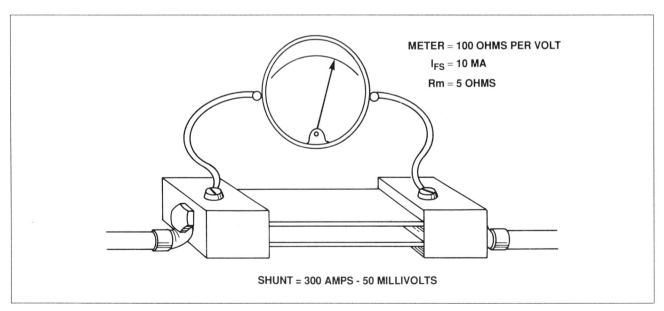

Figure 4-2. The voltage drop across an ammeter shunt is proportional to the amount of current flowing through it.

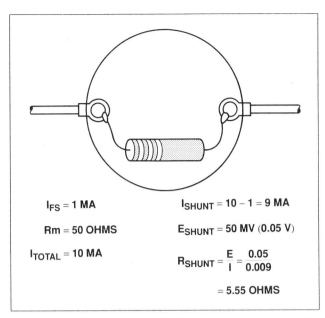

Figure 4-3. Computation of the resistance of a milliameter shunt.

milliamp flows through the meter when they are connected in parallel.

$$R_{shunt} = \frac{E}{I} = \frac{0.050}{0.009} = 5.55 \text{ ohms}$$

According to the Ohm's law formula ($R = E/I$) we find that if the meter is connected in parallel with a 5.55-ohm shunt, one milliamp will flow through the meter when nine milliamps flow through the shunt. In other words, the meter will deflect full scale when ten milliamps flow through the circuit.

C. Voltmeters

A D'Arsonval meter can be used to measure voltage by using enough resistance in series with the meter movement to limit the current to the value for which the meter will give full scale deflection. The meter sensitivity is rated in ohms per volt, which is the number of ohms of resistance we must have in the circuit for each volt we wish to measure. If, for example, our meter has a sensitivity of 1,000 ohms per volt and a resistance of 1,000 ohms, we would need a resistor of 500 ohms in series with the meter to limit the current to one milliampere when the meter is placed across a 1.5-volt battery.

The resistors that are placed in series with the meter movement are called multiplier resistors, or simply multipliers, because they are used to multiply the range of the basic meter.

Multi-range voltmeters use one meter movement with several different multipliers. These multipliers

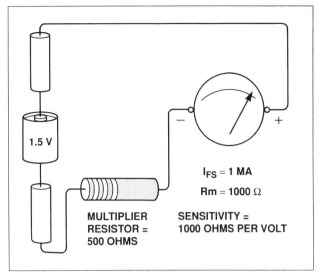

Figure 4-4. Circuit of a voltmeter.

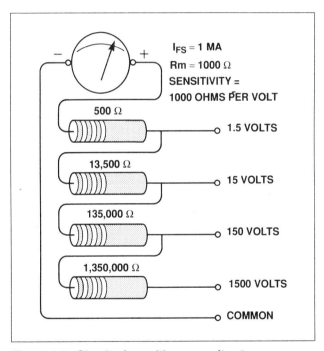

Figure 4-5. Circuit of a multi-range voltmeter.

are usually arranged so the current for each succeedingly higher range flows through the multipliers for all of the lower ranges. When the meter in figure 4-5 is used to measure 1.5 volts, the test leads are connected to the common terminal and the 1.5-volt terminal. Current then flows through the meter and the 500-ohm resistor. To measure 15 volts, connect to the common and the 15-volt terminals and the current will flow through the meter, the 500-ohm, and the 13,500-ohm multiplier resistors. The total resistance in the circuit is 15,000 ohms. To measure voltages as high as 1,500, the current must flow through all of the resistors in the voltmeter circuit.

Most multi-range meters use a selector switch rather than separate terminals. When using this type of meter, the range is set for a voltage higher than that anticipated. After the meter is connected and the needle is deflected, reduce the range with the selector switch to get a needle deflection in the center third of the scale, where the meter is most accurate.

D. Ohmmeter

Resistance is most easily determined by measuring the current through the unknown resistor when a known voltage is placed across it.

1. Series Ohmmeter

The series ohmmeter uses small flashlight or penlight batteries in series with a fixed and an adjustable resistor and the meter.

For the meter to deflect full-scale, the total resistance must equal 4,500 ohms, since we are using a 4.5-volt battery, and the meter has a sensitivity of 1,000 ohms per volt. Because the battery voltage changes with use, a variable resistor is included in the circuit to "zero" or standardize the meter before each use. To set the meter up for use, short the test leads together and turn the zero adjusting knob until the meter reads full-scale. Here, the scale is marked zero ohms, indicating that there is no resistance between the test leads. When the leads are separated, the needle will drop back to the left side of the scale where the dial is marked "∞". This is the symbol for infinity, and indicates that there is an infinite resistance between the test leads, and no current is flowing.

In this particular circuit, if a resistance of 4,500 ohms, the same as the total circuit resistance, is placed between the test leads, the current will drop to one-half of its original value. At this point, the dial is marked 4.5k ohms. The scale on a series ohmmeter is highly non-linear, meaning that there is no uniform distance between the graduations. The numbers are fairly widely separated at the low-range end, on the right side of the dial, but are crowded very tightly at the high end, the left side. For most accurate measurement of resistance, you should use a scale that will give a pointer deflection in the center third of the dial.

Different resistance ranges are selected by using different values of battery voltage and of the fixed resistor.

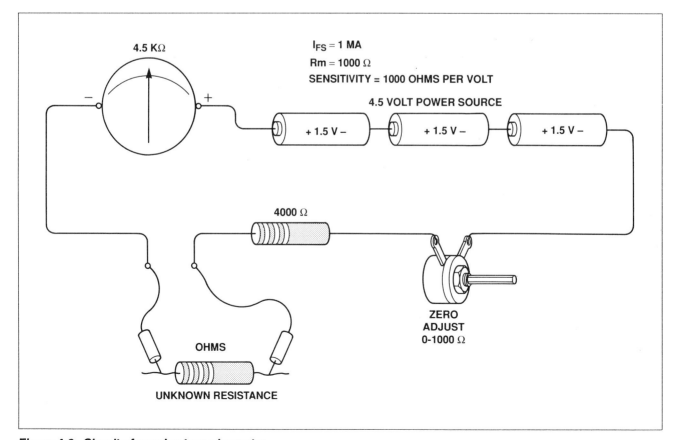

Figure 4-6. Circuit of a series-type ohmmeter.

2. Potentiometer-Type Ohmmeter

The simple series-type ohmmeter has a problem of seriously crowding the resistances on the high end of the scale, and this problem is solved to some extent by the potentiometer-type ohmmeter. The scale is still non-linear, but it is not crowded nearly so badly.

A low resistance standard resistor is in series with the battery and the resistance to be measured. This sets up a voltage divider circuit, and the voltage across the standard resistor is proportional to the current through the unknown resistance.

A sensitive meter having a high internal resistance is used to measure the voltage drop across the standard resistor; hence the name potentiometer-type ohmmeter. Rather than being calibrated in volts or millivolts, the meter dial is calibrated in ohms. When the test leads in figure 4-7 are shorted together, all of the voltage of the 1.5-volt battery is dropped across the standard resistor, and the meter can be adjusted with the zero-set variable resistor to give full-scale deflection. At this point, the dial is marked zero, since there is no resistance between the leads. When the leads are separated, the battery circuit is open and no current can flow. The point at which the needle rests is marked "∞" indicating that there is an infinite resistance between the leads.

3. Shunt-Type Ohmmeter

It is sometimes necessary to measure very low resistances; for example, the resistance of the primary winding of a magneto coil. Neither a series nor a potentiometer-type ohmmeter is suitable for this kind of measurement, but a shunt-type ohmmeter may be used.

The shunt-type ohmmeter in figure 4-8 uses a meter having a very low internal resistance of four ohms in series with a switch, a 4,000-ohm fixed resistor, and a 600-ohm variable resistor. All of this is in series with three 1.5-volt batteries. When the switch is closed and the test leads are open, the variable resistor is adjusted until the meter reads infinity. You will notice that the infinity mark is on the right side, or the full scale side, of the

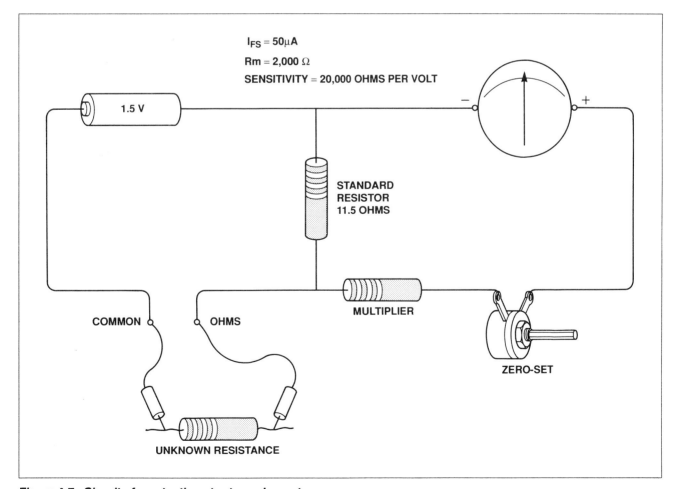

Figure 4-7. Circuit of a potentiometer-type ohmmeter.

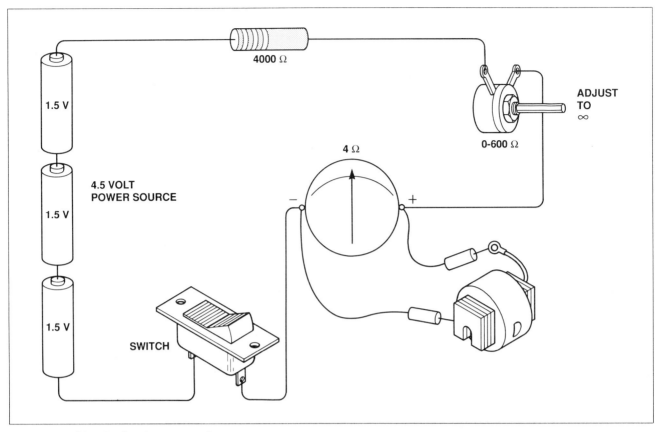

Figure 4-8. Circuit of a shunt-type ohmmeter.

dial which is opposite that of the other types of ohmmeters.

When the test leads are shorted, all of the current will flow through them and bypass the meter, so the meter will read zero. If a resistance of 4 ohms is placed between the leads, half of the current will flow through the meter and half through the resistance. The meter will then read half-scale.

4. Megohmmeter

We sometimes need to measure very high resistance values that require a voltage in excess of that provided in any of the standard ohmmeters. When we have a requirement of this nature, the megohmmeter (or "megger") is used. A hand-cranked generator with a slip clutch allows the operator to produce a voltage of several hundred volts. When the leads are separated and the crank is turned, the pointer will deflect fully to the left, indicating that there is an infinite resistance between the leads. When a high resistance is placed between the leads, a second coil within the meter will pull the needle away from the infinity mark, and it will come to rest at the proper resistance measurement. Meggers are most often used for measuring insulation resis-

Figure 4-9. A megohmmeter measures high resistances by applying a high voltage across the resistance being measured.

tance in ignition systems and other high voltage circuits.

It is important to exercise caution when testing resistance with a megger. The high voltage generated by the megger may arc to ground through defective insulation in a wire being tested and, if conditions

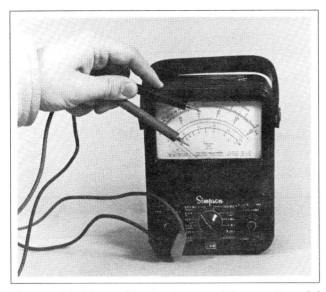

Figure 4-10. The multimeter is one of the most useful electrical measuring instruments used by the A&P technician.

are right, cause damage to equipment or injury to personnel. Some maintenance organizations or companies may limit or prohibit the use of meggers.

E. Multimeters

The most versatile electrical measuring instrument used by the aircraft technician is the multimeter. This handy tool has a single meter movement and a selector switch which may be used to select any of a number of circuits.

1. Analog Multimeters

Analog multimeters have voltage ranges from 0 to 2.5, 10, 50, 250, 1,000, and 5,000 volts for both AC and DC. They can measure direct current in ranges of 100 microamperes, 10, 100, and 500 milliamps, and 10 amps. Resistance can be measured in ranges from 0-2,000 ohms, 0-200,000 ohms and 0-20 megohms. The meter has a sensitivity of 20,000 ohms per volt for measuring DC, but because of the rectifier circuit, its sensitivity for AC is 1,000 ohms per volt.

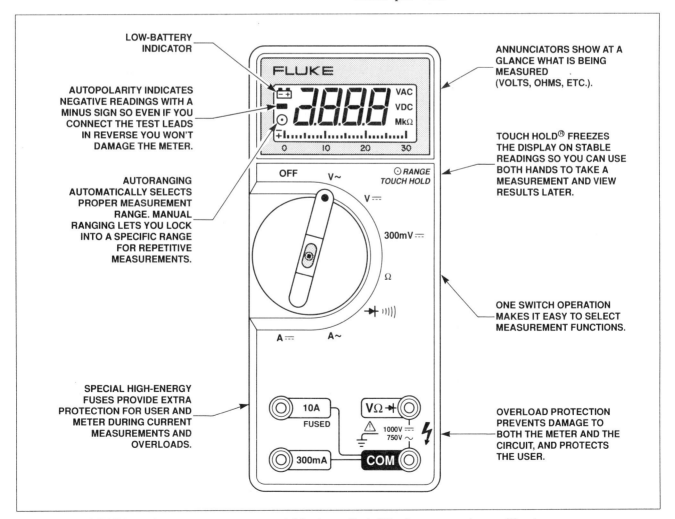

Figure 4-11. A DMM can have a great many special features that differ from an analog multimeter.

All analog multimeters have a "zero adjust" knob to reset the scale as the internal battery discharges with time and usage.

2. Digital Multimeters

A digital multimeter (DMM) is simply an electronic ruler for making electrical measurements. It may have any number of special features, but mainly, a DMM measures volts, ohms, and amperes.

a. Resolution, Digits and Counts

Resolution refers to how small or fine a measurement the meter can make. By knowing the resolution of a DMM, you can determine whether the meter could measure down to only a volt or a millivolt ($1/1{,}000$ of a volt).

The terms bits and counts are used to describe a meter's resolution. DMMs are grouped by the number of counts or digits they display.

A 3-½ digit meter, for example can display three full digits ranging from 0 to 9, and one "half" digit which displays only a one or is left blank. A 3-½ digit meter will display up to 1,999 counts of resolution.

It is more precise to describe a meter by counts of resolution than by digits. Today's 3-½ digit meters may have enhanced resolution of up to 3,200 or 4,000 counts.

b. Accuracy

Accuracy is the largest allowable error that will occur under specific operating conditions. In other words, it is an indication of how close the DMMs displayed measurement is to the actual value of the signal being measured.

Accuracy for a DMM is usually expressed as a percent of reading. An accuracy of ±1% of reading means that for a displayed reading of 100.0 volts,

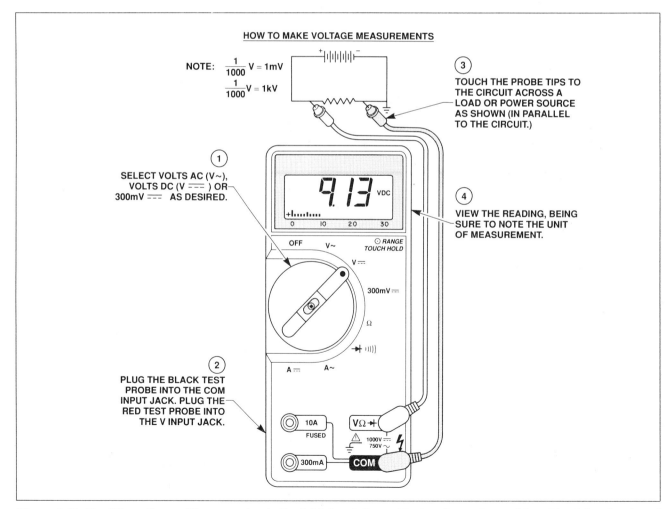

Figure 4-12. For DC readings of the correct polarity (+/–), touch the red test probe to the positive side of the circuit and the black probe to the negative side or to the circuit ground. If you reverse the connections, a DMM with auto polarity will merely display a minus sign indicating negative polarity. With an analog meter you risk damaging the meter.

the actual value of the voltage could be anywhere between 99.0V and 101.0V.

Specifications may also include a range of digits added to the base accuracy specification. This indicates how many counts the digit to the extreme right of the display may vary. So the accuracy example above might be stated as ±1% +2. Therefore, a display reading of 100.0 volts could actually represent a voltage between 98.8 volts and 101.2 volts.

For high accuracy and resolution, the digital display excels, displaying three or more digits for each measurement. The analog needle display is less accurate and has lower effective resolution, since you have to estimate values between the lines.

Some DMMs have a bar graph display. A bar graph shows changes and trends in a signal just like an analog needle, but is more durable and less prone to damage.

c. Measuring Voltage

One of the most basic tasks of a DMM is measuring voltage. A typical DC voltage source is a battery, like the one used in you car. The wall outlets in your home are common sources for AC voltage.

Some devices convert AC to DC. For example, electronic equipment such as TVs, stereos, VCRs and computers that you plug into the wall outlet use devices called rectifiers to convert AC into DC. This DC voltage is used to power the electronic circuits in these devices.

Testing for proper supply voltage is usually the first thing measured when troubleshooting a circuit. If there is no voltage present, or if it is too high or too low, the voltage problem should be corrected before investigating further.

The waveforms associated with AC voltages are either sinusoidal (sine waves) or non-sinusoidal (sawtooth, square, ripple, etc.). DMMs display the root-mean-square (RMS) value of these voltage waveforms. The RMS value is the effective or equivalent DC value of the AC voltage.

Most meters, called "average responding", give accurate RMS readings if the AC voltage signal is

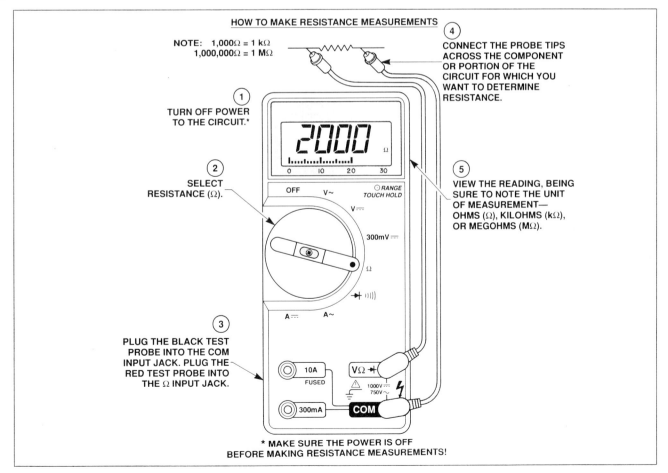

Figure 4-13. Using a DMM, resistance measurements should never be made with the circuit energized. Damage to the meter is not only possible, but probable.

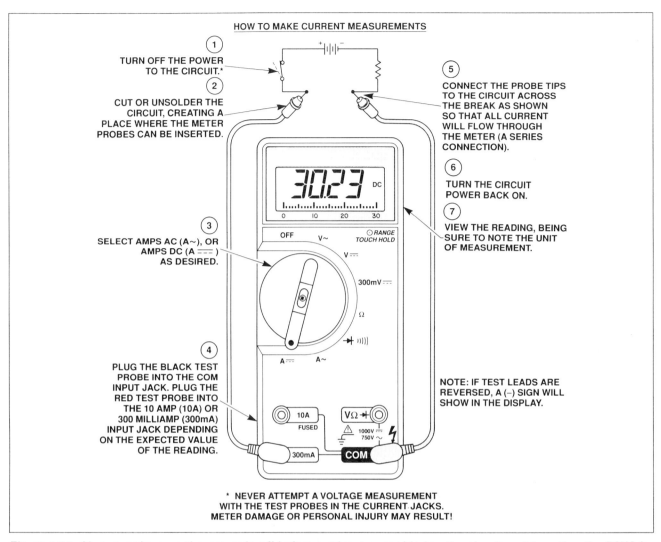

Figure 4-14. Always make sure the power is off before cutting or unsoldering the circuit and inserting the DMM for current measurement. Even small amounts of current can be dangerous.

a pure sine wave. Averaging meters are not capable of measuring non-sinusoidal signals accurately. Special DMMs called "true-RMS" meters will accurately measure the correct RMS value, regardless of the waveform.

A DMM's ability to measure AC voltage can be limited by the frequency of the signal. Most DMMs can accurately measure AC voltages with frequencies from 50 Hz to 500 Hz. DMM accuracy specifications for AC voltage and AC current should state the frequency range of a signal that the meter can accurately measure.

d. Measuring Resistance

Resistance is measured in ohms. Resistance values may vary from a few milliohms for contact resistance to billions of ohms for insulators. Most DMMs measure as little as 0.1 ohm, and may measure as high as 300 Megohms. Infinite resistance is read as "OL" on some displays, and means that the resistance is greater than the meter can measure. Open circuits will also read OL on some meter displays. For accurate low resistance measurements, resistance in the test leads must be subtracted from the total resistance measurement.

Resistance measurements must be made with the circuit power off, otherwise damage to the meter and the circuit may result. Some DMMs provide protection in the ohms mode in case of accidental contact with voltages. The level of protection may vary greatly between different makes and models of meter.

If the DMM supplies less than 0.3 volts DC test voltage for measuring resistances, it will be able to measure the values of resistors that are isolated in a circuit by diodes from semiconductor junctions. This often allows you to test resistors on a circuit board without unsoldering them.

e. Continuity

Continuity is a quick go/no-go resistance test that distinguishes between an open and closed circuit.

A DMM with a continuity beeper allows you to complete many continuity tests quickly and easily. The meter beeps when is detects a closed circuit, so you don't have to look at the meter as you test. The level of resistance required to trigger the beeper varies from model to model.

f. Measuring Current

Current measurements are different from other measurements made with a DMM. Current measurements are made with the meter connected in series with the circuit. The entire current being measured flows through the meter. Also, the test leads must be plugged into a different set of input jacks on the meter.

F. Electrodynamometer

1. Electrodynamometer Wattmeter

An electrodynamometer operates in a manner similar to a D'Arsonval meter, except that an electromagnet is used instead of the permanent magnet for the fixed field.

A large coil having a few turns of heavy wire is connected in series with the load, and the strength of the magnetic field is proportional to the amount of current flowing through the load. The movable voltage coil is connected across the load, and its magnetic strength is proportional to the amount of voltage dropped across the load. The magnetic fields caused by the current and the voltage react with each other to move the pointer an amount that is proportional to the power dissipated by the load.

Electrodynamometer wattmeters may be used in either DC or AC circuits. In an AC circuit, they measure the true power, because if the current and voltage are out of phase in the circuit, they will also be out of phase with each other in the coils of the instrument, and the resultant field will cause the pointer to deflect an amount proportional to the true power rather than the apparent power.

The apparent power in an AC circuit may be found by measuring the current with an AC ammeter and the voltage with an AC voltmeter. The product of these two values is the apparent power.

Apparent Power (volt–amps) = volts amps

The power factor of the circuit can be found as the quotient of the true power divided by the apparent power.

$$\text{Power factor} = \frac{\text{True power (watts)}}{\text{Apparent power (volt–amps)}}$$

2. Electrodynamometer Voltmeters and Ammeters

Electrodynamometers may be used for voltmeters and ammeters to measure AC as well as DC values, because the polarity of both the fixed and movable fields reverse at the same time. The sensitivity of this type of meter is considerably lower than that of the D'Arsonval-type meter.

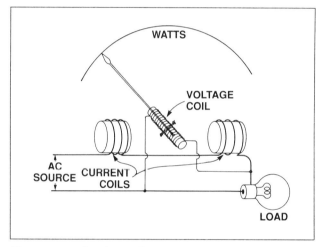

Figure 4-15. An electrodynamometer-type wattmeter.

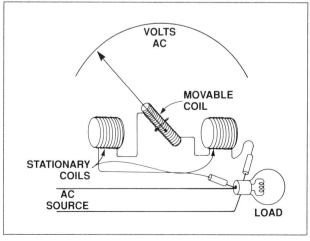

Figure 4-16. An electrodynamometer-type voltmeter can be used in either AC or DC circuits.

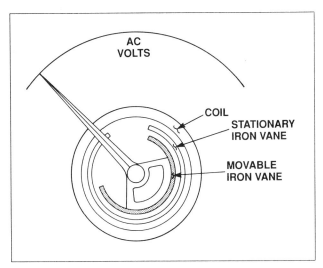

Figure 4-17. A repulsion-type, moving-vane meter can be used in either AC or DC circuits. This principle may be used for ammeters or voltmeters.

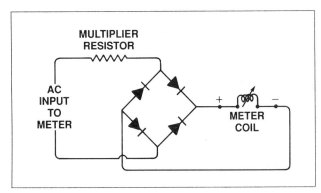

Figure 4-18. D'Arsonval meters must use a rectifier such as this 4-diode, bridge-type, full-wave rectifier when they are used in an AC circuit.

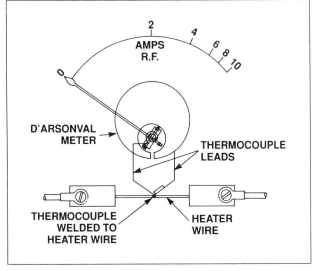

Figure 4-19. Thermocouple-type ammeters are useful for measuring high-frequency AC.

G. Repulsion-Type Moving Vane Meters

The repulsion type moving vane meter, like the electrodynamometer, can be used to measure either AC or DC voltage or current.

If the instrument is an ammeter, its coil has a relatively few turns of heavy wire, but if it is designed as a voltmeter, the coil will have many turns of fine wire. Inside the coil there are two vanes, one fixed and the other attached to the pointer staff and free to move beside the fixed vane, but restrained by a calibrated hairspring. When current flows in the coil, both the fixed and moving vanes are magnetized with the same polarity, and they repel each other, driving the pointer over the scale. The greater the amount of current, the farther the pointer will deflect. It makes no difference in which direction the current flows, the pointer will always deflect upscale. It is because of this feature that moving-vane meters are usable on AC without the need of a rectifier.

H. D'Arsonval Meters with Rectifiers

D'Arsonval meter movements may be adapted for use in AC circuits by using a rectifier to change the AC into DC before it flows through the meter coil. After the AC is rectified, it operates the meter as DC.

Most D'Arsonval meters used in AC circuits use a 4-diode, full-wave, bridge-type rectifier, connected in the circuit as we see in figure 4-18.

I. Thermocouple-Type Ammeters

Low-frequency AC can be measured with an electrodynamometer or by a repulsion type moving vane meter, but when the frequency is in the range of kilohertz or megahertz, this type of meter is not usable, and the thermocouple-type indicator is used.

The AC to be measured passes through a small piece of resistance wire inside the meter case. The greater the amount of current, the more the wire will be heated. A thermocouple, made of two dissimilar metals welded together, is welded to the resistance wire. The other ends of the thermocouple are connected to the moving coil of a D'Arsonval-type meter movement. A voltage is generated in a thermocouple that is proportional to the difference in the temperature between the two junctions. And since the junction at the meter movement has a relatively constant temperature, the amount of voltage and therefore the current through the meter is proportional to the temperature of the resistance wire. This temperature is proportional to the amount of the current.

The meter scale is calibrated in amperes, and since the amount of heat produced in the wire is a function of the square of the current (P = $I^2 \times R$). The scale is not uniform, but the numbers are bunched up on the low end of the scale and spread out as the current increases. When the current doubles, there is four times as much deflection. Thermocouple-type ammeters are usable for DC or audio-frequency or radio-frequency AC.

J. Vibrating Reed Frequency Meters

For precise frequency measurement, integrated circuit chips having clock circuits are used to actually count the cycles in a given time period and display the frequency as a digital display. But a much simpler type of frequency meter is used for determining the frequency of the AC produced by aircraft alternators.

These frequency meters use a series of metal reeds of different lengths. The center reed has a resonant frequency of exactly 400 Hz, and the reeds on one side are progressively longer, while those on the other side are progressively shorter.

A coil through which alternating current flows is wound around the fixture that holds these reeds, and the magnetic fields from the AC cause the fixture to vibrate at the frequency of the AC. The reed whose natural resonant frequency is that of the AC will vibrate with a large amplitude and will show up as a blur, while the other reeds will be stationary or moving with far less amplitude.

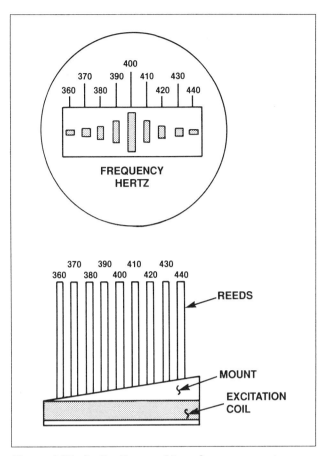

Figure 4-20. A vibrating reed-type frequency meter.

Chapter V
Aircraft Batteries

A battery is a device composed of two or more cells in which chemical energy is converted into electrical energy. The chemical nature of the battery components provides an excess of electrons at one terminal and a deficiency at the other. When the two terminals are joined by a conductor, electrons flow, and as they flow, the chemical composition of the active material changes. In a primary cell, the active elements become exhausted and cannot be restored, while in a secondary cell, electricity from an external source will restore the active material to its original, or charged, condition.

A. Primary Cells
1. Carbon-Zinc Cells

The most commonly used primary cell, or dry cell, as it is called, is of the carbon-zinc type. In this cell, a carbon rod is supported in a zinc container by a moist paste containing ammonium chloride, manganese dioxide, and granulated carbon. When a conductor joins the zinc case and the carbon rod, electrons leave the zinc and flow to the carbon. The zinc is left with positive ions which attract negative chlorine ions from the ammonium

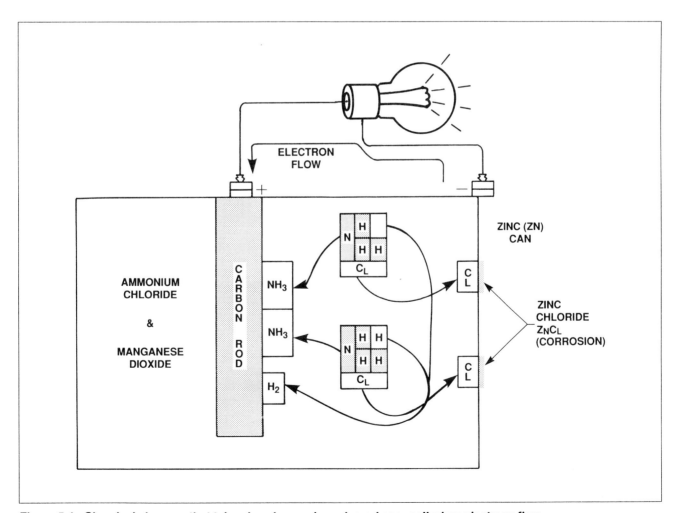

Figure 5-1. Chemical changes that take place in a carbon-zinc primary cell when electrons flow.

chloride electrolyte and form zinc chloride on the inside of the can. Zinc chloride is actually corrosion, and it eats away the zinc, eventually causing the battery to leak.

The positive ammonium ions that are left in the electrolyte are attracted to the carbon rod, where they accept the electrons that have just arrived from the zinc. As they are neutralized, they break down into ammonia and hydrogen gasses which are absorbed into the moist manganese dioxide.

Carbon-zinc cells produce one and a half volts regardless of their size, but the size of the cell determines the amount of current it can supply. Leakage, which at one time was a problem with carbon-zinc batteries, is minimized by the use of the effective seals and by enclosing the zinc can inside a steel jacket.

2. Alkaline Cells

When longer life is needed, alkaline cells are used, rather than the less expensive carbon-zinc variety. In the modern alkaline cell, the center electrode is a zinc rod supported in a carbon container by a moist electrolyte of potassium hydroxide. All of this is housed inside a steel case, with an insulating disc isolating the center electrode from the case.

Potassium hydroxide has a lower resistance that ammonium chloride, and these cells will produce more current than carbon-zinc cells and so are excellent for tape recorders and other devices which contain motors. In order for alkaline cells to be interchangeable with carbon-zinc cells, their polarity has to be reversed. This is done by mounting the cell in an insulated steel outer case. The negative center terminal of the cell bears against the outer case by a spring, and the case of the cell contacts only the center conductor of the outer shell. By doing this, both types of cells have negative outer cases and positive center terminals.

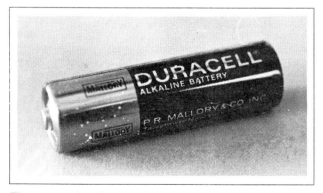

Figure 5-2. An alkaline cell uses a carbon container and a zinc center electrode with potassium hydroxide as an electrolyte.

3. Mercury Cells

Another type of alkaline cell is the mercury cell. It is used in some hearing aids, cameras, and for other applications where a high capacity for small size overcomes their higher cost.

A pellet of mercuric oxide is placed inside a steel container, and a porous insulator or separator is placed over the pellet. A roll of extremely thin corrugated zinc is placed over the paper which is saturated by a solution of potassium hydroxide which serves as the electrolyte. The cell is then closed with a steel cap that contacts the zinc, and is insulated from the container. The container is the positive terminal, and the cap is the negative. In applications where it is necessary to have a negative case, the same procedure is used to reverse the polarity as was used by the other alkaline cell.

B. Secondary Cells

It is possible to construct a cell which, instead of consuming or destroying one of its elements as it is used, will convert the element into a chemical that can be changed back into its original material by flowing electricity into it. Cells of this type are called secondary cells. They do not produce electrical energy, but merely store it in chemical form. It is for this reason that batteries of these cells are called storage batteries.

1. Lead-Acid Batteries

The most commonly used storage battery for aircraft is the lead-acid battery.

a. Construction

The positive plate is made up of a grid of lead and antimony filled with lead peroxide. The negative

Figure 5-3. A mercury cell is used when a high capacity is needed in a very small cell.

Figure 5-4. A typical lead-acid aircraft battery.

plate uses a similar grid, but its open spaces are filled with spongy lead in which an expander material is used to prevent its compacting back into a dense form of lead, losing much of its surface area.

The cells are made up of a group of positive plates, joined together and interlaced between a stack of negative plates. Porous separators keep the plates apart and hold a supply of electrolyte in contact with the active material.

The battery container was formerly made of hard rubber, but is now made of a high-impact plastic and has individual compartments for the cells.

Connector straps join the cells and provide the external terminals of the battery. A cover seals the cells in the case, and holes in the cover provide access to the cells for servicing them with water and for checking the condition of the electrolyte.

In aircraft batteries, these cell openings are closed with vented screw-in-type caps which have lead weights inside them to close the vent when the battery is tipped. This prevents the electrolyte spilling in unusual flight attitudes.

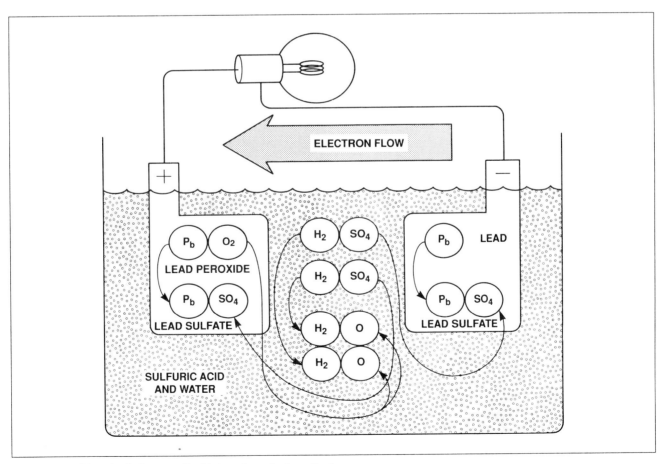

Figure 5-5. Chemical changes that take place during discharge.

An electrolyte consisting of a mixture of sulfuric acid and water covers the plates and takes an active part in the charging and discharging of the cell.

b. Chemical Changes During Discharge

When a conductor connects the positive and negative terminals of the battery, electrons flow from the lead to the lead peroxide. When electrons leave the lead, it leaves behind positive ions which attract the negative sulfate radicals from the sulfuric acid in the electrolyte. This combination forms lead sulfate on the negative plate.

The electrons arriving at the positive plate drive the negative oxygen radicals from the lead peroxide. This oxygen joins up with the hydrogen in the electrolyte that had lost its sulfate radical, and this now becomes water (H_2O). The lead that was left on the positive plate attracts sulfate radicals from the electrolyte and becomes lead sulfate.

Now, with lead sulfate on both the positive and negative plates, and with the electrolyte diluted by the water that has formed in it, the battery is discharged and electrons no longer flow.

c. Chemical Changes During Charge

If a discharged battery is attached to a source of DC having the proper voltage, and the positive plates of the battery connected to the positive terminal of the source, electrons will be drawn from the positive plate and force into the negative plates.

Electrons arriving at the negative plates drive the negative sulfate radicals out of the lead sulfate back into the electrolyte, where they join with the hydrogen from the water to form sulfuric acid (H_2SO_4).

When the electrons flowed from the positive plates, they left behind positively charged lead atoms which attract oxygen from the water in the electrolyte to form lead peroxide (PbO).

Now when the battery is fully charged, the positive plate has again become lead peroxide, the negative plate has become lead, and the electrolyte again has a high concentration of sulfuric acid. All during the charging process, as the electrolyte is being changed back into sulfuric acid, hydrogen gas is released in the form of bubbles. As the charge is completed, the bubbling increases.

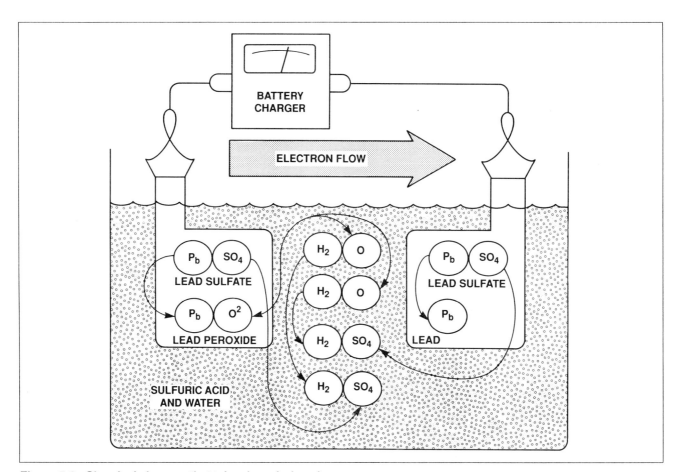

Figure 5-6. Chemical changes that take place during charge.

d. Determining the Condition of Charge

The open-circuit voltage of a lead-acid battery remains relatively constant, at about 2.1 volts per cell, and so does not tell us the state of charge of the battery. But since the concentration of acid in the electrolyte changes as the battery is used, its specific gravity gives us a good indication of the state of charge.

When electrolyte is prepared for a battery, sulfuric acid having a specific gravity of 1.835 is mixed with pure water to get a mixture having a specific gravity of between 1.265 and 1.275.

The electrolyte of a fully charged battery will have a specific gravity of between 1.275 and 1.300 with an electrolyte temperature of 80°F. When the battery is discharged until its specific gravity is down to 1.150, there is not enough chemical strength in the electrolyte to convert the active materials into lead sulfate, and the battery is considered to be discharged.

e. Ratings

1) Voltage

The open-circuit voltage of a lead-acid battery is 2.10 volts per cell when the electrolyte has a specific gravity of 1.265. The physical size of the cell or the number of plates has no effect on this voltage.

When a load is placed on the battery, the active material begins to convert into lead sulfate which has a higher resistance than the fully charged plates. This increased internal resistance will cause the closed-circuit terminal voltage to drop, and when it is down to about 1.75 volts per cell, the battery is, for all practical purposes, discharged.

2) Capacity

The capacity of a battery is its ability to produce a given amount of current for a specified time, and is expressed in ampere-hours. One ampere-hour of capacity is the amount of electricity that is put into or taken from a battery when a current of one ampere flows for one hour. Any combination of flow and time that moves the same amount of electricity is also one ampere-hour. For example, a flow of one-half amp for two hours or two amps for one-half hour is each one ampere-hour. In theory, a 100-ampere-hour battery will be able to produce 100 amps for one hour, 50 amps for two hours, or 20 amps for five hours.

The amount of active material, the area of the plates, and the amount of electrolyte determine the capacity of a battery. But also of importance in rating a battery is the rate at which the current flows.

Figure 5-7. The specific gravity of the electrolyte is measured with a hydrometer.

BATTERY VOLTAGE	PLATES PER CELL	DISCHARGE RATE*					
		5-HOUR		20-MINUTE		5-MINUTE	
		A.H.	AMPS	A.H.	AMPS	A.H.	AMPS
12	9	25	5	16	48	12	140
24	9	17	3.4	10.3	31	6.7	80
* Battery is considered discharged when closed-circuit voltage drops to 1.2 volts per cell.							

Figure 5-8. Relationship between ampere-hour capacity and discharge rate.

a) Standard Rate—Five-Hour Discharge

The standard rating used to specify the capacity of a battery is the five-hour discharge rating. This is the number of ampere-hours of capacity of the battery when there is sufficient current flow to drop the voltage of a fully charged battery to 1.75 volts per cell at the end of five hours.

b) High Rate—Twenty-Minute Discharge and Five-Minute Discharge

Two ratings are used to specify the capacity of a battery when the current is taken out at a high rate. Both ratings will be lower than the five-hour rating. Figure 5-8 shows the relationship between the capacity at the five-hour rating and that for the twenty-minute and the five-minute rating for typical 12- and 24-volt aircraft batteries.

f. Battery Testing

1) Specific Gravity

As we have previously stated, the specific gravity of the electrolyte is one of the best indicators of the state of charge of a lead-acid battery.

We determine the specific gravity with a hydrometer, an instrument which measures the depth to which the calibrated stem of a float sinks into the electrolyte. A sample of electrolyte is drawn from the cell by the syringe, and the float riding on the surface of the liquid is visible in the glass tube. The graduation on the stem which is even with the liquid shows the specific gravity of the electrolyte. The more dense the liquid, the less the float will sink, and the less dense, the deeper the float will be able to go.

The temperature of the electrolyte affects its specific gravity, and so a correction must be made if the temperature is not standard. 80°F is used as the reference, and if the electrolyte is warmer, it is less dense; so the float will sink deeper into it and will indicate that the state of charge is lower than it actually is. To correct for temperature effect. We use a correction chart similar to the one in figure 5-9.

If the specific gravity read by the hydrometer is 1.240, and the electrolyte temperature is 60°F, we know that the corrected specific gravity will be lower than is indicated because the lower temperature has made the electrolyte more dense. According to the chart, a correction of eight points should be made. The indication was 1.240, and after we subtract 0.008, we have a corrected specific gravity of 1.232.

2) Cell Test

If the battery construction is such that the voltage of the individual cells can be measured, we can get a good indication of the behavior of the cells under load.

We must first determine that the electrolyte level is proper, and if it is not correct, it should be adjusted.

Apply a heavy load to the battery for about three seconds by cranking the engine with the ignition switch off.

CAUTION: Be sure that the propeller is clear before engaging the starter.

Now, turn on the landing lights and taxi lights to draw about ten amps, and while the load current is flowing, measure the voltage of each cell.

(A) Electrolyte Temperature		Correction Points Add or Subtract	(B) Specific Gravity	Freezing Point	
°C	°F			°C	°F
60	140	+ 24			
55	130	+ 20	1.300	– 70	– 90
49	120	+ 16			
43	110	+ 12	1.275	– 62	– 80
38	100	+ 8			
33	90	+ 4	1.250	– 52	– 62
27	80	0			
23	70	– 4	1.225	– 37	– 35
15	60	– 8			
10	50	– 12	1.200	– 26	– 16
5	40	– 16			
– 2	30	– 20	1.175	– 20	– 4
– 7	20	– 24			
– 13	10	– 28	1.150	– 15	+ 5
– 18	0	– 32			
– 23	– 10	– 36	1.125	– 10	+ 13
– 28	– 20	– 40			
– 35	– 30	– 44	1.100	– 8	+ 19

Figure 5-9.
(A) Specific gravity correction for non-standard temperature.
(B) Freezing point of lead-acid battery with various electrolyte specific gravities.

A fully charged battery with all of the cells in good condition should have a voltage of 1.95 volts for each cell, and all of the cells should be within 0.05 volt of each other.

If some of the cells are below 1.95 volts, but all are within 0.05 volt of each other, the cells are in good condition but the battery is low. If any of the cells read higher than 1.95 volts and there is more than a 0.05-volt difference between any of them, there is a defective cell in the battery.

g. Servicing and Charging

1) Inspection and Servicing

One of the most important aspects of battery servicing is keeping the battery clean and all of the terminals tight and free of corrosion. Any corrosion on the battery terminals or the cable should be removed, and the battery box and the top of the battery should be scrubbed with a soft bristle brush and a solution of bicarbonate of soda (baking soda) and water. When washing the top of the battery, be sure that you do not get any soda in the cells, as it will neutralize the electrolyte. After the battery and box are perfectly clean, rinse them with clean water and dry thoroughly, then coat the battery terminals with petroleum jelly or general purpose grease, and touch up any paint damage to the box or adjacent area with an acid-proof paint.

The electrolyte in each cell should just cover the plates, and most batteries have an indicator to show you the correct level. If it is low, add distilled or demineralized water to the level specified. You should never add acid to the battery unless it has been spilled, and then, *follow the recommendations of the battery manufacturer in detail*. The normal loss of liquid in a battery is the loss of water caused by its decomposition during charging.

Most new batteries are received in a dry-charged state with the cells dry and sealed. When putting one of these batteries into service, remove the cell seals and pour in the electrolyte that is shipped with the battery. The battery is now ready for installation. In order to have it fully charged, it must be given a slow freshening, or boost, charge. Allow the battery to sit for an hour or so after the boost charge and adjust the level of the electrolyte.

It is normally not necessary for you to mix electrolyte, as properly diluted acid is readily available. But if it should ever become necessary to dilute acid, *it is extremely important that the acid be poured into the water, and never the other way*

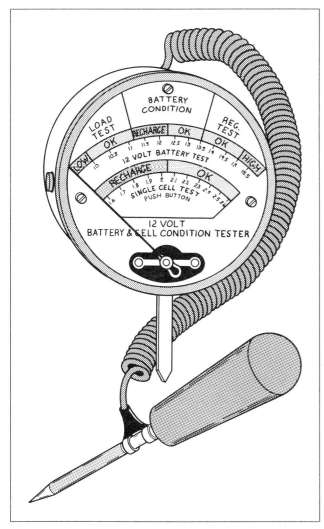

Figure 5-10. A low-range voltmeter is needed to determine the condition of the cells under load.

Figure 5-11. Electrolyte level should come up to the indicator inside the cell.

around. If water is poured into the acid, the water, being less dense, will float on top of the acid, and a chemical action will take place along the surface where they meet. This action will generate enough heat to boil the water, and the boiling will splash acid out the container and can cause serious injury if it gets on the skin or in the eyes.

If acid should get into your eyes, flush them with generous amounts of clean water and get medical attention as soon as possible.

When acid is poured into the water, it will mix with the water and distribute the heat throughout the entire container, raising the temperature of the water, but not causing any violent reaction or boiling.

There are two different kinds of electrolyte, automotive and aircraft; with automotive having a lower specific gravity when charged. *Do not use automotive electrolyte in an aircraft battery.* It may never reach full charge with automotive electrolyte.

2) Constant-Current Charging

The most effective way of charging a battery is by putting current back into it at a constant rate, and for the most efficient charging, this rate is usually no more than seven percent of the ampere-hour rating of the battery. For example, if you are charging a 35-ampere-hour battery, and do not have specific information from the battery manufacturer stating otherwise, you should charge it at a rate not exceeding two and a half amperes.

Constant-current chargers have a voltage adjustment so that the voltage can be increased as the charge progresses. In this way, the charger voltage can be kept just enough above the rising voltage of the battery to maintain a constant current of the correct rate.

When charging batteries with a constant-current charger, the batteries are connected in series across the charger, and the charging rate is controlled by varying the voltage that is produced by the charger.

The batteries being charged may be of different voltages, but they should all require the same charging rate. Begin the charge cycle with the maximum recommended current for the battery with the lowest capacity, and when the cells are gassing freely, decrease the current and continue the charge until the proper number of ampere hours of charge have been put into them.

3) Constant-Voltage Charging

The generating system in an aircraft charges the battery by the constant voltage method. A fixed voltage, slightly higher than the battery voltage, is connected across the battery.

The low voltage of a discharged battery will allow a large amount of current to flow into it when the charge is first begun, but as the charge continues, and the voltage rises, the current will decrease.

The voltage of the charger is sufficiently high that even when the battery is fully charged, about one ampere of current will flow into it.

Constant-voltage chargers are often used as shop chargers, but care should be exercised when using them. The battery can be overheated by the high charging rate when the charger is first connected across a fully discharged battery. The boost charge provided by a constant-voltage charger will not fully charge a battery, but it will usually supply enough charge to start the engine and allow the aircraft generating system to complete the charge.

h. Battery Installation

Before installing any battery in an aircraft, be sure you know that the battery is correct for the aircraft, that the voltage and ampere-hour ratings are as specified, and that the battery fits the battery box properly.

Some aircraft use two batteries connected in parallel to provide a reserve of current for starting and for extra-heavy electrical loads. Be sure that the batteries installed in this type of arrangement are the batteries specified in the aircraft service manual.

Most aircraft use a single-wire electrical system, with the negative terminal of the battery connected to the aircraft structure. *NOTE: When installing the battery, connect the "hot" lead first. If you should short-circuit between the battery and the aircraft with your wrench, you will not cause a spark if the ground lead has not been connected. When removing a battery, always disconnect the ground lead first for the same reason.*

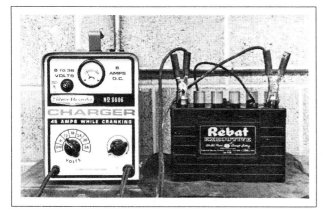

Figure 5-12. The constant-current charger is the most effective type of battery charger for shop use.

Be sure that the battery box is properly vented, if a vent is required, and that the battery box drain extends through the aircraft skin. Some batteries are of the manifold type, which do not require a separate battery box. There is a cover over the cells, and the area above the cells is vented to the outside of the aircraft structure.

The fumes emitted from storage batteries as they are charged are highly corrosive to the metals of which aircraft are made, and they must be neutralized before they are released into the atmosphere. Many battery installations have vent sump jars containing absorbent pads moistened with a solution of bicarbonate of soda and water.

No battery installation is complete until you know that the battery will supply enough current to crank the engine, and that the aircraft generating system will keep the battery charged.

2. Nickel-Cadmium Batteries

Turbine engines, with their demand for extremely high current for starting, have encouraged the development of an alkaline battery for aircraft use.

One of the disadvantages of the use of a lightweight lead-acid battery for high current requirements, is its buildup of internal resistance and subsequent voltage drop during high-rate discharges. The nickel-cadmium battery has a very distinct advantage in this regard. Its internal resistance is very low, and so its voltage remains constant until it is almost totally discharged. This low resistance is also an advantage in recharging, as it will accept high charging rates without damage.

While high charging currents are possible, there are dangers involved. This dangers begins with a breakdown of the cellophane-like material that separates the plates in the nickel cadmium cell. This breakdown is usually the result of high temperatures resulting from high rates of charge. The breakdown of the cell separator creates a short circuit, and current flow increases. The increased current flow creates heat, the heat will cause further breakdown of the separator material, and the cycle continues. This process is known as vicious-cycling or thermal runaway.

New cell separator materials, and advanced on-board charging equipment have reduced the likelihood of thermal runaway. Nickel-cadmium battery installations are also required to have temperature monitoring equipment that will enable the flight crew to recognize an overheat condition that may lead to thermal runaway.

a. Construction

Most nickel-cadmium batteries are made up of individual removable cells. The positive plates are made of powdered nickel (plaque) fused (or sintered) to a nickel wire-mesh screen. This porous material is impregnated with nickel hydroxide.

The negative plates are made of the same type of plaque as was used on the positive plates, but the negative plates are impregnated with cadmium hydroxide.

Separators of nylon and cellophane prevent the plates touching each other, and the stack-up of plates and separators is assembled into a polystyrene or nylon cell case, and the case is sealed.

A thirty-percent, by weight, solution of potassium hydroxide and water is used as the electrolyte. The specific gravity of this liquid is between 1.24 and 1.30 at 80°F and since it acts only as a conductor during charging and discharging, its specific gravity is no indication of the state of charge of the battery.

The individual cell produces an open circuit voltage of between 1.55 and 1.80 volts, and batteries for 12-volt aircraft may use either nine or ten cells, while batteries for 24-volt aircraft are made up of 19 or 20 individual cells.

Figure 5-13. Nickel-cadmium batteries are made up of individual cells in transparent cases.

b. Chemical Changes During Charge

When charging current flows into the battery, oxygen is driven from the cadmium oxide on the negative plate and leaves metallic cadmium. The nickel hydroxide on the positive plate accepts some of the released oxygen and becomes more highly oxidized. The charge continues until all of the oxygen is removed from the negative plate, and then gassing occurs in the cell as the electrolyte is decomposed by electrolysis. When the battery is fully charged, the electrolyte is driven from the plates, and immediately after the charge is completed, will be at its highest level.

c. Chemical Changes During Discharge

As electrons flow from the negative terminal of the nickel-cadmium battery, oxygen is driven from the positive plate and is recovered by the negative plate.

d. Servicing Nickel-Cadmium Batteries

The electrolytes used by nickel-cadmium and lead-acid batteries are chemically opposite, and either type of battery can be contaminated by fumes from the charging of the other. For this reason, it is extremely important that separate facilities be used for servicing nicad batteries, completely away from the area used for lead-acid batteries.

The alkaline electrolyte used in nickel-cadmium batteries is corrosive, and it can burn your skin or cause severe injury if it gets into your eyes. Be careful when handling this liquid, and if any of it is spilled, neutralize it with vinegar or boric acid, and flush the area with clean water.

Nickel-cadmium battery manufacturers supply detailed service information for each of their products, and these directions must be followed closely. Each nickel-cadmium battery should have a service record that will follow the battery to the service facility each time that it is removed for service or testing. It is very important to perform service in accordance with the manufacturer's instructions, and to record all work on the battery service record.

Most nickel-cadmium batteries will get an accumulation of potassium carbonate on top of the cells. This white powder forms when electrolyte spewed from the battery combines with carbon dioxide. The amount of this deposit is increased by charging the battery too fast, or by the electrolyte level being too high. If there is an excessive amount of potassium carbonate, check the voltage regulator and the level of electrolyte in the cells.

Scrub all of the deposits off of the top of the cells with a nylon or other type of non-metallic bristle brush, and dry the battery thoroughly with a soft flow of compressed air.

Check for electrical leakage between the cells and the steel case by using a milliammeter between the positive terminal of the battery and the case. If there is more than about 100 milliamps of leakage, the battery should be disassembled and thoroughly cleaned.

Check all of the hardware in the cell connectors for their condition and to be sure that there is no trace of corrosion.

Dirty contacts or improperly torqued nuts will cause overheating and burned hardware.

The only way of actually determining the condition of a nickel-cadmium battery is to fully charge it, and then discharge it at a specified rate and measure its ampere-hour capacity. When charging, use the

Figure 5-14. Cell-to-case leakage should be measured with a milliammeter.

Figure 5-15. Short all of the cells to assure that they will all be exactly the same before recharging the battery during a deep-cycle operation.

OBSERVATION	PROBABLE CAUSE	CORRECTIVE ACTION
High-trickle charge—When charging at constant voltage of 28.5 (±0.1) volts, current does not drop below 1 amp after a 30-minute charge.	Defective cells.	While still charging, check individual cells. Those below .5 volts are defective and should be replaced. Those between .5 and 1.5 volts may be defective or may be unbalanced, those above 1.5 volts are alright.
High-trickle charge after replacing defective cells, or battery fails to meet amp-hour capacity check.	Cell imbalance.	Discharge battery and short out individual cells for 8 hours. Charge battery using constant-current method. Check capacity and if OK, recharge using constant-current method.
Battery fails to deliver rated capacity.	Cell imbalance or faulty cells.	Repeat capacity check, discharge and constant-current charge a maximum of three times. If capacity does not develop, replace faulty cells.
No potential available.	Complete battery failure.	Check terminals and all electrical connections. Check for dry cell. Check for high-trickle charge.
Excessive white crystal deposits on cells. (There will always be some potassium carbonate present due to normal gassing.)	Excessive spewage.	Battery subject to high charge current, high temperature, or high liquid level. Clean battery, constant-current charge and check liquid level. Check charger operation.
Distortion of cell case.	Overcharge or high heat.	Replace cell.
Foreign material in cells—black or gray particles.	Impure water, high heat, high concentration of KOH, or improper water level.	Adjust specific gravtiy and electrolyte level. Check battery for cell imbalance or replace defective cell.
Excessive corrosion of hardware.	Defective or damaged plating.	Replace parts.
Heat or blue marks on hardware.	Loose connections causing overheating of intercell connector or hardware.	Clean hardware and properly torque connectors.
Excessive water consumption. Dry cell.	Cell imbalance.	Proceed as above for cell imbalance.

Figure 5-17. Nickel-cadmium troubleshooting chart.

five-hour rate, and charge it until the cell voltage is that specified by the battery manufacturer. When it is fully charged, and immediately after it is taken off of the charger, measure the level of the electrolyte. If it is low, adjust it by adding distilled water. If the level is not checked immediately after the charge is completed, the level will drop so the battery should be allowed to sit for about three hours and then the electrolyte adjusted to a lower specified level. When water is added, the amount and cell location must be recorded on the battery service record.

When the battery is fully charged and the electrolyte adjusted, it must be discharged at a specified rate and its ampere-hour capacity measured. If the capacity is less than it should be, it is an indication that some of the cells are

Figure 5-16. Complete the deep-cycle operation by charging the battery to 140% of its ampere-hour capacity.

unbalanced, and they must be equalized by a process known as deep-cycling.

To deep-cycle the battery, continue to discharge it at a rate somewhat lower than that used for the capacity test. When the cell voltage is down to around 0.2 volt per cell, short across each cell with a shorting strap. Leave the strap across the cells for three to eight hours, to be sure that all of the cells are *completely* discharged. This process is known as equalization.

After equalization the battery is ready to charge. Nickel-cadmium batteries may be charged using either the constant-voltage or constant-current methods. The constant-voltage method will result in a faster charge, but the constant-current is most widely used. For either system, *the battery manufacturer's service instructions must be followed exactly.*

Monitor the battery during charge, and measure individual cell voltages. The manufacturer will specify a maximum differential between cells during the charging process. If a cell exceeds the specification, it must be replaced. Battery manufacturers will specify the maximum number of cells that may be replaced, before the battery must be retired. Cell replacement should be entered in the battery service record.

As the battery nears completion of the charge, gassing of the cells will occur. This is normal, and in fact must occur before the cell is fully charged. It is normal to overcharge the nicad battery to 140% of its amp-hour capacity.

If the battery has been properly serviced and is in good condition, each cell should have a voltage of between 1.55 and 1.80 at a temperature between 70° and 80°F. If the voltage of any cell is not within this range, the cell must be replaced.

Chapter VI
Electrical Generators and Motors

A. DC Generators

Energy for the operation of most electrical equipment in an airplane depends upon the electrical energy supplied by a generator. A generator is any machine which converts mechanical energy into electrical energy by electromagnetic induction. Generators designed to produce alternating current are called AC generators; generators which produce direct current energy are called DC generators.

For airplanes equipped with direct-current electrical systems, the DC generator is the regular source of electrical energy. One or more DC generators, driven by the engine(s), supply electrical energy for the operation of all units in the electrical system, as well as energy for charging the battery. In most cases only one generator is driven by each engine, but in some large airplanes, two generators may be driven by a single engine. Aircraft equipped with alternating-current systems use electrical energy supplied by AC generators, also called alternators.

1. Theory of Operation

After the discovery that an electric current flowing through a conductor creates a magnetic field around the conductor, there was considerable scientific speculation about whether a magnetic field could create a current flow in a conductor. In 1831, the English scientist Michael Faraday demonstrated this could be accomplished. This discovery is the basis for the operation of the generator, which signaled the beginning of the electrical age.

To show how an electric current can be created by a magnetic field, a demonstration similar to that illustrated in figure 6-1 can be used. Several turns of wire are wrapped around a cardboard tube, and the ends of the conductor are connected to a galvanometer. If a simple bar magnet is plunged into the cylinder, the galvanometer can be observed to deflect in one direction from its zero (center) position (figure 6-1(A)). When the magnet is at rest inside the cylinder, the galvanometer shows a reading of zero, indicating that no current is flowing (figure 6-1(B)). In figure 6-1(C), the galvanometer indicates a current flow in the opposite direction when the magnet is pulled from the cylinder.

The same results may be obtained by holding the magnet stationary and moving the cylinder over the magnet. This indicates that a current flows when there is relative motion between the wire coil and the magnetic field.

When a conductor is moved through a magnetic field (figure 6-2) an electromotive force (emf) is induced in the conductor. The direction (polarity) of

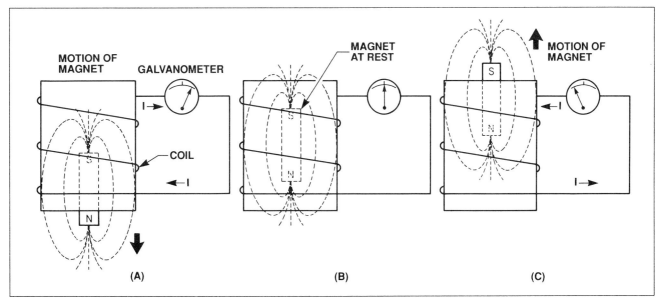

Figure 6-1. Inducing a current flow.

the induced emf is determined by the magnetic lines of force and the direction the conductor is moved through the magnetic field.

The left-hand rule for generators (not to be confused with the left-hand rules used with a coil) can be used to determine the direction of the induced emf, as shown in figure 6-3. The first finger of the left hand is pointed in the direction of the magnetic lines of force (north to south), the thumb is pointed in the direction of movement of the conductor through the magnetic field, and the second finger points in the direction of the induced emf. When two of these three factors are known, the third may be determined by the use of this rule.

When a loop conductor is rotated in a magnetic field (figure 6-4), a voltage is induced in each side of the loop. The two sides cut the magnetic field in opposite directions, and although the current flow is continuous, it moves in opposite directions with respect to the two sides of the loop. If sides A and B of the loop are rotated half a turn, so that the sides of the conductor have exchanged positions, the induced emf in each wire reverses its direction. This is because the wire formerly cutting the lines of force in an upward direction is now moving downward.

The value of an induced emf depends upon three factors:

1. The number of wires moving through the magnetic field.
2. The strength of the magnetic field.
3. The speed of rotation.

A simple generator is illustrated in figure 6-5, together with the components of an external generator circuit which collect and use the energy produced by the generator. The loop of wire (figure 6-5, A and B) is arranged to rotate in a magnetic field. When the plane of the loop of wire is parallel to the magnetic lines of force, the voltage induced in the loop causes a current to flow in the direction indicated by the arrows in figure 6-5. The voltage

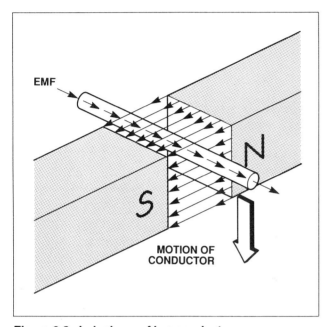

Figure 6-2. Inducing emf in a conductor.

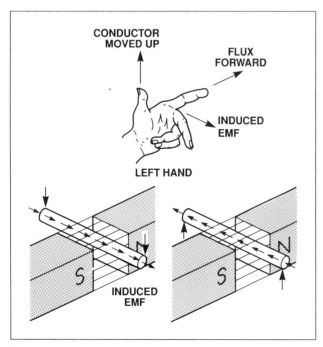

Figure 6-3. An application of the generator left-hand rule.

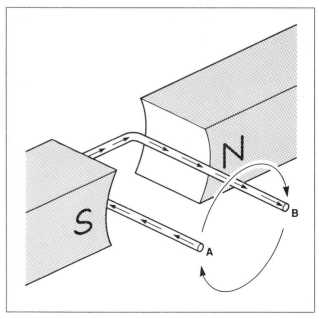

Figure 6-4. Voltage induced in a loop.

induced at this position is maximum, since the wires are cutting the lines of force at right angles and thus are cutting more of the lines of force per second than in any other position relative to the magnetic field.

As the loop approaches the vertical position shown in figure 6-6, the induced voltage decreases because both sides of the loop (A and B) are approximately parallel to the lines of force and the rate of cutting is reduced. When the loop is vertical, no lines of force are cut since the wires are momentarily traveling parallel to the magnetic lines of force, and there is no induced voltage.

The rotation of the loop continues and the number of lines of force cut increases until the loop has rotates an additional 90° to a horizontal plane. As shown in figure 6-7, the number of lines of force cut and the induced voltage once again are maximum. The direction of cutting, however, is in the opposite direction to that occurring in figures 4-5 and 4-6, so the direction (polarity) of the induced voltage is reversed.

As rotation of the loop continues, the number of lines of force having been cut again decreases, and the induced voltage becomes zero at the position shown in figure 6-8, since the wires A and B are again parallel to the magnetic lines of force.

If the voltage induced throughout the entire 360° of rotation is plotted, the curve shown in figure 6-9 results. As you can see from the illustration, the output of a loop rotating in a magnetic field is alternating current. By replacing the slip rings of the basic AC generator with two half-cylinders, called a commutator, a basic DC generator (figure 6-10) is obtained. In this illustration, the black side of the coil is connected to the black segment and the white side of the coil to the white segment. The segments are insulated from each other. The two stationary brushes are placed on opposite sides of the commutator and are so mounted that each brush contacts each segment of the commutator as it revolves simultaneously with the loop. The rotating parts of a DC generator (coil and commutator) are called an armature.

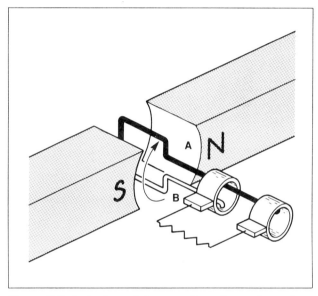

Figure 6-6. Inducing minimum voltage in an elementary generator.

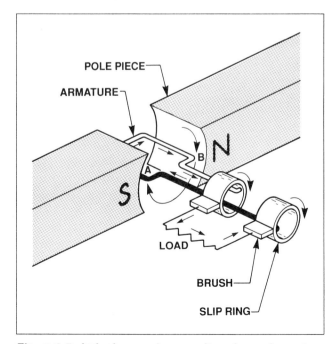

Figure 6-5. Inducing maximum voltage in an elementary generator.

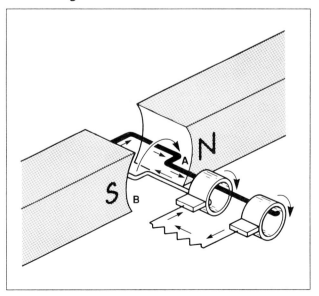

Figure 6-7. Inducing maximum voltage in the opposite direction.

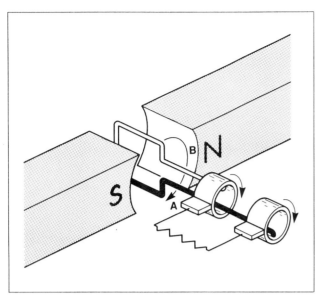

Figure 6-8. Inducing a minimum voltage in the opposite direction.

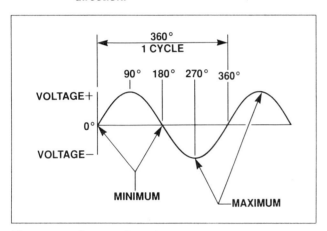

Figure 6-9. Output of an elementary generator.

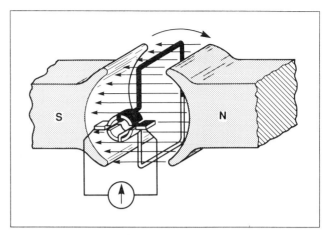

Figure 6-10. Basic DC generator.

Figure 6-11 illustrates the generation of a DC voltage using a commutator. The loop in position (A) is rotating clockwise, but no lines of force are cut by the coil sides and no emf is generated. The black brush is shown coming into contact with the black segment of the commutator, and the white brush is just coming into contact with the white segment.

In position (B) of the figure, the flux is being cut at a maximum rate and the induced emf is maximum. At this time, the black brush is contacting the black segment and the white brush is contacting the white segment. The deflection of the meter is toward the right, indicating the polarity of the output voltage.

At position (C), the loop has completed 180° of rotation. Again, no flux lines are being cut and the output voltage is zero. The important condition to observe at position (C) is the action of the segments and brushes. The black brush at the 180° angle is contacting both black and white segments on one side of the commutator, and the white brush is contacting both segments on the other side of the commutator. After the loop rotates slightly past the 180° point, the black brush is contacting only the white segment and the white brush is contacting only the black segment.

Because of this switching of commutator electrodes, the black brush is always in contact with the coil side moving downward, and the white brush is always in contact with the coil side moving upward. Though the current actually reverses its direction in the loop in exactly the same way as in the AC generator, commutator action causes the current to flow always in the same direction through the external circuit or meter.

The voltage generated by the basic DC generator in figure 6-11 varies from zero to its maximum twice for each revolution of the loop. This variation of DC voltage is called "ripple," and may be reduced by using two or more loops, or coils as shown in figure 6-12(A). As the number of loops is increased, the variation between maximum and minimum values of voltage is reduced, and the output voltage of the generator approaches a steady DC value.

The number of commutator segments is increased in direct proportion to the number of loops; that is, there are two segments for one loop, four segments for two loops, and eight segments for four loops.

The voltage induced in a single-turn loop is small. Increasing the number of loops does not increase the maximum value of the generated voltage, but increasing the number of turns in each loop will increase this value.

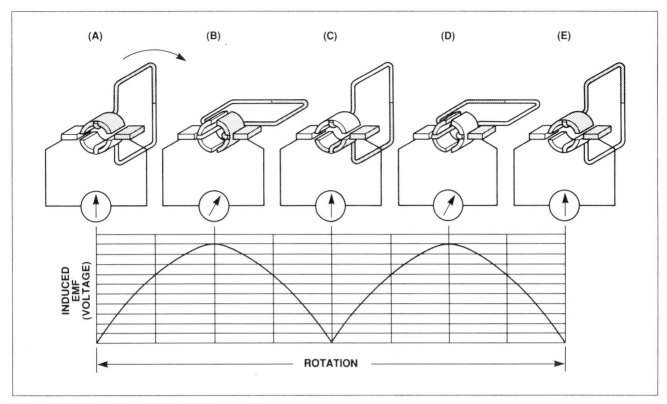

Figure 6-11. Operation of a basic DC generator.

2. Construction Features of DC Generators

Generators used on aircraft may differ somewhat in design, since they are made by various manufacturers. All, however, are of the same general construction and operate similarly. The major parts, or assemblies of a DC generator are a field frame, a rotating armature, and a brush assembly. The parts of a typical aircraft generator are shown in figure 6-13.

a. Field Frame

The field frame (also called the yoke) constitutes the foundation for the generator. The frame has two functions: It completes the magnetic circuit between the poles and acts as a mechanical support for the other parts of the generator. In figure 6-14(A), the frame for a two-pole generator is shown in cross-section. A four-pole generator frame is shown in (B) of the same figure.

In small generators, the frame is made of one piece of iron, but in larger generators, it is usually made up of two parts bolted together. The frame has high magnetic properties and, together with the pole pieces, forms the major part of the magnetic circuit. The field poles (pole shoes) (figure 6-14) are bolted to the inside of the frame and form a core on which the field coil windings are mounted. The

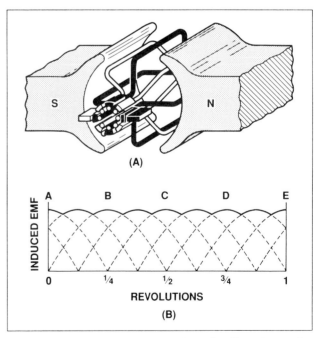

Figure 6-12. Increasing the number of coils reduces the ripple in the voltage.

poles are usually laminated to reduce eddy current losses and serve the same purpose as the iron core of an electromagnet; that is, they concentrate the lines of force produced by the field coils. The

101

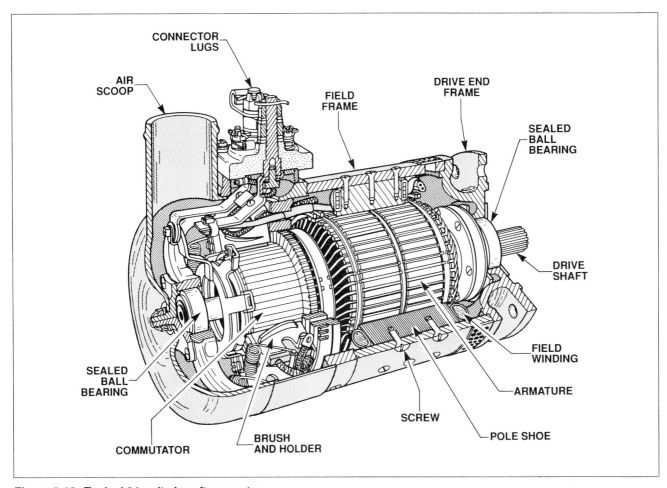

Figure 6-13. Typical 24-volt aircraft generator.

entire frame, including the field poles, is made from high quality magnetic iron or sheet steel.

A practical DC generator uses electromagnets instead of permanent magnets. To produce a magnetic field of the necessary strength with permanent magnets would greatly increase the physical size of the generator.

The field coils are made up of many turns of insulated wire and are usually wound on a form which fits over the iron core of the pole shoe to which it is securely fastened (figure 6-15). The exciting current, which is used to produce the magnetic field and which flows through the field coils, is obtained from an external source or from the current generated by the unit itself. No electrical connection exists between the windings of the field coils and the pole pieces.

Most field coils are connected in such a manner that the poles show alternate polarity. Since there is always one north pole for each south pole, there must always be an even number of poles in any generator.

Note that the pole pieces in figure 6-14 project from the frame. Because air offers a great amount of reluctance to the magnetic field, this design reduces the length of the air gap between the poles and the rotating armature and increases the efficiency of the generator. When the pole pieces are made to project as shown in figure 6-14, they are called salient poles.

b. Armature

The armature assembly consists of armature coils wound on an iron core, a commutator, and associated mechanical parts. Mounted on a shaft which rotates in bearings located in the end frames of the generator, it rotates through the magnetic field produced by the field coils. The core of the armature acts as an iron conductor in the magnetic field and, for this reason, is laminated to prevent the circulation of eddy currents.

A drum-type armature (figure 6-16) has coils placed in slots in the core, but there is no electrical connection between the coils and core. The use of slots increases the mechanical safety of the armature. Usually, the coils are held in place in the slots by means of wooden or fiber wedges. The coil ends are brought out to individual segments of the commutator.

c. Commutators

Figure 6-17 shows a cross-sectional view of a typical commutator. The commutator is located at the end of the armature and consists of wedge-shaped segments of hard-drawn copper, insulated from each other by thin sheets of mica. The segments are held in place by steel V-rings or clamping flanges fitted with bolts. Rings of mica insulate the segments from the flanges. The raised portion of each segment is called a riser, and the leads from the armature coils are soldered to the risers. When the segments have no risers, the leads are soldered to short slits in the ends of the segments.

One end of a coil attaches to one bar and the other end is soldered to the adjacent bar. Each coil laps over the preceding one. This method is called lap winding and is illustrated in figure 6-18.

The brushes ride on the surface of the commutator, forming the electrical contact between the armature coils and the external circuit. A flexible, braided-copper conductor, commonly called a pigtail, connects each brush to the external circuit. The brushes, usually made of high-grade carbon and held in place by brush holders insulated from the frame, are free to slide up and down in their holders in order to follow any irregularities in the surface of the commutator. The brushes are usually adjustable so that the pressure of the brushes on the commutator can be varied and the position of the brushes with respect to the segments can be adjusted.

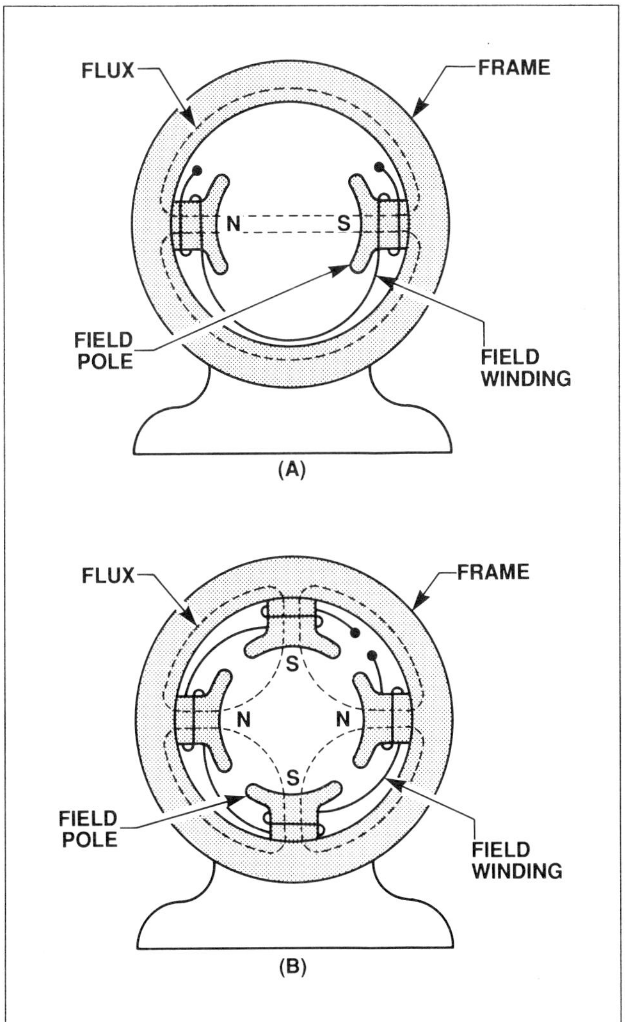

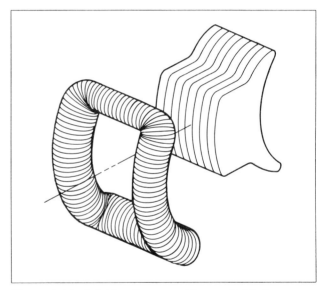

Figure 6-15. A field coil removed from a field shoe.

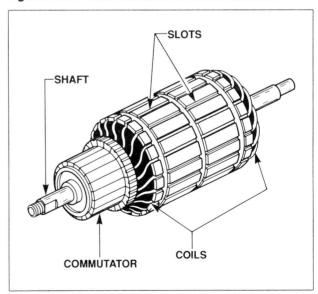

Figure 6-16. A drum-type armature.

Figure 6-14. A two-pole and a four-pole frame assembly.

103

The constant making and breaking of connections to the coils in which voltage is being induced necessitates the use of material for brushes which has a definite contact resistance. Also, this material must be such that the friction is low, to prevent excessive wear. The high-grade carbon used in the manufacture of brushes must be soft enough to prevent undue wear of the commutator and yet hard enough to provide reasonable brush life, and since the contact resistance of carbon is fairly high, the brush must be quite large to provide a large area of contact. The commutator surface is highly polished to reduce friction as much as possible. Oil or grease must never be used on a commutator, and extreme care must be used when cleaning it to avoid marring or scratching the surface.

3. Types of DC Generators

There are three types of DC generators: series-wound, shunt-wound, and shunt-series or compound-wound. The difference in type depends on the relationship of the field winding to the external circuit.

a. Series-Wound

The field winding of a series generator is connected in series with the external circuit, called the load (figure 6-20). The field coils are composed of a few turns of large wire; the magnetic field strength depends here more on the current flow than the number of turns in the coil.

Series generators have very poor voltage regulation under changing load, since the greater the current through the field coils to the external circuit, the greater the induced emf and the greater the terminal voltage. Therefore, when the load is increased, the voltage increases; likewise, when the load is decreased, the voltage decreases.

The output voltage of a series-wound generator may be controlled by a rheostat in parallel with the field windings, as shown in figure 6-20(A). Since the series-wound generator has such poor regulation, it is never employed as an airplane generator. Generators in airplanes have field windings which are connected either in shunt or in compound.

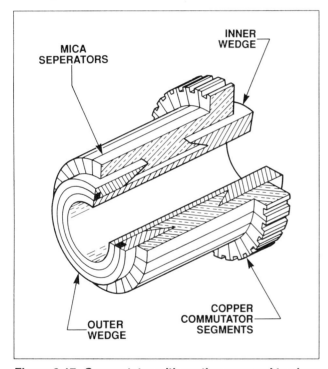

Figure 6-17. Commutator with portion removed to show construction.

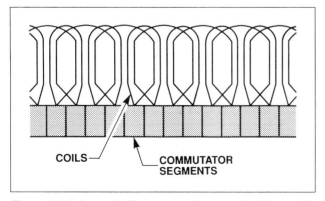

Figure 6-18. Lap winding connects one end of two coils to each commutator segment and the ends of each coil to adjacent segments.

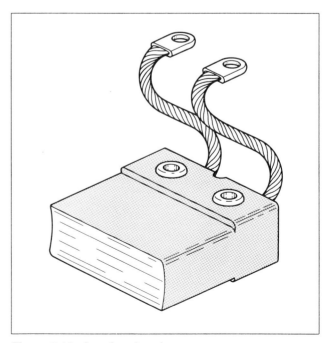

Figure 6-19. A carbon brush.

b. Shunt-Wound

A generator having a field winding connected in parallel with the external circuit is called a shunt generator and is shown in figure 6-21. The field coils of a shunt generator contain many turns of a small wire; the magnetic strength is derived from the large number of turns rather than the current strength through the coils.

If a constant voltage is desired, the shunt-wound generator is not suitable for rapidly fluctuating loads. Any increase in load causes a decrease in the terminal or output voltage, and any decrease in the load causes an increase in the terminal voltage. This is because the armature and the load are connected in series, and all current flowing in the external circuit passes through the armature winding. Because of the resistance in the armature winding, there is a voltage drop (IR drop = current × resistance). As the load increases, the armature current increases and the IR drop in the armature increases. The voltage delivered to the terminals is the difference between the induced voltage and the voltage drop; therefore there is a decrease in terminal voltage. This decrease in voltage causes a decrease in field strength, because the current in the field coils decreases in proportion to the decrease in terminal voltage; with a weaker field, the voltage is further decreased.

When the load decreases, the output voltage increases accordingly, and a larger current flows in the windings. This action is cumulative, so the output voltage continues to rise to a point called field saturation, after which there is no further increase in output voltage.

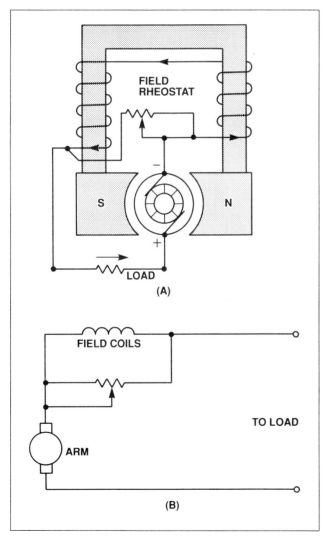

Figure 6-20. Diagram and schematic of a series-wound generator.

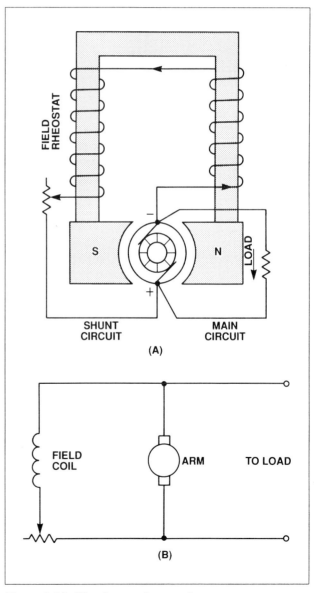

Figure 6-21. Shunt-wound generator.

The terminal voltage of a shunt generator can be controlled by means of a rheostat inserted in series with the field winding as shown in (A) of figure 6-21. As the resistance is increased, the field current is reduced; consequently, the generated voltage is reduced also. For a given setting of the field rheostat, the terminal voltage at the armature brushes will be approximately equal to the generated voltage minus the IR drop produced by the load current in the armature; thus, the voltage at the terminals of the generator will drop as the load is applied. Certain voltage-sensitive devices are available which automatically adjust the field rheostat to compensate for variations in load. When these devices are used, the terminal voltage remains essentially constant.

The output and voltage regulation capabilities of shunt-type generators make them suitable for light to medium duty use on aircraft. While once popular, these units have largely been replaced by DC alternators.

c. Compound-Wound

A compound-wound generator combines a series winding and a shunt winding in such a way that the characteristics of each are used to advantage. The series field coils are made of a relatively small number of turns of large copper conductor, either circular or rectangular in cross section, and are connected in series with the armature circuit. These coils are mounted on the same poles which the shunt field coils are mounted and, therefore, contribute a magnetomotive force which influences the main field flux of the generator. A diagrammatic and a schematic illustration of a compound-wound generator is shown in figure 6-22.

If the ampere-turns of the series field act in the same direction as those of the shunt field, the combined magnetomotive force is equal to the sum of the series and shunt field components. Load is added to a compound generator in the same manner in which load is added to a shunt generator, by increasing the number of parallel paths across the generator terminals. Thus, the decrease in total load resistance with added load is accompanied by an increase in armature circuit and series-field circuit current.

The effect of the additive series field is that of increased field flux with increased load. The extent of the increased field flux depends on the degree of saturation of the field as determined by the shunt field connection. Thus, the terminal voltage of the generator may increase or decrease with load, depending on the influence of the series field coils.

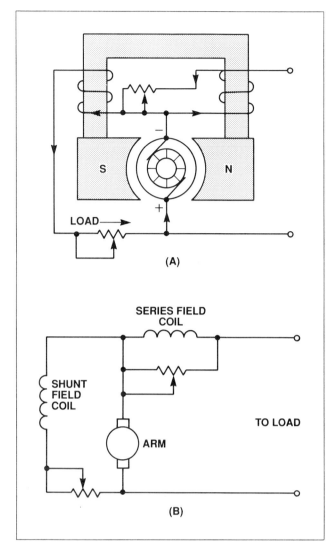

Figure 6-22. Compound-wound generator.

This influence is referred to as the degree of compounding.

Changes in terminal voltage with increasing load depends upon the degree of compounding. A flat-compound generator is one in which the no-load and full-load voltages have the same value. The under-compound generator has a full-load voltage less than the no-load value, and an over-compound generator has a full-load voltage which is higher than the no-load value.

Compound generators are usually designed to be overcompounded. This feature permits varied degrees of compounding by connecting a variable shunt across the series field. Such a shunt is sometimes called a diverter. Compound generators are used where voltage regulation is of prime importance.

If the series field aids the shunt field, the generator is said to be cumulative-compounded (figure 6-

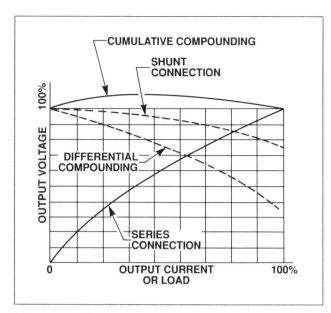

Figure 6-23. Generator characteristics.

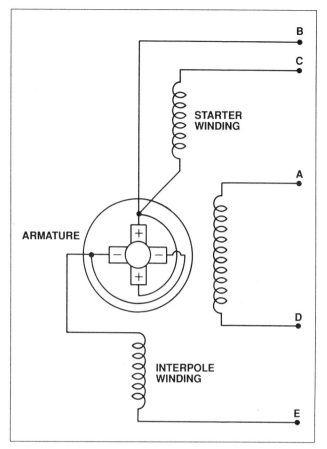

Figure 6-24. Internal wiring diagram of a typical starter-generator.

22(B)). If the series field opposes the shunt field, the generator is said to be differentially compounded.

Differential generators have somewhat the same characteristics as series generators in that they are essentially constant-current generators. However, they generate rated voltage at no load, the voltage dropping materially as the load current increases. Constant-current generators are ideally suited as power sources for electric arc welders and are used almost universally in electric arc welding.

If the shunt field of a compound generator is connected across both the armature and the series field, it is known as a long-shunt connection. If the shunt field is connected across the armature alone, it is called a short-shunt connection. These connections produce essentially the same generator characteristics.

A summary of the characteristics of the various types of generators discussed is shown graphically in figure 6-23.

d. Starter-Generators

Most small turbine engines are equipped with starter-generators rather than separate starters and generators. This effects an appreciable weight saving, as both starters and generators are quite heavy and they are never used at the same time.

The armature of a starter-generator is splined to fit into a drive pad on the engine, rather than being connected through a clutch and drive jaws as starters are.

Starter-generators are equipped with two or three sets of field winding. In figure 6-24, we have a schematic of a typical starter-generator. The generator circuit consists of the armature, a series field around the interpoles and a shunt field for generator control. A series motor field is wound around the pole shoes inside the field frame, and the end of this winding is connected to the C terminal.

For starting, current flows from the battery or external power unit through the series winding and the armature. As soon as the engine starts, the start relay disconnects this winding and connects the generator circuit to the aircraft electrical system.

4. Armature Reaction

Current flowing through the armature sets up electromagnetic fields in the windings. These new fields tend to distort or bend the magnetic flux between the poles of the generator from a straight line path. Since armature current increases with the load, the distortion becomes greater as the load increases. This distortion of the magnetic field is called armature reaction (in figure 6-25).

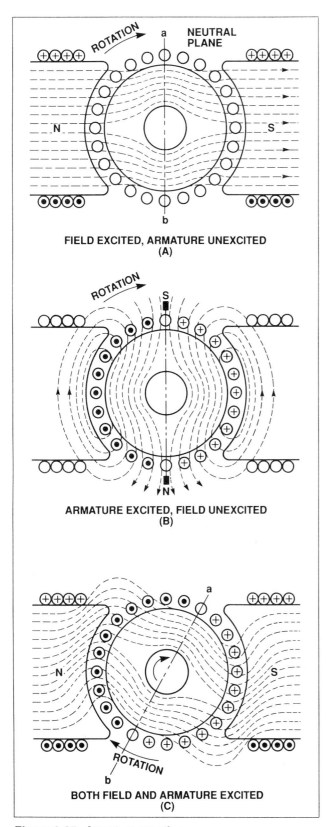

Figure 6-25. Armature reaction.

Armature windings of a generator are spaced in such a way that, during rotation of the armature, there are certain positions when the brushes contact two adjacent segments, thereby shorting the armature windings to these segments. Usually, when the magnetic field is not distorted, there is no voltage being induced in the shorted windings, and, therefore, no harmful results occur from the short circuit. However, when the field is distorted by armature reaction, a voltage is induced in these shorted windings and sparking takes place between the brushes and the commutator segments. Consequently, the commutator becomes pitted, the wear on the brushes becomes excessive, and the output of the generator is reduced.

To correct this condition, the brushes are set so that the plane of the coils which are shorted by the brushes is perpendicular to the distorted magnetic field. This is accomplished by moving the brushes forward in the direction of rotation. This operation is called shifting the brushes to the neutral plane, or plane of commutation.

The neutral plane is the position where the plane of the two opposite coils is perpendicular to the magnetic field in the generator. On a few generators, the brushes can be shifted manually ahead of the normal neutral plane to the neutral plane caused by field distortion. On nonadjustable brush generators, the manufacturer sets the brushes for minimum sparking.

Interpoles may be used to counteract some of the effects of field distortion, since shifting the brushes is inconvenient and unsatisfactory, especially when the speed and load of the generator are changing constantly. An interpole is a pole placed between the main poles of a generator. A four-pole generator with interpoles is shown in figure 6-26. This generator has four interpoles, which are north and south poles, alternately as are the main poles.

An interpole has the same polarity as the next main pole in the direction of rotation. The magnetic flux produced by an interpole causes the current in the armature to change direction as an armature winding passes under it. This cancels the electromagnetic fields about the armature windings. The magnetic strength of the interpoles varies with the load on the generator; and since field distortion varies with the load, the magnetic field of the interpoles counteracts the effects of the field set up around the armature windings and minimizes field distortion. The interpoles tend to keep the neutral plane in the same position for all loads of the generator.

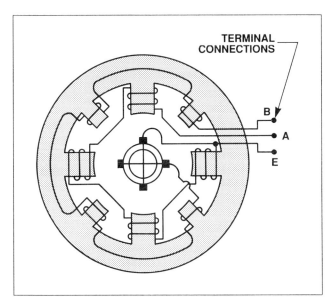

Figure 6-26. Generator with interpoles.

5. Generator Ratings

A generator is rated in power output. Since a generator is designed to operate at a specified voltage, the rating is usually given as the number of amperes the generator can safely supply at its rated voltage. Generator rating and performance data are stamped on the name plate attached to the generator. When replacing a generator, it is important to choose one of the proper rating.

The rotation of generators is termed either clockwise or counterclockwise, as viewed from the driven end. Usually, the direction of rotation is stamped on the data plate. If no direction is stamped on the data plate, the rotation may be marked by an arrow on the cover plate of the brush housing. It is important that a generator with the correct direction of rotation be used; otherwise the polarity will be reversed.

The speed of an aircraft engine varies from idle RPM to takeoff RPM; however, during the major portion of a flight, it is at a constant cruising speed. The generator drive is usually geared to revolve the generator between 1-1/8 and 1-1/2 times the engine crankshaft speed. Most aircraft generators have a speed at which they begin to produce their normal voltage. Termed the "coming-in" speed, it is usually about 1,500 RPM.

6. Generator Terminals

On most large 24-volt generators, electrical connections are made to terminals marked B, A, and E (see figure 6-27). The positive armature lead in the generator connects to the B terminal. The negative armature lead connects to the E terminal. The positive end of the shunt field winding connects to terminal A, and the opposite end connects to the negative terminal brush. Terminal A receives current from the negative generator brush through the shunt field winding. This current passes through the voltage regulator and back to the armature through the positive brush. Load current, which leaves the armature through the negative brush, comes out of the E lead and passes through the load before returning through the positive brush.

7. Regulation of Generator Voltage

Efficient operation of electrical equipment in an airplane depends on a constant voltage supply from the generator. Among the factors which determine the voltage output of a generator, only one, the strength of the field current, can be conveniently controlled. To illustrate this control, refer to the diagram in figure 6-27 showing a simple generator with a rheostat in the field circuit.

If the rheostat is set to increase the resistance in the field circuit, less current flows through the field winding and the strength of the magnetic field in which the armature rotates decreases. Consequently, the output of the generator decreases. If the resistance in the field circuit is decreased with the rheostat, more current flows through the field windings, the magnetic field becomes stronger, and the generator produces a greater voltage.

This principle may be further developed by the addition of a solenoid which will electrically connect or remove the field rheostat from the circuit as the voltage varies. This arrangement is illustrated in figure 6-28. With the generator running at normal

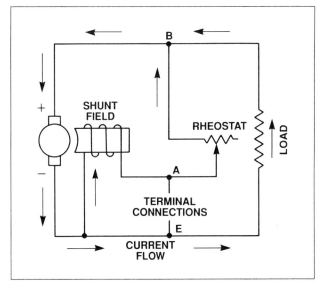

Figure 6-27. Regulation of generator voltage by field rheostat.

speed and switch K open, the field rheostat is adjusted so that the terminal voltage is about 60% of normal. Solenoid S is weak and contact B is held closed by the spring. When K is closed, a short circuit is placed across the field rheostat. This action causes the field current to increase and the terminal voltage to rise.

When the terminal voltage rises above a certain critical value, the solenoid's downward pull exceeds the spring tension and contact B opens, reinserting the field rheostat in the field circuit. This additional resistance reduces the field current and lowers terminal voltage.

When the terminal voltage falls below a certain value, the solenoid armature contact B is closed again by the spring, the field rheostat is now shorted, and the terminal voltage starts to rise. Thus, an average voltage is maintained with or without load changes.

The dashpot P provides smoother operation by acting as a dampener to prevent hunting. The capacitor C across contact B eliminates sparking.

Certain light aircraft employ a three-unit regulator for their generator systems. This type of regulator includes a current limiter and a reverse current cutout in addition to a voltage regulator. A three-unit regulator is seen in figure 6-29.

The action of the voltage regulator unit is similar to the vibrating-type regulator described above. The second of the three units is a current regulator to limit the output current of the generator. The third unit is a reverse-current cutout that disconnects the battery from the generator. If the battery

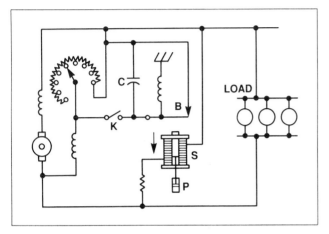

Figure 6-28. Vibrating-type voltage regulator.

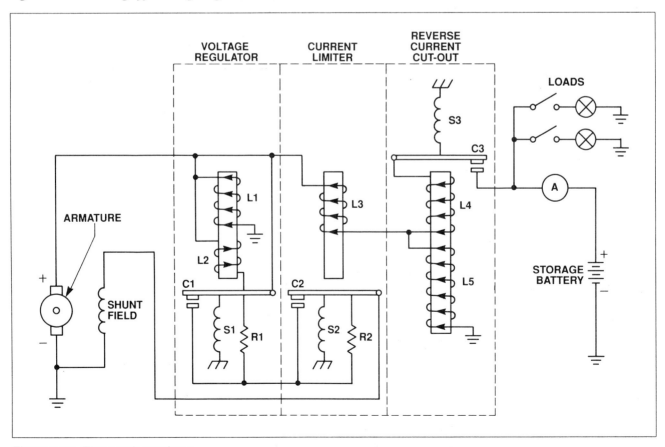

Figure 6-29. Three-unit regulator.

is not disconnected, it will discharge through the generator armature when the generator voltage falls below that of the battery. When this occurs, the battery will attempt to drive the generator as a motor. This action is called "motoring" the generator and, unless it is prevented, it will discharge the battery in a short time.

Vibrating-type regulators cannot be used with generators which require a high field current since the contacts will pit or burn. Heavy-duty generator systems require a different type of regulator, such as the carbon-pile voltage regulator.

The carbon-pile voltage regulator depends on the resistance of a number of carbon disks arranged in a pile or stack. The resistance of the carbon stack varies inversely with the pressure applied. When the stack is compressed under appreciable pressure, the resistance in the stack is less. When the pressure is reduced the resistance of the carbon stack increases because there is more air space between the disks, and air has a high resistance.

Pressure on the carbon pile depends upon two opposing forces: a spring and an electromagnet. The spring compresses the carbon pile, and the electromagnet exerts a pull which decreases the pressure.

Whenever the generator voltage varies, the pull of the electromagnet varies. If the generator voltage rises above a specific amount, the pull of the electromagnet increases, decreasing the pressure exerted on the carbon pile and increasing its resistance. Since this resistance is in series with the field, less current flows through the field winding, there is a corresponding decrease in field strength, and the generator voltage drops. On the other hand, if the generator output drops below the specified value, the pull of the electromagnet is decreased and the carbon pile places less resistance in the field winding circuit. In addition, the field strength increases and the generator output increases. A small rheostat provides a means of adjusting the current flow through the electromagnet coil. Figure 6-30 shows a typical 24-volt voltage regulator with its internal circuits.

8. DC Generator Service and Maintenance

Because of the relative simplicity and durable construction generators will operate many hours without a hint of trouble. The routine inspection and service at each 100-hour or annual inspection interval is generally all that is required. Generator overhaul is often accomplished at the same time as engine overhaul, minimizing aircraft down time and increasing the likelihood of trouble free operation when the aircraft is placed back in service.

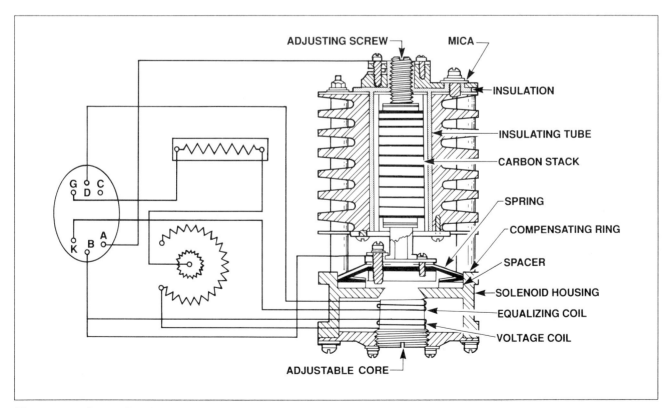

Figure 6-30. A 24-volt voltage regulator showing internal circuits.

111

a. Routine Inspection and Servicing

The 100-hour and/or annual inspection of the generator should include the following items:

1. Inspect the generator for security of mounting, checking the mounting flange for cracks and looseness of mounting bolts.
2. Inspect the mounting flange area for oil leaks.
3. Inspect generator electrical connections for cleanliness and security of attachment.
4. Remove the band covering the brushes and commutator. Use compressed air to blow out accumulated dust. Inspect the brushes for wear, and freedom of movement. Check the tension of the brush springs, using a spring scale.
5. Inspect the commutator for cleanliness, wear, and pitting.
6. Inspect the area around the commutator and brush assemblies for any particles of solder. This indicates that overheating has occurred, and the generator should be removed.

If a DC generator is unable to keep the aircraft battery charged, and if the ammeter does not show the proper rate of charge, you should first check the aircraft electrical system that is associated with the battery and with the generator. Physically check every connection in the generator and battery circuit and electrically check the condition of all of the fuses and circuit breakers. Check the condition of all ground connections for the battery, battery contactor, and the generator control units. When you have determined that there is no obvious external problem and the generator armatures does turn when the engine is cranked, you should next determine which is at fault, the generator or the generator control.

One of the easiest ways of determining which unit is not operating is to connect a voltmeter between the G terminal of the voltage regulator and ground to check the voltage output of the generator. Because this check requires that the generator be turning it must be accomplished with the engine running, or on an appropriate test stand. In either case, observe the proper safety precautions. Even if the field winding is open, or the voltage regulator is malfunctioning, the generator should produce residual voltage; that is, the voltage produced by the armature cutting across the residual magnetic field in the generator frame. This should be about one or two volts.

If there is no residual voltage, it is possible that the generator need only have the residual magnetism restored. Residual magnetism is restored by an operation known as "flashing the field". This is accomplished by momentarily passing current through the field coils in the same way that it normally flows.

The method of accomplishing this task varies with the internal connections of the generator, and with the type of voltage regulator used. Be certain to follow the specific instructions of the equipment manufacturer as failure to do so could result in damage to the generator and/or voltage regulator.

If the generator produces residual voltage but no output voltage, the trouble could lie with the generator or with the regulator. To determine which, operate the engine at a speed high enough for the generator to produce an output, and bypass the voltage regulator with a jumper wire. Again, this connection will vary with the type of generator and regulator being used, and this operation should be performed in accordance with the manufacturer's recommendations.

If the generator produces voltage with the regulator shorted, the problem is with the voltage regulator. Be sure that the regulator is properly grounded, as a faulty ground connection will prevent its functioning properly.

It is possible to service and adjust some vibrator-type generator controls but, because of the expense of the time involved and the test equipment needed to do the job properly, most servicing is done by replacing a faulty unit with a new one.

If the generator does not produce an output voltage when the regulator is bypassed, remove the generator from the engine and overhaul it or replace it with an overhauled unit.

b. Generator Overhaul

Generator overhaul is normally accomplished at any time the generator has been determined to be inoperative, or at the same time as the aircraft's engine is overhauled. The generator overhaul may be accomplished in some aircraft repair facilities, but is nowadays more often the job of an FAA Certified Repair Station licensed for that operation.

The steps involved in the overhaul of a generator are the same as for the overhaul of any unit: (1) disassembly, (2) cleaning, (3) inspection and repair, (4) reassembly, and (5) testing.

1) Disassembly

The disassembly instruction for specific units will be covered in the overhaul instructions supplied by the manufacturer, and must be followed exactly. Some specialized tools may be required for removing the pole shoes. The screws holding these in place are usually staked to prevent their accidently backing out, and some force will be required to remove them.

Care must be taken and the manufacturer's instructions followed when removing bearings as incorrect procedures or tools could result in damage to the bearings or their seating area.

2) Cleaning

Care must be taken in cleaning electrical parts. The proper solvents must be used, and generally parts are not submerged in solvent tanks. The use of a wrong solvent could remove some of the lacquer-type insulation used on field coils and armatures and cause short circuits.

3) Inspection and Repair

Inspect components for physical damage, corrosion, and wear, and repair or replace as required. Testing of electrical components is usually accomplished using a growler and an electrical multimeter. The growler is a specially designed test unit for DC generators and motors. A variety of tests on the armature and field coils may be performed using this equipment. Growlers consist of a laminated core wound with many turns of wire and is connected to 110-volts AC. The top of the core forms a vee into which the armature of a DC generator will fit. The coil and laminated core of the growler form the primary of a transformer, of which the generator's armature becomes the secondary. Also included on most growlers is a 110-volt test lamp.

This is a simple series circuit with a light bulb which will illuminate when the circuit is completed. A growler may be seen in figure 6-31.

Because the growler forms the primary of a transformer, and the armature the secondary, it stands to reason that when a voltage is applied to the primary a voltage will appear in the secondary. If there are any open circuits in the secondary (armature) no voltage will be present.

To test an armature for opens, it should be placed on the growler and the growler energized. Using the probes attached to a test lamp, test each coil of the armature by placing the probes on adjacent segments. The lamp should light with each set of commutator bars. Failure of the test lamp to illuminate indicates an open in that coil, and replacement of the armature is called for. This test is illustrated in figure 6-32.

Armatures may be tested for shorts by placing them on a growler, energizing the unit and holding a thin steel strip (hacksaw blade) slightly above the armature. Slowly rotate the armature on the growler, if there are any shorts in the armature windings, the blade will vibrate vigorously.

A third test for armatures that may be accomplished using the 110-volt test lamp is for grounds. If a ground exists between the windings and the core of the armature exists, it will be indicated by placing one lead of the test lamp on the armature shaft and moving the other probe

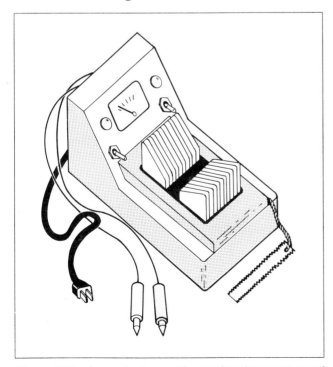

Figure 6-31. A growler is used for testing the armature of a DC generator.

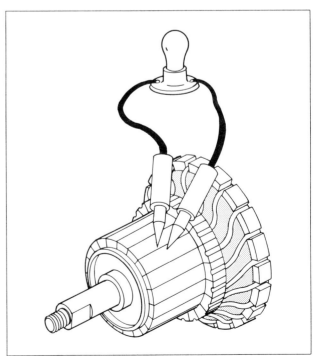

Figure 6-32. Testing an armature for open coils.

around all of the commutator bars. This test may also be accomplished using an ohmmeter.

The field coils may be tested for grounds between the coils and the field frame using the 110-volt test light. The test lamp should be connected in series with the field frame and one terminal of the field coil. If the lamp lights, there is a ground and the coil is defective. This test may also be accomplished using an ohmmeter. Infinite resistance between the windings and the field frame should be indicated.

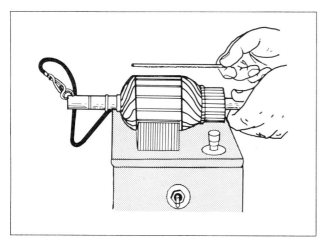

Figure 6-33. Testing an armature for shorts using a growler.

To test the field coil for continuity using an ohmmeter, select the low ohms scale. A shunt field coil should indicate between 2 and 30 ohms, depending on the specific coil. A series-type field coil will show almost no resistance.

A current draw test may be specified by the manufacturer. This test is accomplished by connecting a battery of the proper voltage across the field coils and measuring the current flow. This value must be within the limits specified in the manufacturer's test specifications. Figure 6-36 illustrates how this test may be accomplished.

Commutators may be resurfaced to remove any irregularities or pitting. This is normally accomplished by turning the armature in a special armature lathe or using an engine lathe equipped with a special holding fixture.

Remove only enough metal to smooth the surface of the commutator. If too much material is removed, the security of the coil ends may be jeopardized. If the commutator is only slightly roughened, it may be smoothed using No. 000 sandpaper. Never use emery cloth or other conductive abrasives as shorting between commutator segments may result.

Following the most recent instructions provided by the manufacturer, the mica insulation between the commutator bars may need to be undercut.

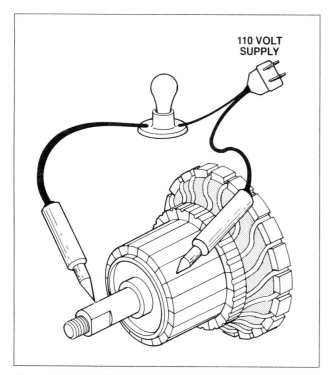

Figure 6-34. Testing for a ground in the armature windings.

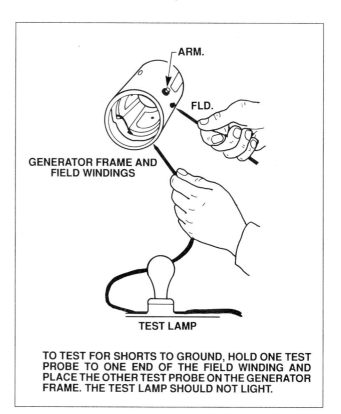

Figure 6-35. Testing the field coils for shorts to ground.

This operation can be accomplished using a special attachment for the armature lathe or simply using a hack-saw blade. When specified, the mica is usually undercut about the same depth as the thickness of the mica or approximately 0.020".

4) Reassembly

Prior to reassembly the painted finish on the exterior of the frame may be restored, and in certain cases the special insulated coatings on the interior surfaces are renewed. Replace all defective parts, and follow the reassembly order specified by the manufacturer.

Make certain that all internal electrical connections are properly made and secured, and that the brushes are free to move in their holders. Check the pigtails on the brushes for freedom. The pigtails should never be permitted to alter or restrict free motion of the brushes.

The purpose of the pigtail is to conduct the current, rather than subjecting the brush spring to currents which would alter its spring action by overheating. The pigtails also eliminate possible sparking of the brush guides cause by movement of the brush within the holder, thus minimizing side wear of the brush.

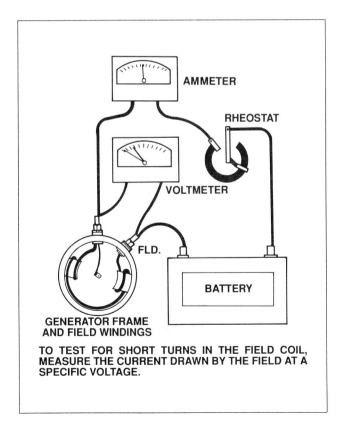

Figure 6-36. Current draw test for generators.

Generator brushes are normally replaced at overhaul, and the spring tension is checked using a spring scale. A carbon, graphite, or light metalized brush should exert a pressure of 1½ to 2½ PSI on the commutator. The spring tension must be within the limits set by the manufacturer, or the springs will have to be replaced.

When a spring scale is used, the measurement of the pressure which a brush exerts is read directly on the scale. The scale is applied at the point of contact between the spring arm and the top of the brush, with the brush installed in the guide. The scale is drawn up until the arm just lifts of the brush surface. At this instant, the force on the scale should be read.

When new brushes are installed they must be seated so that the face of the brush will match the contour of the commutator. This is to maximize the contact area. Seating may be accomplished using a strip of No. 000 or finer sandpaper.

The brush should be lifted slightly to permit the insertion of the sandpaper strip under the brush rough side out (figure 6-37). Pull the sandpaper in the direction of armature rotation, being careful to keep the ends of the sandpaper as close to the commutator as possible in order to avoid rounding the edges of the brush. When pulling the sandpaper back to the starting point, the brush should be raised so it does not ride on the sandpaper. The brush should be sanded only in the direction of rotation.

After the generator has run for a short period, brushes should be inspected to make sure that pieces of sand have not become embedded in the brush and are collecting copper.

Under no circumstances should emery cloth or similar abrasives be used for seating brushes (or smoothing commutators), since they contain conductive materials.

5) Testing

Operational testing of generators may be accomplished on test benches built for that purpose. Bench testing allows an opportunity for the technician to flash the field, and assure proper operation of the unit before installation on the aircraft. Generator manufacturers supply test specifications in their overhaul instructions, follow them exactly.

B. DC Alternators and Their Controls

Two of the limitations of DC generators for aircraft installations are the limited number of pairs of poles that can be used, and the fact that the load current is produced in the rotating member and must be brought out of the generator through the brushes.

DC alternators solve these two problems, and since they produce three-phase AC and convert it into direct current with built-in solid-state rectifiers, their output at low engine speed allows them to keep the battery charged even when the aircraft is required to operate on the ground with the engine idling—as it must often do when waiting for clearance to takeoff.

1. DC Alternators

DC alternators do exactly the same thing as DC generators. They produce AC and convert it into DC before it leaves the device. The difference being that in an alternator, field current is taken into the rotor through brushes which ride on smooth slip rings. The AC load current is produced in the fixed windings of the stator, and after it is rectified by six solid-state diodes it is brought out of the alternator through solid connections.

a. The Rotor

The rotor of an alternator consists of a coil of wire wound on an iron spool between two heavy iron segments that have interlacing fingers around their periphery. Some rotors have four fingers and others have as many as seven. Each finger forms one pole of the rotating magnetic field.

The two ends of the coil pass through one of the segments and each end of the coil is attached to an insulated slip ring. The slip rings, segments, and coil spool are all pressed onto a hardened steel rotor shaft which is either splined or has a key slot machined in it. This shaft can be driven from an engine accessory pad or fitted with a pulley and belt driven from an engine drive. The slip-ring end of the shaft is supported in the housing with a needle bearing and the drive end with a ball bearing.

Two carbon brushes ride on the smooth slip rings to bring current into the field and carry it back out to the regulator.

b. The Stator

The stator coils in which the load current is produced are wound in slots around the inside periphery of the stator frame, which is made of thin laminations of soft iron. There are three sets of coils in the stator and these sets are joined into a Y-connection with one end of each set of windings brought out of the stator and attached to the rectifier.

With the stator wound in the three-phase configuration, there is a peak of current produced in each set of windings every 120° of rotor rotation. The AC produced in the stator looks much like what we see in figure 6-41(A), and after it is rectified by the three-phase, full-wave rectifier, the DC output is much like that in figure 6-41(B).

The large number of field poles and the equally large number of coils in the stator cause the al-

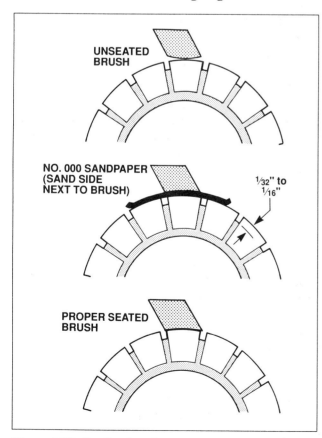

Figure 6-37. Seating brushes with sandpaper.

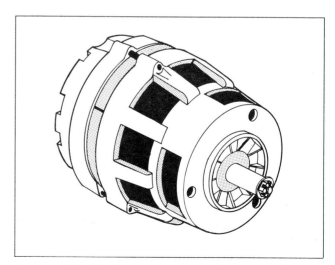

Figure 6-38. DC alternators are used in almost all of the modern aircraft that require a low or medium amount of electrical power.

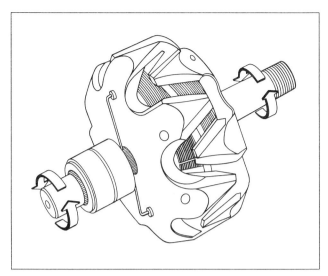

Figure 6-39. DC alternator rotor.

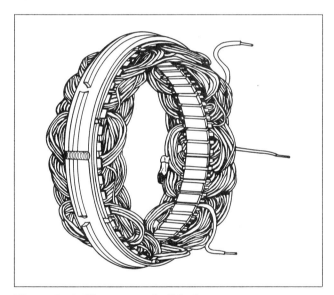

Figure 6-40. The stator of a DC alternator.

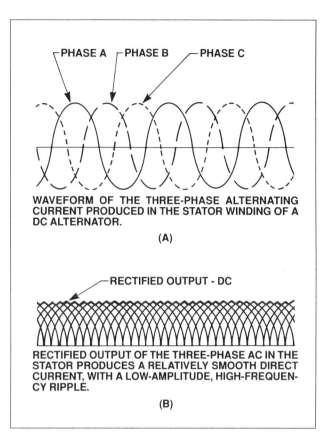

Figure 6-41. Output of a DC alternator.

ternator to put out its rated current at a low alternator RPM.

c. The Rectifier

The three-phase, full wave rectifier is made up of six heavy-duty silicon diodes. Three of them are pressed into the slip-ring end frame, and the other three are pressed into a heat sink that is electrically insulated from the end frame.

In figure 6-42, we see the way this rectifier converts the three-phase AC output of the stator into direct current. Since we are dealing with solid-state devices whose symbols use arrowheads, it is easier to follow the action if we think in terms of conventional current that follows the direction in which the arrows point. This imaginary flow is from positive to negative.

At the instant the output terminal of winding A is positive with respect to the output end of winding C, current flows through diode 1 in the heat sink, through the load, and back through diode 2 that is pressed into the alternator end frame. From this diode, it flows back through winding C. As the rotor continues to turn, winding B becomes positive with respect to winding A and the current flows through diode 3, the load, and back through diode 4 and winding A. Next, C becomes positive with respect to B and current flows out through diode 5 and back through diode 6. In this way, the current flows through the load in the same direction all of the time. The terminal connected to the heat sink is the positive terminal of the alternator and the end frame into which the anodes of diodes 2, 4, and 6 are pressed becomes the negative terminal of the alternator.

2. Alternator Controls

The voltage produced by an alternator is controlled in exactly the same way it is controlled in a generator, by varying the field current. When the voltage rises above the desired value, the field current is decreased and when the voltage drops too low, the field current is increased.

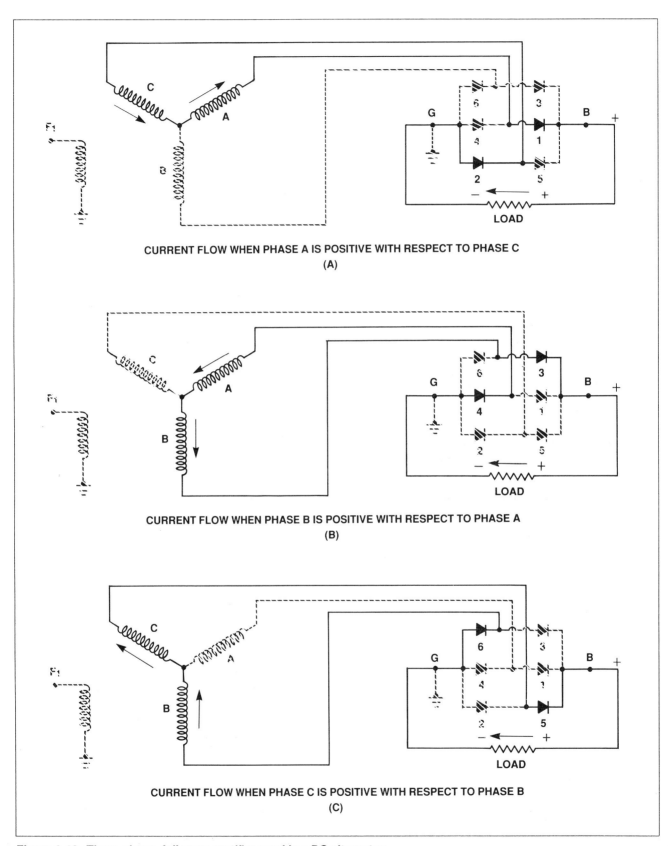

Figure 6-42. Three-phase, full-wave rectifier used in a DC alternator.

This action may be accomplished in low output alternators with vibrator-type controls which interrupt the field current by opening the contacts. A much more efficient means of voltage control has been devised which uses a transistor to control flow of field current.

The transistorized voltage regulator utilizes both vibrating points and transistors for voltage control. The vibrating points operate in exactly the same way they do in a normal vibrator-type voltage regulator, but instead of the field current flowing through the contacts, only the transistor base current flows through them. This is so small compared with the field current which flows through the emitter-collector portion of the transistor that there is no arcing at the contacts. A simplified schematic of this circuit is shown in figure 6-43.

The transistorized voltage regulator is a step in the right direction, but semiconductor devices may be used to replace all of the moving parts and a completely solid-state voltage regulator may be built. These units are very efficient, reliable, and generally have no serviceable components. If the unit is defective, it will be removed and replaced with a new one.

Alternator control requirements are different from those of a generator for several reasons. An alternator uses solid-state diodes for its rectifier and since current cannot flow from the battery into the alternator, there is no need for a reverse-current cutout relay. The field of an alternator is excited from the system bus, whose voltage is limited either by the battery or by the voltage regulator, so there is no possibility of the alternator putting out enough current to burn itself out, as a generator with its self-excited field can do. Because of this, there is no need for a current limiter. There must be one control with an alternator, however, that is not needed with a generator, and that is some means of shutting off the flow of field current when the alternator is not producing power. This is not needed in a generator since its field is excited by its own output. An alternator uses either a field switch or a field relay.

Most modern aircraft alternator circuits employ some form of overvoltage protection to remove the alternator from the bus if it should malfunction in such a way that its output voltage rises to a dangerous level.

3. DC Alternator Service and Maintenance

If an alternator fails to keep the battery charged, you should first determine that all of the alternator

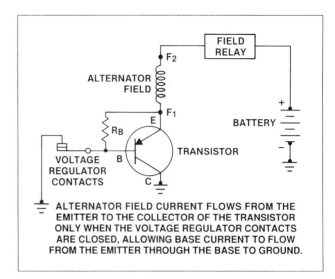

Figure 6-43. Transistorized voltage regulator.

and battery circuits in the electrical system are properly connected and that there are no open fuses or circuit breakers. There should be battery voltage at the B terminal of the alternator and at the Batt or (+) terminal of the voltage regulator.

It is extremely important in an alternator installation that the battery be connected with the proper polarity, and that anytime an external power source is connected to the aircraft it must have the correct polarity. Connecting a battery or APU with the poles reversed can burn out the rectifying diodes.

The solid-state diodes in an alternator are quite rugged and have a long life when they are properly used, but they can be damaged by excessive voltage or by reverse current flow. For this reason, an alternator must never be operated without being connected to an electrical load, as the voltage can rise high enough to destroy the diodes.

Alternators receive their field current from the aircraft bus and do not depend upon residual magnetism to get them started. Since there is no need for residual magnetism, alternators must *never* have their field flashed (or polarized) as this action is sometimes called.

To aid in systematic troubleshooting, some manufacturers have made test equipment available that can be plugged into the aircraft electrical system between the voltage regulator and the aircraft system to indicate by use of indicator lights whether the trouble is in the voltage regulator, the overvoltage sensing circuit, or in the alternator field or output circuit. By using this type of test equipment, much time can be saved and unnecessary replacement of good components can be avoided. If systematic troubleshooting indicates that the alternator is at fault, it can be disassembled and repaired.

There are basically two problems that prevent an alternator from producing electrical power. The most likely is a shorted or open diode in the rectifying circuit and there is a possibility of an open circuit in the field.

To check for a shorted circuit, measure the resistance between the B terminal of the alternator and ground. Set the ohmmeter on the R × 1 scale and measure the resistance. The reverse the ohmmeter leads and measure the resistance again. With one measurement, the batteries in the ohmmeter forward bias the diodes and you should get a relatively low resistance reading. When you reverse the leads, the batteries reverse bias the diodes and you should get an infinite or a very high reading. If you do not get an infinite or very high reading, one or more of the diodes are shorted.

You cannot detect an open diode with this type of ohmmeter test, because the diodes are connected in parallel and the ohmmeter cannot detect one diode that is open when it is in parallel with other diodes that are good. But, if the alternator does not produce sufficient output voltage and everything else appears in good condition, you should disconnect the diodes from the circuit and test them individually with an ohmmeter. A good diode will have an infinite or very high resistance in one direction and a relatively low resistance when the ohmmeter leads are reversed. An open diode will give a high resistance reading with both positions of the ohmmeter leads, and the diode will have to be replaced.

C. Alternators

DC is normally used as the main electrical power for aircraft, because it can be stored and the aircraft engines can be started using battery power. Large aircraft require elaborate ground service facilities and require external power sources for starting, they can take advantage of the weight saving provided by using alternating current for their main electrical power.

AC has the advantage over DC in that its voltage can be stepped up or down. If we need to carry current for a long distance, we can pass the AC through a step-up transformer to increase the voltage and decrease the current. The high voltage AC can be conducted to the point it will be used through a relatively small conductor, and at its destination it is passed through a step-down transformer where its voltage is lowered and its current is stepped back up to the value we need.

It is an easy matter to convert AC into DC when we need direct current to charge batteries or to operate variable speed motors. All we need to do is pass the AC through a series of semiconductor diodes. This changes the AC into DC with relatively little loss.

1. Types of Alternators

Alternators are classified in several ways in order to distinguish properly the various types. One means of classification is by the type of excitation system used. In alternators used on aircraft, excitation can be affected by one of the following methods:

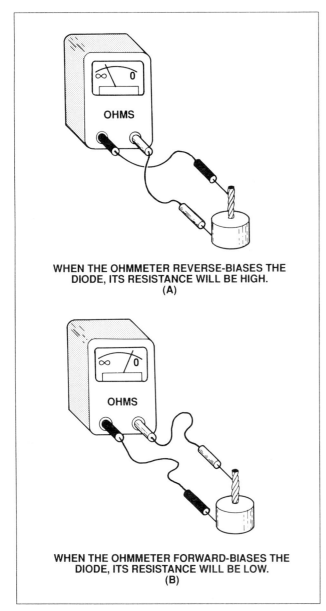

Figure 6-44. Method of checking an alternator rectifier diode.

a. *A direct-connected, DC generator.* This system consists of a DC generator fixed on the same shaft with the AC generator. A variation of this system is a type of alternator which uses DC from the battery for excitation, after which the alternator is self-excited.

b. *By transformation and rectification from the AC system.* This method depends on residual magnetism for initial AC voltage build-up, after which the field is supplied with rectified voltage from the AC system.

c. *Integrated brushless-type.* This arrangement has a DC generator on the same shaft with an AC generator. The excitation circuit is completed through silicon rectifiers rather than a commutator and brushes. The rectifiers are mounted on the generator shaft and their output is fed directly to the AC generator's main rotating field.

Another method of classification is by the number of phases of output voltage. AC generators may be single-phase, two-phase, three-phase, or even six-phase and more. In the electrical systems of aircraft, the three-phase alternator is by far the most common.

Still another means of classification is by the type of stator and rotor used. From this standpoint, there are two types of alternators: the revolving-armature and the revolving-field. The revolving-armature alternator is similar in construction to the DC generator in that the armature rotates through a stationary magnetic field. The revolving-armature alternator is found only in alternators of low power rating and generally is not used.

The revolving-field alternator (figure 6-45) has a stationary armature winding (stator) and a rotating-field winding (rotor). The advantage of having a stationary armature winding is that the armature can be connected directly to the load without having sliding contacts in the load circuit. The direct connection to the armature circuit makes possible the use of large cross-section conductors adequately insulated for high voltage.

Since the rotating-field alternator is used almost universally in aircraft systems, this type will be explained, in detail, as a single-phase, two-phase, and three-phase alternator.

a. Single-Phase Alternator

Since the emf induced in the armature of a generator is alternating, the same sort of winding can be used on an alternator as on a DC generator. This type of alternator is known as a single-phase alternator, but since the power delivered by a single-phase circuit is pulsating, this type of circuit is

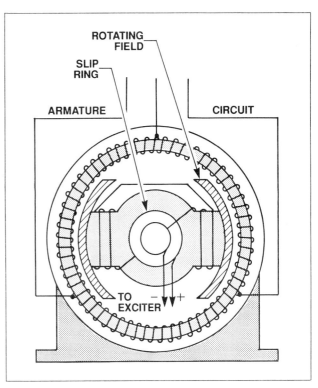

Figure 6-45. Alternator with stationary armature and rotating field.

objectionable in many applications. Figure 6-46 illustrates a schematic diagram of a single-phase alternator having four poles.

The four stator windings are connected to each other so that the AC voltages are in phase or "series adding". All four stator coil groups are connected in series so that the voltages induced in each winding will add to give a total voltage that is four times the voltage in any one winding.

b. Two-Phase Alternator

Two-phase alternators have two or more single-phase windings spaced symmetrically around the stator. In a two-phase alternator, there are two single-phase windings spaced physically so that the AC voltage induced in one is 90° out of phase with the voltage induced in the other. The windings are electrically separate from each other. When one winding is being cut by maximum flux, the other is being cut by no flux. This condition establishes a 90° relation between the two phases.

c. Three-Phase Alternator

A three-phase or polyphase circuit, is used in most aircraft alternators, instead of a single or two-phase alternator. The three-phase alternator has three single-phase windings spaced so that the voltage induced in each winding is 120° out of phase with the voltage in the other two windings.

A simplified schematic diagram, showing each of the three phases, is illustrated in figure 6-47. The rotor is omitted for simplicity. The waveforms of the voltage are shown to the right of the schematic. The three voltages are 120° apart and are similar to the voltages that would be generated by three single-phase alternators, whose voltages are out of phase by 120°. The three phases are independent of each other.

Rather than having six leads from the three-phase alternator, one of the leads from each phase may be connected to form a common junction. The stator is then called Y- or star-connected, and is illustrated in figure 6-48(A). The common lead may or may not be brought out of the alternator. If it is brought out it is called the neutral lead.

A three-phase stator can also be connected so that the phases are connected end-to-end as shown in figure 6-48(B). This arrangement is called a delta connection.

d. Brushless Alternators

Most of the AC generators used in the large jet-powered aircraft are of the brushless type and are usually air-cooled. Since the brushless alternators have no current flow between brushes or slip rings, they are quite efficient at high altitudes where brush arcing could be a problem.

In figure 6-49, we have a schematic of a brushless alternator. The exciter field current is brought into the alternator from the voltage regulator. Here it produced the magnetic field for the three-phase exciter output. Permanent magnets furnish the magnetic flux to start the generator producing an output before field current flows. The voltage produced by these magnets is called residual voltage. The output from the exciter is rectified by six silicon diodes, and the resulting DC flows

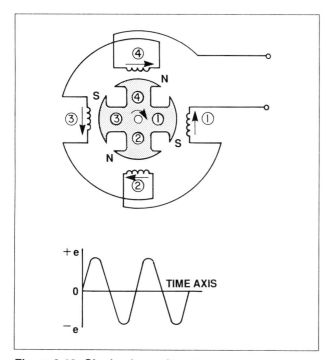

Figure 6-46. Single-phase alternator.

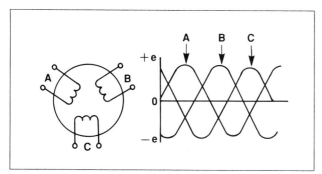

Figure 6-47. Simplified schematic of three-phase alternator with output waveforms.

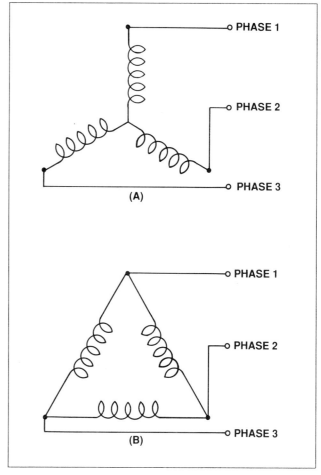

Figure 6-48. Schematic diagrams of Y and delta connections.

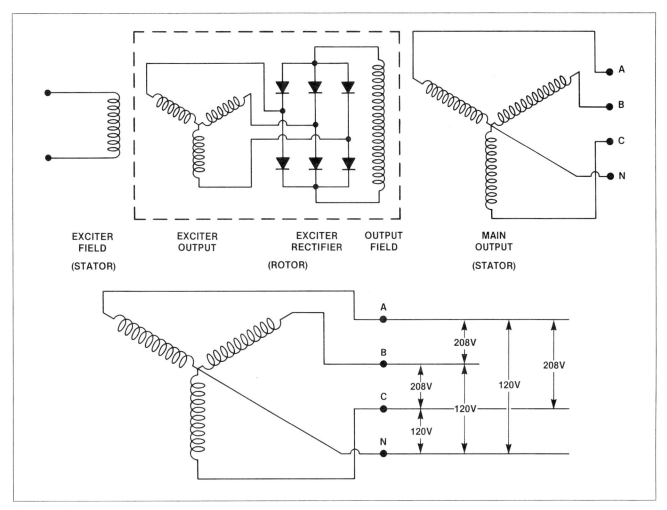

Figure 6-49. Brushless alternator.

through the output field winding. The exciter output winding, the six diodes, and the output field winding are all mounted on the generator shaft and rotate as a unit. The three-phase output stator windings are wound in slots in the laminated frame of the alternator housing, and their ends are connected in the form of a Y with the neutral and the three-phase windings brought out to terminals on the outside of the housing. These alternators are usually designed to produce 120 volts between any of the phase terminals and the neutral terminal and 208 volts between any of the phase terminals.

2. Alternator Ratings

The true power produced in an AC generator is the product of the voltage and that portion of the current that is in phase with the voltage, and is expressed in watts or kilowatts. It is this power that determines the amount of useful work the electricity can do.

a. KVA

AC generators are rated, however, not in watts, but in volt-amps, which is a measure of the apparent power being produced by the generator. Because of the outputs of most large aircraft alternators the ratings will generally be expressed in KVA (kilo volt-amperes). The reason for using this rating is that it is the heating effect of the current in the generator windings that limits generator output, and this current flows in the windings whether it is producing power or not.

b. Frequency

The frequency of the AC produced by an AC generator is determined by the number of poles and the speed of the rotor. The faster the speed, the higher the frequency will be; the lower the speed, the lower the frequency becomes. The more poles on the rotor, the higher the frequency will be for any given speed. The frequency of the alternator

123

in cycles per second (hertz) is related to the number of poles and the speed, as expressed by the equation:

$$F = \frac{P}{2} \times \frac{N}{60} = \frac{PN}{120}$$

where

P = the number of poles
N = the speed in RPM

For example, a two-pole, 3,600 RPM alternator has a frequency of:

$$\frac{2 \times 3,600}{120} = 60 \text{ cycles per second (hertz)}$$

Inductance is a characteristic of a conductor that produces a back voltage when current changes its rate of flow or direction. Since AC is constantly changing its rate and periodically changing direction, there is always a back voltage being produced. This back voltage causes an opposition to current flow that we call inductive reactance. The higher the frequency of AC, the greater this opposition will be. Commercial AC (house current) in the United States has a frequency of 60 Hz, but most aircraft systems use 400-Hz AC. At this higher frequency the inductive reactance is high and current is low. Motors can produce their torque when wound with smaller wire and transformers can be made much smaller and lighter for use with this higher frequency.

To provide a constant frequency as the engine speed varies, many engine-driven aircraft AC generators are connected to the engine through a hydrostatically operated constant speed drive unit, a CSD.

These drive units normally consist of an axial-piston variable-displacement hydraulic pump driven by the engine, supplying fluid to an axial-piston hydraulic motor which drives the generator. The displacement of the pump is controlled by a governor which senses the rotational speed of the AC generator. This governor action holds the output speed of the generator constant and maintains the frequency of the AC at 400-Hz, plus or minus established tolerances.

Some of the modern jet aircraft produce their alternating current with a generator similar to the one we see in figure 6-51. This unit is called an Integrated Drive Generator (IDG) and it includes a constant speed drive unit in the housing with the generator.

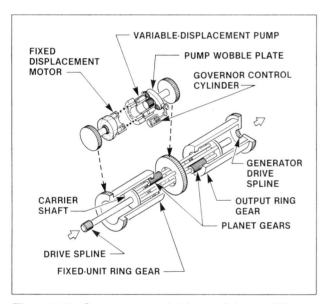

Figure 6-50. Constant-speed drive, axial-gear differential used in the Sunstrand Integrated Drive Generator.

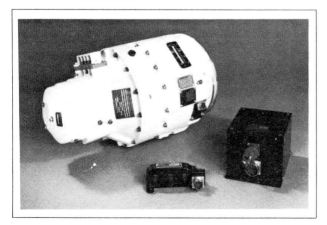

Figure 6-51. A Sunstrand Integrated Drive Generator with its control unit and a current transformer assembly.

3. Alternator Maintenance

Maintenance and inspection of alternator systems is similar to that of DC systems. Proper maintenance of an alternator requires that the unit be kept clean and that all electrical connections are tight and in good repair. Because alternators and their drive systems differ in design and maintenance requirements, no attempt will be made here to detail those procedures. Specific information may be found in the manufacturer's service publications and in the maintenance program approved for the particular aircraft.

D. DC Motors

Many devices in an airplane, from the starter to the automatic pilot, depend upon the mechanical energy furnished by direct-current motors. A DC motor is a rotating machine which transforms DC electrical energy into mechanical energy.

1. Motor Theory

Whenever a current-carrying wire is placed in the field of a magnet, a force acts on the wire. The force is not one of attraction or repulsion; however, it is at right angles to the wire and also at right angles to the magnetic field set up by the magnet.

The action of the force upon a current-carrying wire placed in a magnetic field is shown in figure 6-52. A wire is located between two permanent magnets. The lines of force in the magnetic field are from the north pole to the south pole. When no current flows (diagram A) no force is exerted on the wire. When current flows through the wire, a magnetic field is set up about it (diagram B). The direction of the field depends upon the direction of current flow. Current in one direction creates a clockwise field about the wire, and current in the other direction, a counter clockwise field.

Since the current-carrying wire produces a magnetic field, a reaction occurs between the field about the wire and the magnetic field between the magnets. When the current flows in a direction to create a counterclockwise magnetic field about the wire, this field and the field between the magnets add or reinforce at the bottom of the wire because the lines of force are in the same direction. At the top of the wire, they subtract or neutralize, since the top lines of force in the two fields are opposite in direction. Thus, the resulting field at the bottom is strong and the one at the top is weak. Consequently, the wire is pushed upward as shown in diagram C of figure 6-52. The wire is always pushed away from the side where the field is the strongest.

If current flow through the wire were reversed in direction, the two fields would add at the top and subtract at the bottom. Since a wire is always pushed away from the strong field, the wire would be pushed down.

a. Force Between Parallel Conductors

Two wires carrying current in the vicinity of one another exert a force on each other because of their magnetic fields. An end view of two conductors is shown in figure 6-53. In view A, electron flow in both conductors is toward the reader, and the magnetic fields are clockwise around the conductors. Between the wires, the fields cancel each other because the directions of the two fields oppose each other. The wires are forced in the direction of the weaker field, toward each other. This force is one of attraction.

In view B, the electron flow in the two wires is in opposite directions. The magnetic fields are, therefore, clockwise in one and counterclockwise in the other, as shown. The fields reinforce each other between the wires, and the wires are forced in the direction of the weaker field, away from each other. This force is one of repulsion.

To summarize: Conductors carrying current in the same direction tend to be drawn together; conductors carrying current in opposite directions tend to be repelled from each other.

b. Developing Torque

If a coil in which current is flowing is placed in a magnetic field, a force is produced which will cause the coil to rotate. In the coil shown in figure 6-54, current flows inward on side A and outward on side B. The magnetic field about B is clockwise and that about A, counterclockwise. As previously explained, a force will develop which pushes side

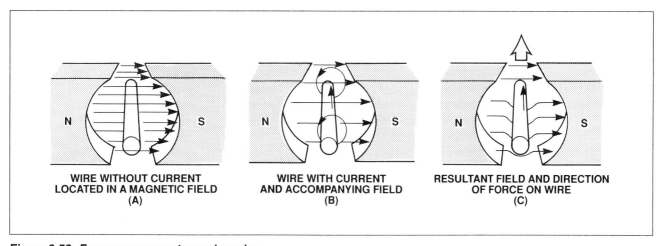

Figure 6-52. Force on a current-carrying wire.

B downward. At the same time, the field of the magnets and the field about A, in which the current is inward, will push side A upward. The coil will thus rotate until its plane is perpendicular to the magnetic lines between the north and south poles of the magnet, as indicated in figure 6-54 by the white coil at right angles to the black coil.

The tendency of a force to produce rotation is called torque. Torque is developed by the reacting magnetic fields about the current-carrying coil just described. This is the torque which turns the coil.

The right-hand motor rule can be used to determine the direction a current-carrying wire will move in a magnetic field. As illustrated in figure 6-55, if the index finger of the right hand is pointed in the direction of the magnetic field and the second finger in the direction of current flow, the thumb will indicate the direction the current carrying conductor will move.

The amount of torque developed in a coil depends upon several factors: the strength of the magnetic field, the number of turns in the coil, and the position of the coil in the field.

c. Basic DC Motor

A coil of wire, through which current flows, will rotate when placed in a magnetic field. This is the technical basis governing construction of a DC motor. Figure 6-56 shows a coil mounted in a magnetic field in which it can rotate. However, if the connecting wires from the battery were permanently connected to the terminals of the coil and there was flow of current, the coil would only rotate until it lined itself up with the magnetic field. Then, it would stop, because the torque at that point would be zero.

A motor, of course, must continue rotating. It is necessary, therefore, to design a device that will reverse the current in the coil just at the time

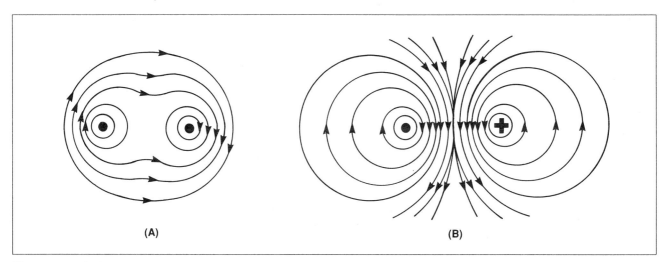

Figure 6-53. Fields surrounding parallel conductors.

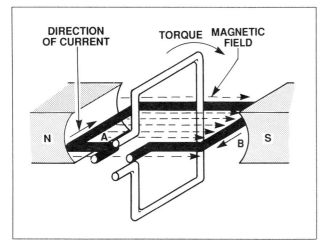

Figure 6-54. Developing a torque.

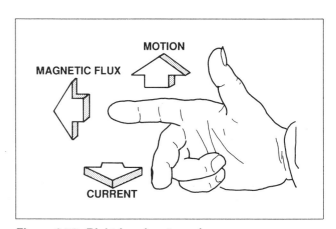

Figure 6-55. Right-hand motor rule.

the coil becomes parallel to the lines of force. This will create torque again and cause the coil to rotate. If the current-reversing device is set up to reverse the current each time the coil is about to stop, the coil can be made to continue rotating as long as desired.

One method of doing this is to connect the circuit so that, as the coil rotates, each contact slides off the terminal to which it connects and slides onto the terminal of opposite polarity. In other words, the coil contacts switch terminals continuously, as the coil rotates, preserving the torque and keeping the coil rotating.

In figure 6-56, the coil terminal segments are labeled A and B. As the coil rotates, the segments slide onto and past the fixed terminals or brushes. With this arrangement, the direction of current in the side of the coil next to the north seeking pole flows toward the reader, and the force acting on that side of the coil turns it downward. The part of the motor which changes the current from one wire to another is called the commutator.

The torque in a motor containing only a single coil is neither continuous nor very effective, because there are two positions where there is actually no torque at all. To overcome this, a practical DC motor contains a large number of coils wound on the armature. These coils are spaced so that, for any position of the armature, there will be coils near the poles of the magnet. This makes the torque both continuous and strong. The commutator, likewise, contains a large number of segments instead of only two.

The armature in a practical motor is not placed between the poles of a permanent magnet but between those of an electromagnet, since a much stronger magnetic field can be furnished. The core is usually made of a mild or annealed steel, which can be magnetized strongly by induction. The current magnetizing the electromagnet is from the same source that supplies the current to the armature.

2. DC Motor Construction

The major parts in a practical motor are the armature assembly, the field assembly, the brush assembly, and the end frames.

a. Armature Assembly

The armature assembly contains a laminated, soft-iron core, coils, and a commutator, all mounted on a rotatable steel shaft. Laminations made of stacks

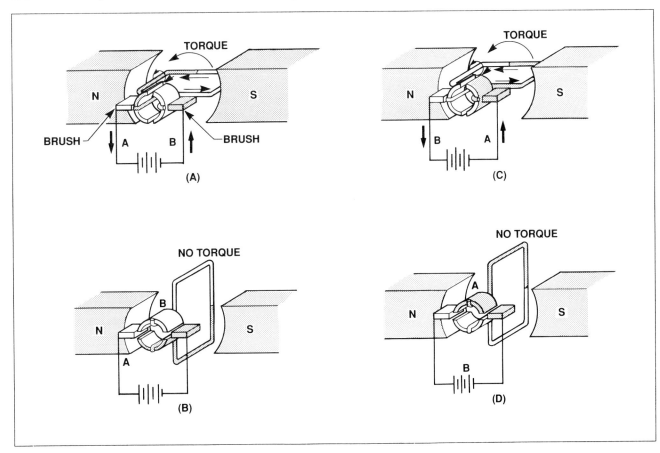

Figure 6-56. Basic DC motor operation.

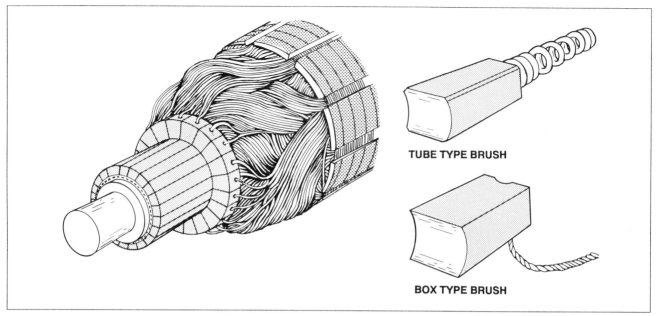

Figure 6-57. Commutator and brushes.

of soft iron, insulated from each other, form the armature core. Solid iron is not used, since a solid-iron core revolving in a magnetic field would heat and use energy needlessly. The armature windings are insulated copper wire, which are inserted in slots insulated with fiber paper (fish paper) to protect the windings. The ends of the windings are connected to the commutator segments. Wedges or steel bands hold the windings in place to prevent them from flying out of the slots when the armature is rotating at high speeds. The commutator consists of a large number of copper segments insulated from each other and the armature shaft by pieces of mica. Insulated wedge rings hold the segments in place.

b. Field Assembly

The field assembly consists of the field frame, the pole pieces, and the field coils. The field frame is located along the inner wall of the motor housing. It contains laminated soft steel pole pieces on which the field coils are wound. A coil, consisting of several turns of insulated wire, fits over each pole piece and, together with the pole, constitutes a field pole. Some motors have as few as two poles, others as many as eight.

c. Brush Assembly

The brush assembly consists of the brushes and their holders. The brushes are usually small blocks of graphitic carbon, since this material has a long service life and also causes minimum wear to the commutator. The holders permit some play in the brushes so they can follow any irregularities in the surface of the commutator and make good contact. Springs hold the brushes firmly against the commutator. A commutator and two types of brushes are shown in figure 6-57.

d. End Frame

The end frame is the part of the motor opposite the commutator. Usually, the end frame is designed so that it can be connected to the unit to be driven. The bearing for the drive end is also located in the end frame. Sometimes the end frame is made a part of the unit driven by the motor. When this is done, the bearing on the drive end may be located in any of a number of places.

3. Types of DC Motors

DC motors may be identified by two major factors. Motors are classified by the type of field-armature connection used, and by the type of duty they are designed for.

a. Field-Armature Connections

There are three basic types of DC motors: series motors, shunt motors, and compound motors. They differ largely in the method in which their field and armature are connected.

1) Series DC Motor

In the series motor, the field windings, consisting of a relatively few turns of heavy wire, are connected in series with the armature winding. This connection is illustrated in figure 6-58. The same current flowing through the field winding also flows through the armature winding. Therefore, any increase in

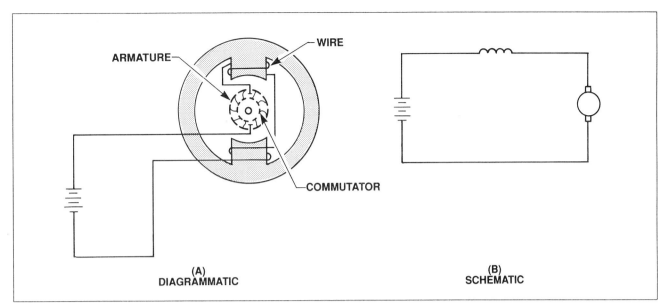

Figure 6-58. Series motor.

current strengthens the magnetism of both the field and the armature.

Because of the low resistance in the windings, the series motor is able to draw large current in starting. This starting current, in passing through both the field and armature windings, produces a high starting torque, which is the series motor's principal advantage.

The speed of a series motor is dependent upon the load. Any change in the load is accompanied by a substantial change in speed. A series motor will run at high speed when it has a light load and at low speed with a heavy load. If the load is removed entirely, the motor may operate at such a high speed that the armature will fly apart. If high starting torque is needed under heavy load conditions, series motors have many applications. Series motors are often used in aircraft as engine starters and for raising and lowering landing gear, cowl flaps, and wing flaps.

2) Shunt DC Motor

In the shunt motor the field winding is connected in parallel, or in shunt with the armature winding (figure 6-59). The resistance in the field winding is high. Since the field winding is connected directly across the power supply, the current through the field is constant. The field current does not vary with motor speed, as in the series motor and, therefore, the torque of the shunt motor will vary only with the current through the armature. The torque developed at starting is less than that developed by a series motor of equal size.

The speed of the shunt motor varies little with changes in load. When all load is removed, it assumes a speed slightly higher than the loaded speed. This motor is particularly suitable for use when constant speed is desired and when high starting torque is not needed.

3) Compound DC Motor

The compound motor is a combination of the series and shunt motors. There are two windings in the field: a shunt winding and a series winding. A schematic of a compound motor is shown in figure 6-60. The shunt winding is composed of many turns of fine wire and is connected in parallel with the armature winding. The series winding consists of a few turns of large wire and is connected in series with the armature winding. The starting torque is higher than in the shunt motor but lower than in the series motor. Variation of speed with load is less than in a series-wound motor, but greater than in a shunt motor. The compound motor is used whenever the combined characteristics of the series and shunt motors are desired.

b. Type of Duty

Electric motors are called upon to operate under various conditions. Some motors are used for intermittent operations; others may operate continuously. Motors built for intermittent duty can be operated for short periods only and, then, must be allowed to cool before being operated again. If such a motor is operated for long periods under full load, the motor will be overheated. Motors built for continuous duty may be operated at rated power for long periods.

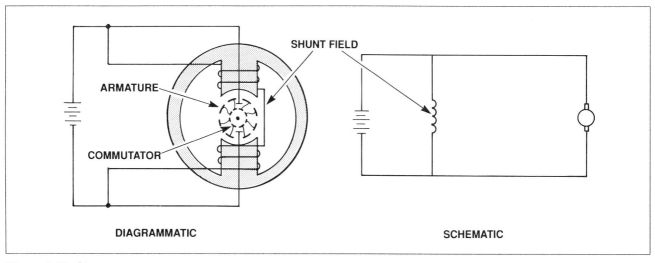

Figure 6-59. Shunt motor.

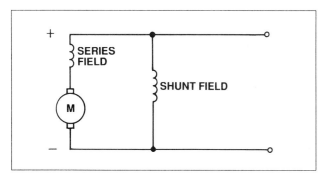

Figure 6-60. Compound motor.

4. Motor Speed and Direction

Certain applications may call for motors whose speed or direction are changed. This may include variable speeds for windshield wipers, or changing the direction of the landing gear retraction motor. Certain internal or external changes may be made in the motor design to allow these operations.

a. Reversing Motor Direction

By reversing the direction of current flow in either the armature or the field windings, the direction of a DC motor's rotation may be reversed. This will reverse the magnetism of either the armature or the magnetic field in which the armature rotates. If the wires connecting the motor to an external source are interchanged, the direction of rotation will not be reversed, since changing these wires reverses the magnetism of both field and armature and leaves the torque in the same direction as before.

One method for reversing the direction of rotation employs two field windings wound in opposite directions on the same pole. This type of motor is called a split field motor. Figure 6-61 shows a series motor with a split field winding. The single-pole, double-throw switch makes it possible to direct current to either of the two windings. When the switch is placed in the lower position, current flows through the lower field winding, creating a north pole at the lower field winding and at the lower pole piece, and a south pole at the upper pole piece. When the switch is placed in the upper position, current flows through the upper field winding, the magnetism of the field is reversed, and the armature rotates in the opposite direction. Some split field motors are built with two separate field windings wound on alternate poles. The armature in such a motor, a four-pole reversible motor, rotates in one direction when current flows through the windings of one set of opposite pole pieces, and in the opposite direction when current flows through the other set of windings.

Another method of reversal, called the switch method, employs a double-pole, double-throw switch which changes the direction of current flow in either the armature or the field. In the illustration of the switch shown in figure 6-62, current direction may be reversed through the field but not through the armature.

When the switch is in the "up" position, current flows through the field windings to establish a north pole at the right side of the motor and a south pole at the left side of the motor. When the switch is moved to the "down" position, this polarity is reversed and the armature rotates in the opposite direction.

b. Changing Motor Speed

Motor speed can be controlled by varying the current in the field windings. When the amount of current flowing through the field windings is increased, the field strength increases, but the motor slows down since a greater amount of counter emf is generated in the armature windings. When the field current is decreased, the field strength decreases, and the motor speeds up because the counter emf is reduced. A motor in which the speed can be controlled is called a variable speed motor. It may be either a shunt or series motor.

In the shunt motor, speed is controlled by a rheostat in series with the field winding (figure 6-63). The speed depends on the amount of current which flows through the rheostat to the field windings. To increase the motor speed, the resistance in the rheostat is increased, which decreases the field current. As a result, there is a decrease in the strength of the magnetic field and in the counter emf. This momentarily increases the armature current and the torque. the motor will then automatically speed up until the counter emf increases and causes the armature current to decrease to its former value. When this occurs, the motor will operate at a higher fixed speed than before.

To decrease motor speed, the resistance of the rheostat is decreased. More current flows through the field windings and increases the strength of the field; then, the counter emf increases momentarily and decreases the armature current. As a result, the torque decreases and the motor slows down until the counter emf decreases to its former value; then the motor operates at a lower fixed speed than before.

In the series motor (figure 6-64), the rheostat speed control is connected either in parallel or in series with the motor field, or in parallel with the motor armature. When the rheostat is set for maximum resistance, the motor speed is increased in the parallel armature connection by a decrease in current. When the rheostat resistance is maximum in the series connection, motor speed is reduced by a reduction in voltage across the motor. For

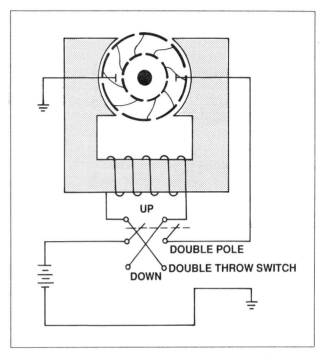

Figure 6-62. Switch method of reversing motor direction.

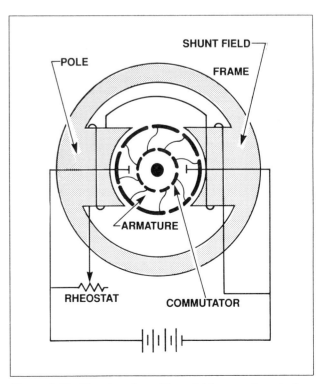

Figure 6-63. Shunt motor with variable speed control.

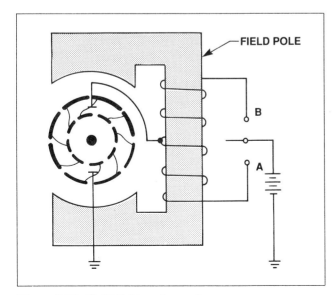

Figure 6-61. Split field series motor.

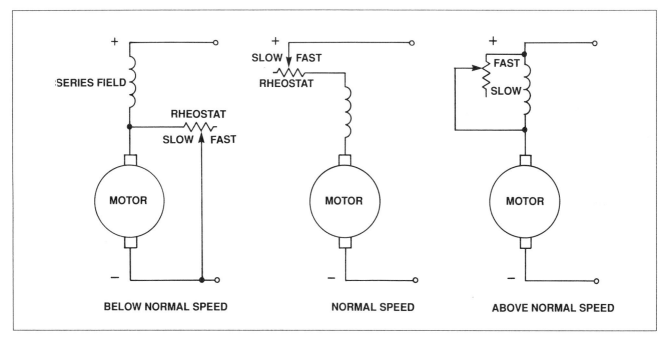

Figure 6-64. Controlling the speed of a series DC motor.

above-normal speed operation, the rheostat is in parallel with the series field. Part of the series field is bypassed and the motor speeds up.

5. Energy Losses in DC Motors

Losses occur when electrical energy is converted to mechanical energy (in the motor), or mechanical energy is converted to electrical energy (in the generator). For the machine to be efficient, these losses must be kept to a minimum. Some losses are electrical, others are mechanical. Electrical losses are classified as copper losses and iron losses; mechanical losses occur in overcoming the friction of various parts of the machine.

Copper losses occur when electrons are forced through the copper windings of the armature and the field. These losses are proportional to the square of the current. They are sometimes called I^2R losses, since they are due to the power dissipated in the form of heat in the resistance of the field and armature windings.

Iron losses are subdivided into hysteresis and eddy current losses. Hysteresis losses are caused by the armature revolving in an alternating magnetic field. It, therefore, becomes magnetized first in one direction and then in the other. The residual magnetism of the iron or steel of which the armature is made causes these losses. Since the field magnets are always magnetized in one direction (DC field), they have no hysteresis losses.

Eddy current losses occur because the iron core of the armature is a conductor revolving in a magnetic field. This sets up an emf across portions of the core, causing currents to flow within the core. These currents heat the core and, if they become excessive, may damage the windings. As far as the output is concerned, the power consumed by eddy currents is a loss. To reduce eddy currents to a minimum, a laminated core usually is used. A laminated core is made of thin sheets of iron electrically insulated from each other. The insulation between laminations reduces eddy currents, because it is "transverse" to the direction in which these currents tend to flow. However, it has no effect on the magnetic circuit. The thinner the laminations, the more effectively this method reduces eddy current losses.

6. Inspection and Maintenance of DC Motors

The inspection and maintenance of DC motors should be in accordance with the guidelines established by the manufacturer. The following is indicative of the types of maintenance checks called for:

a. Check the operation of the unit driven by the motor in accordance with the instructions covering the specific installation.

b. Check all wiring, connections, terminals, fuses, and switches for general condition and security.

c. Keep motors clean and mounting bolts tight.

d. Check brushes for condition, length, and spring tension. Minimum brush lengths, correct spring tension, and procedures for replacing brushes are given in the applicable manufacturer's instructions.

e. Inspect commutator for cleanliness, pitting, scoring, roughness, corrosion or burning. Check for high mica (if the copper wears down below the mica, the mica will insulate the brushes from the commutator). Clean dirty commutators with a cloth moistened with the recommended cleaning solvent. Polish rough or corroded commutators with fine sandpaper (000 or finer) and blow out with compressed air. Never use emery paper since it contains metal particles which may cause shorts. Replace the motor if the commutator is burned, badly pitted, grooved, or worn to the extent that the mica insulation is flush with the commutator surface.

f. Inspect all exposed wiring for evidence of overheating. Replace the motor if the insulation on leads or windings is burned, cracked, or brittle.

g. Lubricate only if called for by the manufacturer's instructions. Most motors used in today's aircraft require no lubrication between overhauls.

h. Adjust and lubricate the gearbox, or unit which the motor drives, in accordance with the applicable manufacturer's instructions covering the unit.

When trouble develops in a DC motor system, check first to determine the source of the trouble. Replace the motor only when the trouble is due to a defect in the motor itself. In most cases, the failure of a motor to operate is caused by a defect in the external electrical circuit, or by mechanical failure in the mechanism driven by the motor.

E. AC Motors

Because of their advantages, many types of aircraft motors are designed to operate on alternating current. In general, AC motors are less expensive than comparable DC motors. In many instances, the AC motors do not use brushes and commutators and, therefore, sparking at the brushes is avoided. They are very reliable and little maintenance is needed. Also, they are well suited for constant-speed applications and certain types are manufactured that have, within limits, variable-speed characteristics. AC motors are designed to operate on poly-phase or single-phase lines and at several voltage ratings.

The subject of AC motors is very extensive, and no attempt has been made to cover the entire field. Only the types of AC motors common to aircraft systems are covered in detail.

The speed of rotation of an AC motor depends upon the number of poles and the frequency of the electrical source of power:

$$RPM = \frac{120 \times Frequency}{Number\ of\ Poles}$$

Since aircraft electrical systems typically operate at 400 Hz, an electric motor at this frequency operates at about seven times the speed of a 60-Hz commercial motor with the same number of poles. Because of this high speed of rotation, 400-Hz AC motors are suitable for operating small high-speed rotors, through reduction gears, in lifting and moving heavy loads, such as the wing flaps, the retractable landing gear, and the starting of engines. The 400-Hz induction type motor operates at speeds ranging from 6,000 RPM to 24,000 RPM.

AC motors are rated in horsepower output, operating voltage, full load current, speed, number of phases, and frequency. Whether the motors operate continuously or intermittently is also considered in the rating.

1. Types of AC Motors

There are three basic types of AC motors. These are the universal motor, the induction motor, and the synchronous motor. Each of the categories may have many variations on the basic operating principle.

a. Universal Motors

Fractional horsepower AC series motors are called universal motors. A universal motor may be operated on either AC or DC, and resembles a DC motor in that it has brushes and a commutator. They are used extensively to operate fans and portable tools, such as drills, grinders, and saws.

b. Induction Motors

The most popular type of AC motor in use is the induction motor. The induction motor needs no electrical connection to the rotating elements, and therefore there are no brushes, commutators, or slip rings to worry about.

Induction motors operate at a fixed RPM that is determined by their design and the frequency of the applied AC. Three-phase induction motors are self-starting and are commonly used when high power is needed. Single-phase induction motors require some form of starting circuit, but once they are started this circuit is automatically disconnected. Single-phase motors operate equally well

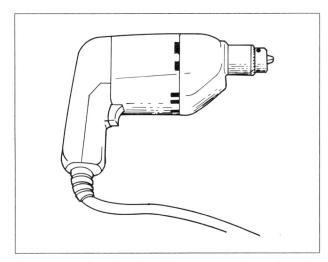

Figure 6-65. An electric drill uses a universal motor which is similar in construction to a series-wound DC motor.

in either direction of rotation, and the direction they turn is determined by the starting circuit.

Many small appliances such as fans and blowers or record players are driven with a small induction motor called a shaded-pole motor because of the way its rotating field is obtained. These small motors have very low starting torque, but their simplicity and low cost make them desirable for application where torque is not important.

1) Construction

An induction motor has a stator which contains a number of electromagnets whose strength and polarity change with the changes in the excitation current. This gives the effect of a rotating magnetic field in the stator, and since the field changes at a specific rate, and induction motor will operate at a fixed speed.

We need two magnetic fields in a motor, and in an induction motor the rotating element is called a squirrel cage because of its construction. A core, made of a stack of thin circular laminations of soft steel, is keyed to the shaft. Slots are cut in the periphery of the core into which bars of copper or aluminum are fitted. End plates of heavy copper or aluminum are fastened to each end of the core, and the bars are welded to these plates.

There is no electrical connection to the rotor, but as soon as current flows in the stator, the lines of magnetic flux produced in the field coils cut across the rotor and induce a voltage in the bars. The rotor has such an extremely low resistance that the induced voltage causes a large current to flow, and this current creates a magnetic field that reacts with the rotating field in the stator.

The steel core of the rotor also has a voltage induced in it, but because it is made up of thin sheets of metal, each covered with an oxide, its resistance is quite high, which keeps the current low. Any current that does flow in the core causes a power loss which is called the iron loss in a motor.

2) Induction Motor Slip

The rotor of an induction motor will assume a position in which the induced voltage is minimized. As a result, the rotor revolves at very nearly the synchronous speed of the stator field, the difference in speed being just sufficient enough to induce the proper amount of current in the rotor to overcome the mechanical and electrical losses in the rotor. If the rotor were to turn at the same speed as the rotating field, the rotor conductors would not be cut by any magnetic lines of force, no emf would be induced in them, no current would flow, and there would be no torque. The rotor would then slow down. For this reason, there must always be a difference in speed between the rotor and the rotating field. This difference in speed is called slip and is expressed as a percentage of the synchronous speed. For example, if the rotor turns at 1,750 RPM and the synchronous speed is 1,800

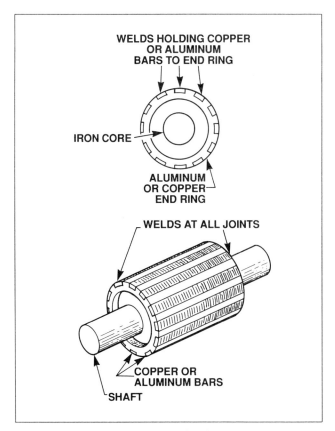

Figure 6-66. The rotor of an induction motor.

RPM, the difference in speed is 50 RPM. The slip is then equal to $50/1{,}800$ or 2.78%.

3) Single-Phase Induction Motor

The previous discussion has applied only to polyphase motors. A single-phase motor has only one stator winding. This winding generates a field which merely pulsates, instead of rotating. When the rotor is stationary, the expanding and collapsing stator field induces currents in the rotor. These currents generate a rotor field opposite in polarity to that of the stator. The opposition of the field exerts a turning force on the upper and lower parts of the rotor trying to turn it 180° from its position. Since these forces are exerted in the center of the rotor, the turning force is equal in each direction. As a result, the rotor does not turn. If the rotor is started turning, it will continue to rotate in the direction in which it is started, since the turning force in that direction is aided by the momentum of the rotor.

4) Shaded-Pole Induction Motor

The first effort in the development of a self-starting, single-phase motor was the shaded-pole induction motor (figure 6-67). A low-resistance, short-circuited coil or copper band is placed across one tip of each small pole, from which, the motor gets the name of shaded-pole. The presence of the ring causes the magnetic field through the ringed portion of the pole face to lag appreciably behind that through the other part of the pole face (figure 6-68). The net effect is the production of a slight component of rotation in the field, sufficient to cause the rotor to revolve. As the rotor accelerates, the torque increases until the rated speed is obtained. Such motors have low starting torque and find their greatest application in small fan motors where the initial torque required is low.

5) Split-Phase Motor

There are various types of self-starting motors known as split-phase motors. Such motors have a starting winding displaced 90 electrical degrees from the main or running winding. In some types, the starting winding has a fairly high resistance, which cause the current in this winding to be out of phase with the current in the running winding. This condition produces, in effect, a rotating field and the rotor revolves. A centrifugal switch disconnects the starting winding automatically after the rotor has attained approximately 25% of its rated speed.

6) Capacitor-Start Motor

With the development of high-capacity electrolytic capacitors, a variation of the split-phase motor,

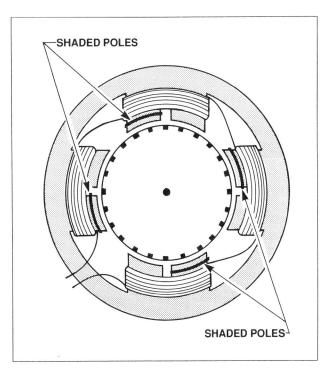

Figure 6-67. Shaded-pole induction motor.

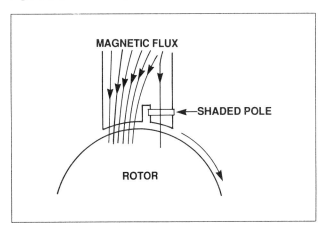

Figure 6-68. Magnetic flux in a shaded-pole motor.

known as the capacitor-start motor, has been made. Nearly all fractional horsepower motors in use today on refrigerators, oil burners, and other similar appliances are of this type (figure 6-69). In this adaptation, the starting winding and running winding have the same size and resistance value. The phase shift between currents of the two windings is obtained by using capacitors connected in series with the starting winding.

Capacitor-start motors have a starting torque comparable to their torque at rated speed and can be used in applications where the initial load is heavy. Again, a centrifugal switch is required for disconnecting the starting winding when the rotor speed is approximately 25% of the rated speed.

Although some single-phase induction motors are rated as high as 2 horsepower (HP), the major field of application is 1 HP, or less, at a voltage rating of 115 volts for the smaller sized and 110 to 220 volts for ¼ HP and up. For even larger power ratings, poly-phase motors generally are used, since they have excellent starting torque characteristics.

7) Direction of Rotation of Induction Motors

The direction of rotation of a three-phase induction motor can be changed by simply reversing two of the leads to the motor. The same effect can be obtained in a two-phase motor by reversing connections to one phase. In a single-phase motor, reversing connections to the starting winding will reverse the direction of rotation. Most single-phase motors designed for general application have provisions for readily reversing connections to the starting winding. Nothing can be done to a shaded-pole motor to reverse the direction of rotation because the direction is determined by the physical location of the copper shading ring.

If, after starting, one connection to a three-phase motor is broken, the motor will continue to run but will deliver only one-third the rated power. Also, a two-phase motor will run at one-half its rated power if one phase is disconnected. Neither motor will start under these abnormal conditions.

c. Synchronous Motors

The synchronous motor is one of the principal types of AC motors. Like the induction motor, the synchronous motor makes use of a rotating magnetic field. Unlike the induction motor, however, the torque developed does not depend on the induction of currents in the rotor. Briefly, the principle of operation of the synchronous motor is as follows: A multiphase source of AC is applied to the stator windings, and a rotating magnetic field is produced. A direct current is applied to the rotor winding, and another magnetic field is produced. The synchronous motor is so designed and constructed that these two fields react to each other in such a manner that the rotor is dragged along and rotates at the same speed as the rotating magnetic field produced by the stator windings.

An understanding of the operation of the synchronous motor can be obtained by considering the simple motor of figure 6-70. Assume that poles A and B are being rotated clockwise by some mechanical means in order to produce a rotating magnetic field, they induce poles of opposite polarity in the soft-iron rotor, and forces of attraction exist between corresponding north and south poles. Consequently, as poles A and B rotate, the rotor is dragged along at the same speed. However, if a load is applied to the rotor shaft, the rotor axis will momentarily fall behind that of the rotating field but, thereafter, will continue to rotate with the field at the same speed, as long as the load remains constant. If the load is too large, the rotor will pull out of synchronism with the rotating field and, as a result, will no longer rotate with the field at the same speed. The motor is then said to be overloaded.

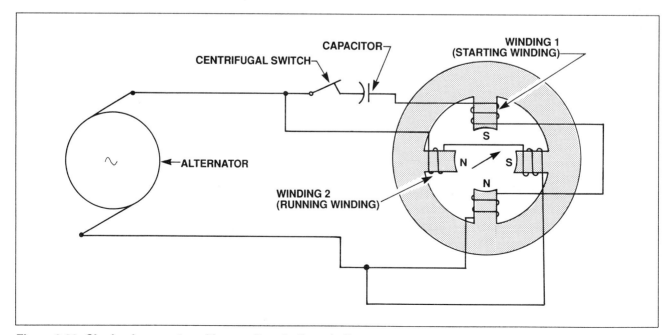

Figure 6-69. Single-phase motor with capacitor starting winding.

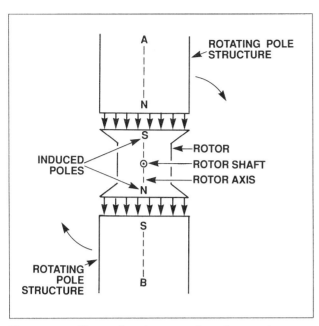

Figure 6-70. Illustrating the operation of a synchronous motor.

Such a simple motor as that shown in figure 6-70 is never used. The idea of using some mechanical means of rotating the poles is impractical because another motor would be required to perform this work. Also, such an arrangement is unnecessary because a rotating magnetic field can be produced electrically by using phased AC voltages. In this respect, the synchronous motor is similar to the induction motor.

The synchronous motor consists of a stator field winding similar to that of an induction motor. The stator winding produces a rotating magnetic field. The rotor may be a permanent magnet, as in small single-phase synchronous motors used for clocks and other small precision equipment, or it may be an electromagnet, energized from a DC source of power and fed through slip rings into the rotor field coils, as in an alternator.

Since a synchronous motor has little starting torque, some means must be provided to bring it up to synchronous speed. The most common method is to start the motor at no load, allow it

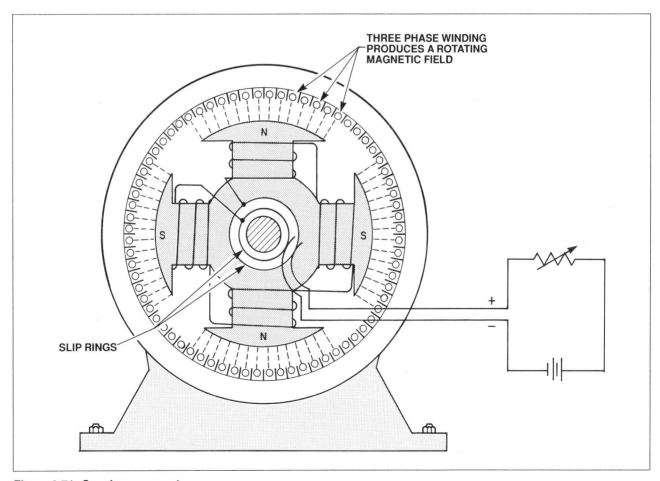

Figure 6-71. Synchronous motor.

137

to reach full speed, and then energize the magnetic field. The magnetic field of the rotor locks with the magnetic field of the stator and the motor operates at synchronous speed.

The magnitude of the induced poles in the rotor shown in figure 6-71 is so small that sufficient torque cannot be developed for most practical loads. To avoid such a limitation on motor operation, a winding is placed on the rotor and energized with DC. A rheostat placed in series with the DC source provides the operator of the machine with a means of varying the strength of the rotor poles, thus placing the motor under control for varying loads.

The synchronous motor is not a self-starting motor. The rotor is heavy and, from a dead stop, it is impossible to bring the rotor into magnetic lock with the rotating magnetic field. For this reason, all synchronous motors have some kind of starting device. One type of simple starter is another motor, either AC or DC, which brings the rotor up to approximately 90% of its synchronous speed. The starting motor is then disconnected, and the rotor locks in step with the rotating field. Another starting method is a second winding of the squirrel-cage type on the rotor. This induction winding brings the rotor almost to synchronous speed, and when the DC is disconnected to the rotor windings, the rotor pulls into step with the field. The latter method is the more commonly used.

2. Maintenance of AC Motors

The inspection and maintenance of AC motors is very simple. The bearings may or may not need frequent lubrication. If they are the sealed type, lubricated at the factory, they require no further attention. Be sure the coils are kept dry and free from oil or other abuse.

The temperature of a motor is usually its only limiting operating factor. A good rule of thumb is that a temperature too hot for the hand is too high for safety.

Next to the temperature, the sound of a motor or generator is the best trouble indicator. When operating properly, it should hum evenly. If it is overloaded, it will "grunt". A three-phase motor with one lead disconnected will refuse to turn and will "growl". A knocking sound generally indicates a loose armature coil, a shaft out of alignment, or armature dragging because of worn bearings.

The inspection and maintenance of all AC motors should be performed in accordance with the applicable manufacturer's instructions.

Chapter VII

Electronic Control Devices

I. BASIC ELECTRONIC THEORY

Electronics is the science of controlling the flow of electrons. When we understand atomic structure and how electrons may move from one atom to the next, we can begin to see how they can be controlled, and their movement used to perform work.

A. Atoms, Crystals, and Energy States

In chapter 1, we discussed the structure of the atom. We saw how the electrons were arranged in "shells" around the nucleus, according to their energy state. We also saw how electrons traveled from atom to atom when an electrical voltage was applied. The outer shell of electrons was capable of accepting extra electrons, or of losing them. Atoms with an excess or deficiency of electrons are known as ions.

When atoms of the same, or different elements combine, they form molecules. For example when one atom of oxygen combines with two atoms of hydrogen, we have one molecule of water. In general, a molecule is stable when each atom in the molecule has access to 8 electrons in its outer shell. Molecules may share electrons to make up the number 8 in the outer shell.

We also learned that some materials which exist in nature have an atomic structure which enables them to easily permit the movement of electrons, and others have a structure which opposes the movement of electrons. Those materials which oppose the movement of electrons are known as insulators. Insulators are often used to prevent the inadvertent flow of electricity. Common insulating materials are plastic, rubber, glass, and ceramics. Materials which permit the flow of electrons are referred to as conductors. The best conductors (gold, silver, copper, and aluminum) are all metals. Metals generally have only one or two electrons in the outer shell, and they are easily removed.

Metals are crystalline structures. The individual atoms are stacked together in a uniform pattern, in such a manner that the outer electrons of each atom are practically touching. This arrangement means that little energy is required to get the electrons moving from atom to atom.

Conductors have two basic ingredients: free electrons, which are easily lured away from their atoms, and resting places in the outer electron shells, which provide stepping stones for electrons moving from atom to atom. Both free electrons and resting places must be present for current to flow.

B. Semiconductors

Halfway between conductors and insulators are semiconductors. The ability to control the conductivity of these materials is the key to the operation of all solid-state electronic equipment from the simplest diode rectifier to the most sophisticated computer system. Semiconductor material can be changed from an insulator to a conductor by applying a small voltage, heat, or light. This ability makes it possible to control the flow of a large electrical current with the application of a small amount of energy.

Like metals, semiconductors are also crystals. They are usually made from atoms which have four electrons in the outer shell. Each atom shares its four outer electrons (valence electrons) with four other atoms in a stable, rigid, crystal.

Because the sharing of electrons in a semiconductor crystal provides access to 8 electrons in

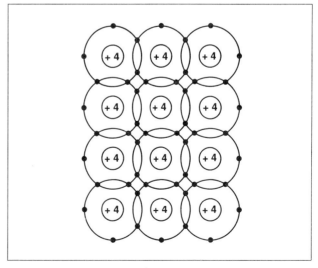

Figure 7-1. All of the valence electrons in an insulator are tightly bound with covalent bonds, and the material will not accept electrons from an outside source.

the outer shell of each atom, these materials are particularly stable and chemically inert. Pure semiconductor crystals are actually good insulators. If we wish to use these materials as conductors, they must be impure or extremely hot. If we trap atoms of another material in this crystal matrix, the impurities (because of their different number of valence electrons) will provide free electrons and resting places, and will turn the semiconductor into a conductor.

Silicon and germanium are two of the more common semiconductor materials. Both have four electrons in their outer shells. If the silicon or germanium are doped—that is if a few parts per million of an element such as arsenic, bismuth, or antimony, which have five valence electrons is mixed in—there will be free electrons in the material after all of the covalent bonds are formed. These electrons are free to move, and so the material is called a donor or an N-type material.

Elements such as boron, indium, and gallium have only three valence electrons, and so when a few parts per million are alloyed with silicon or germanium, there will be an area where covalent bonds have not formed. These are the resting places for electrons, and are referred to as holes. Each hole represents a place where an electron can go, but which is not currently filled. When an electron is displaced it leaves a hole in which another free electron may come to rest. Material doped in this way is called an acceptor or P-type material.

II. DIODES

N- and P-type silicon or germanium may be joined either by a junction or a point contact, to form a semiconductor diode. To understand the way this diode works, let's consider a junction formed by a piece of P-type and a piece of N-type silicon.

The holes in the P-type material attract the electrons in the N-type and, at the junction, they combine. This leaves a depletion area where there are no more free electrons or holes.

When the electrons leave the N-type material, its side of the junction assumes a slight positive charge because of the lack of electrons. As the holes leave the edge of the P material, its side assumes a negative charge. These charges constitute a barrier, or a potential hill. The intensity of this barrier is proportional to the width of the depletion area.

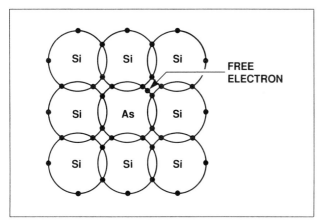

Figure 7-2. When an insulator is doped with an impurity having five valence electrons, it produces an N-type material, one with an excess of electrons.

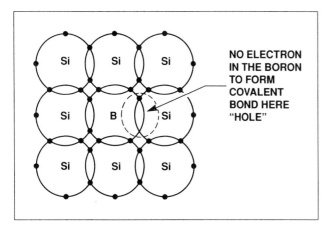

Figure 7-3. When an insulator is doped with an impurity having only three valence electrons, it produces a P-type material which will accept free electrons.

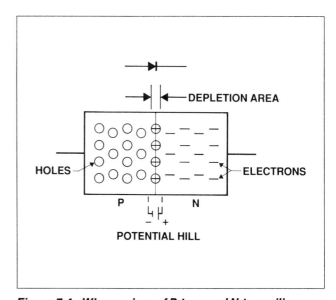

Figure 7-4. When a piece of P-type and N-type silicon are joined, the electrons and holes combine along the junction and form a depletion area or a potential hill.

If a voltage source is attached to a semiconductor diode with its positive terminal connected to the P material, and its negative terminal to the N material, the diode is said to be forward biased. In this state electron scan flow through the diode.

The negative potential of the battery forces the electrons toward the junction, while the positive voltage forces the holes toward the junction. At the junction, the electrons and holes combine, making room for more electrons to enter the N side. In the depletion area, holes and electrons are continually combining and the area becomes extremely narrow. As a result, the barrier, or potential hill, is very small.

When the power source is turned around so that the positive terminal attaches to the N material, and the negative terminal to the P, the electrons and holes will be attracted away from the junction. This will enlarge the depletion area so much, and the potential hill will become so high that no electrons or holes can combine, and there will be no electron flow in the external circuit. The junction is now said to be reversed biased.

A simple way to remember the function of a diode in a circuit is to think of it as an electrical check valve. The diode works much like the check valves used in plumbing systems. Each allows flow only in one direction. This analogy is illustrated in figure 7-7.

If you can identify the side of the diode containing the P-type material you can determine which direction the P-N junction will allow current to flow by remembering "positive to P conducts".

A. Ideal Diodes

In our discussion above, we have looked at diodes as if they were ideal in their operation. With the ideal diode, the current flow is immediately unlimited, even with the smallest forward bias. This is because an ideal diode would have zero resistance. Real diodes have some internal resistance and do

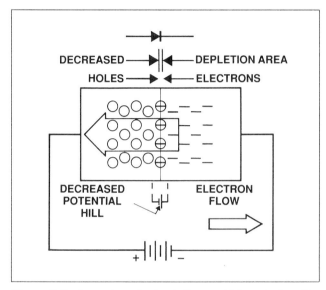

Figure 7-5. A semiconductor is forward-biased when the positive terminal of the power source is attached to the P-type material and the negative terminal is attached to the N-type material.

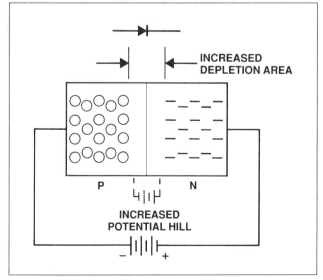

Figure 7-6. When a semiconductor is reverse-biased, no current can flow.

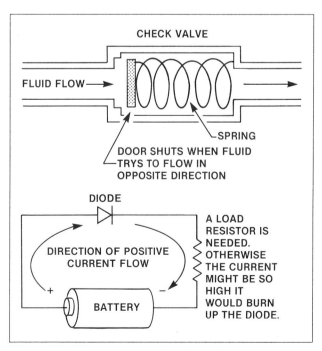

Figure 7-7. Check valve analogy for a diode.

not respond in a linear manner. As we begin to discuss diode applications in aircraft systems, keep in mind that by understanding this non-linear action we can use various types of diodes in tasks far more complicated than a simple check valve.

B. Diode Applications

Diodes are capable of performing many tasks in a circuit. We may find diodes, singly or in combination, used to convert AC electricity to DC, as radio signal detectors, frequency converters, modulators, regulators, clippers, and more.

First on our list above, and probably first in most peoples mind, is the use of diodes as rectifiers. Because an AC signal reverses its polarity, a single diode connected in the line will allow to pass only the half of the signal that forward biases the diode. The result is called half-wave rectified current. A little more complicated circuit using two diodes can result in full-wave rectified current. This combination of two diodes providing full wave rectification is what is used in the alternator on your car to produce a DC current to keep the battery charged.

When looking at diode applications it is important to keep in mind that positive and negative are relative terms. It is a common misconception that the voltage on the P end of the diode must be positive while the voltage on the N material must be negative before the diode will conduct. All that really matters is which end of the diode is *more* positive. This relationship of which is more positive or more negative is very important to understanding the operation of solid state equipment.

C. Special Purpose Diodes

1. Zener Diodes

We have in figure 7-8 a curve representing the characteristics of a semiconductor diode, showing the way the current flow varies as a result of the applied voltage.

When a diode is forward-biased, its current flow is roughly proportional to the applied voltage; that is, the diode is essentially linear. Above its burnout point, however, enough current can pass through it to overheat it and burn it up. When the diode is reverse-biased, very little current can pass, usually only a few milliamperes. This is called leakage current and it is strongly dependent on the temperature of the diode.

In all semiconductor diodes, there is a rather sharply defined point of reverse voltage called the zener voltage, (or avalanche point) at which point the diode stops acting as a check valve and breaks

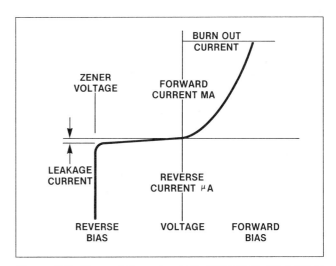

Figure 7-8. Current-flow characteristics through a zener diode.

down, allowing current to pass in the reverse direction. You notice from the curve in figure 7-8 that when the zener voltage is reached, the current increases almost instantaneously until it reaches such a value that it will burn out the diode. Typically, a limiting resistor is installed in the circuit to prevent damage to the diode.

Zener diodes are used in circuits for protection against over-voltage and to prevent spikes of induced voltage damaging sensitive circuits and components.

2. Silicon-Controlled Rectifiers (SCR)

A silicon-controlled rectifier is a special-type of diode having three P-N junctions. It is similar to a silicon diode in its outward appearance, except for its extra terminal, the gate. The gate connection is used to control the current flow through the anode-cathode circuit within the SCR.

In a normal control circuit (figure 7-9) two of the junctions are forward-biased and one is reverse-biased. In this state, no electrons can flow through the load. If the gate is momentarily connected to the positive voltage at the anode, the reverse-biased junction will become forward-biased, and electrons will flow through the SCR. Once this flow starts, it will maintain the forward bias, and the flow will continue until the voltage across the SCR is removed. No further voltage need be applied to the gate. Once an SCR starts conducting, it will continue until the supply voltage is shut off, or in the case of AC, until the next half cycle.

Electrical circuits for lights and for some types of motors may limit the current by dropping the voltage supplied to these devices, but this requires a relatively large resistor to dissipate the heat. It is possible to dim lights and control the speed of universal motors by using a silicon-controlled rectifier, an SCR, rather than a resistor. An SCR decreases the amount of current supplied to the device, not by dropping the voltage and dissipating power, but by controlling the time in the AC cycle that current is allowed to flow.

3. Triacs

One of the limitations of a silicon-controlled rectifier is that it controls only one-half of the cycle of AC. A triac, on the other hand, overcomes this by acting as though it were two SCRs connected side by side in opposite directions. An SCR requires a positive pulse to trigger its gate, but a triac can be triggered by a pulse of either polarity, and the output waveform will appear as that in figure 7-10. For full power, the triac is triggered at the beginning of the cycle and all of the current flows, but if it is triggered later in the cycle, as we see here, only about one-half of the current flows.

4. Photodiodes

As the science of electronics becomes more sophisticated, we are able to use some of the energy found in light for some of our devices and circuits. Light energy is electromagnetic in nature, and one of its characteristics is that it will increase the reverse bias or leakage current in a semiconductor device. A photodiode is a special diode, made in such a way that a light shining through an aperture in its case will release free electrons into the depletion area and cause it to conduct.

A practical circuit using a photodiode is seen in figure 7-11 where we have a photodiode in the coil circuit of a relay. When the photodiode is dark, no current can flow and the relay is open. When the light shines through the aperture in the diode's case, the diode will break down so that current from the battery can flow through the relay coil and close the contacts.

5. Light Emitting Diodes (LED)

When a photon of light strikes a photoconducting material, an electron is freed and, conversely, if

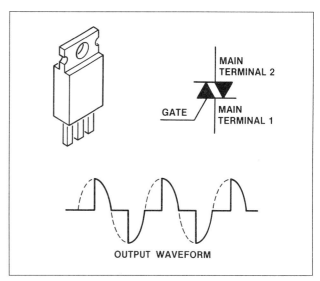

Figure 7-10. A triac may be triggered with a pulse of either polarity and will conduct during the portion of the cycle after it has been triggered.

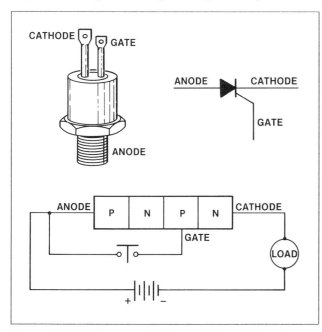

Figure 7-9. A silicon-controlled rectifier will start to conduct when a positive pulse is applied to its gate. It will continue to conduct as long as the anode and the cathode are forward biased.

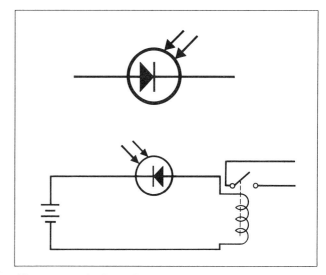

Figure 7-11. A photodiode will conduct when sufficient light strikes its junction to cancel the effect of the reverse bias.

a free electron in a piece of light-emitting semiconductor material falls into a hole, a photon of light is released.

Light emitting diodes find some of their most popular aviation applications in the newer generation of digital display instruments, but calculators, clocks and watches are familiar to all of us as non-aviation applications. In these devices, the LEDs are arranged in a group of seven bars, such as we see in figure 7-12, and an integrated circuit control device provides a forward bias on the segments that are needed to form the digit.

The most common LEDs emit a red light, however, green, yellow and blue LEDs are also available. Light-emitting material has found an almost unlimited number of applications in digital displays for electronic devices. When a light emitting semiconductor is reverse-biased, no current can flow and it remains dark, but when it is forward-biased, it will emit light.

III. TRANSISTORS

Diodes serve well as check valves, but we must also be able to control the amount of flow. We can do this with a three layer semiconductor, called the transistor.

A. Bipolar Transistors

A bipolar transistor is essentially a sandwich of N-type silicon between two pieces of P-type silicon, or a piece of P-type silicon between two pieces of N-type. The design and construction of these miniature electron control valves are quite complex, but here we are concerned only with their operation. The symbol used to designate a transistor and its connections are seen in figure 7-13.

There are three basic components to the common transistor: the emitter (E), the base (B) and the collector (C). The emitter, base or collector can be an N- or P-type material depending on the type of transistor (PNP or NPN). Between each N and P material is a junction, identical to that found

Figure 7-12. Light emitting diodes (LEDs) emit light when they are forward-biased and current flows through them.

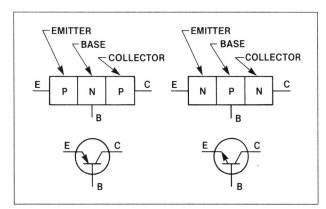

Figure 7-13. Commonly used transistor symbols.

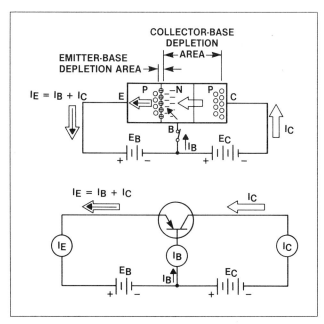

Figure 7-14. The emitter-base junction is forward-biased and base current is flowing. By controlling the small base current, the much larger emitter-collector current may be controlled.

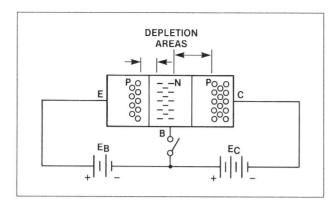

Figure 7-15. When the base circuit is opened, no base current can flow, and therefore, there is no emitter-collector current.

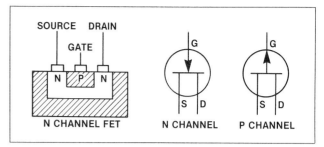

Figure 7-16. A field-effect transistor is a high-impedance semiconductor device that is controlled by varying the voltage on its gate.

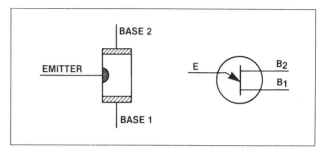

Figure 7-17. A unijunction transistor produces a pulse output when the voltage between its emitter and base rises to a preset value.

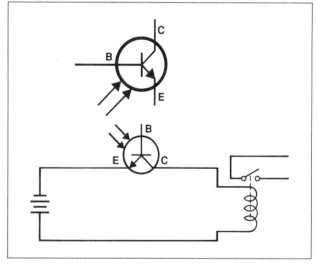

Figure 7-18. A phototransistor will conduct when sufficient light strikes its emitter-base junction to provide a forward bias.

in the diode. The E, B and C typically represent the three connections of a common transistor.

For a transistor to conduct, its emitter-base junction must be forward-biased and the collector-base junction must be reverse-biased (figure 7-14). A small amount of current flows between the emitter and the base, and this causes an extremely narrow depletion area, or a very low resistance. The relatively large reverse-bias between the collector and the base forces electrons into the emitter-base junction where they are attracted to the positive voltage source attached to the emitter. Since the base is so thin, the electrons which leave the collector return to the positive terminal of the emitter source. A very small base current is all that is needed to keep the emitter-collector resistance low enough to allow a large flow of emitter-collector current.

When the base circuit is opened (figure 7-15), there is no longer a force available to keep the emitter-base depletion area reduced; and there will be no attraction for the electrons from the negative terminal of the collector source across both depletion areas to the positive terminal of the emitter source. When there is no base current flow, there can be no collector current flow.

When the base of a PNP transistor is negative with respect to its emitter, base current will flow, and the transistor will conduct between its collector and its emitter. It is sometimes desirable, in a circuit, to have a positive voltage on the base to cause the transistor to conduct, and when this is the case, an NPN transistor can be used. An NPN transistor is similar in almost all ways to a PNP except for the arrangement of the doped areas. When biasing an NPN transistor, the voltage polarities must be exactly opposite those used for a PNP. For maximum condition with both types, the emitter-base junction should be forward-biased and the collector-base junction reverse-biased.

B. Field Effect Transistor

A field effect transistor (FET) is a solid-state device in which the emitter-collector current is controlled by a voltage rather than by a current. In figure

7-16, we see that an FET may be constructed of a channel of either N-type or P-type silicon and, sitting in this channel like a valve, is the gate. One end of the channel is called the source, and the other end, the drain. An N-channel FET has a P-type gate, so when a positive voltage is applied to the gate, the FET is forward-biased, and there will be a flow of electrons between the source and the drain. When a negative voltage is applied to the gate, the FET will be reverse-biased, and the flow between the source and the drain will be pinched off.

C. Unijunction Transistors

A unijunction transistor (UJT) is sometimes called a double-base diode. It is made of a single crystal of uniformly doped N-type silicon and has contacts at each of its ends and a small P-type emitter located near its middle. A UJT acts like a good insulator until the voltage at the emitter becomes sufficiently high to trigger it, and then it can conduct with a minimum of resistance. UJTs are used in such circuits as relaxation oscillators, where it is necessary to provide short, high-intensity current pulses when the control voltage rises to a given value. They may also be used to provide the gate pulses needed for silicon controlled rectifiers or triac circuits.

D. Phototransistors

A phototransistor is activated by light, and is used for applications where the current flow allowed by a diode is not sufficient for the circuit. When light of sufficient intensity strikes the collector-base junction, enough electrons will flow from the base to turn the transistor on, and amplified current can flow from the emitter to the collector and operate any device in the circuit that requires more current than could be supplied through a diode.

Chapter VIII

Aircraft Electrical Systems

I. WIRING INSTALLATION

A. Wiring Diagrams

There are a number of different types of electrical diagrams available to help us understand the electrical systems we are called upon to service. These include the block diagram, the pictorial diagram, and the schematic diagram. We will take a brief look at each of these types.

1. Block Diagrams

The block diagram uses very few component symbols, but rather uses blocks to tell us how a particular portion of the system operates. In figure 8-1, we have a block diagram of the power system of a Boeing 727 aircraft. From this diagram, we can tell nothing about the wires required or the types of components, but the interrelation between the parts of the system is clear. We can tell what each generator does, and visualize the way each of the relays and breakers tie the system together, and the way in which we can isolate portions of the circuit. A block diagram gives us an overview of a system in the simplest way.

2. Pictorial Diagrams

The symbols used on an electrical schematic are actually a language of their own and are often unin-

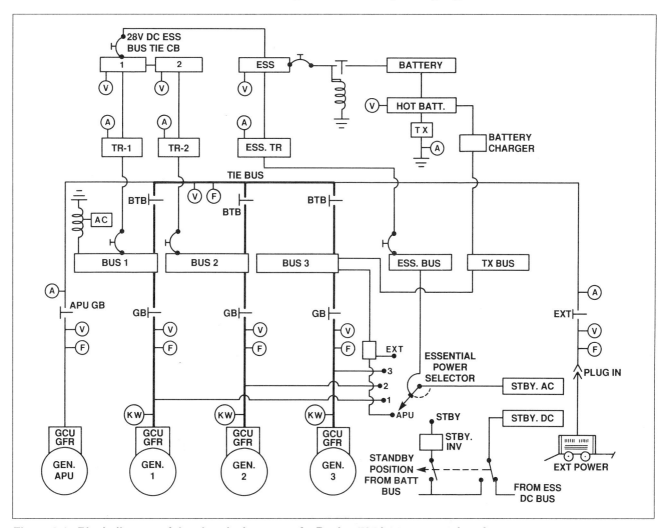

Figure 8-1. Block diagram of the electrical system of a Boeing 727 jet transport aircraft.

147

telligible to persons not schooled in their use. In figure 8-2, we have a pictorial diagram of the electrical system used in a popular single-engine aircraft. You will notice that this is a sort of mixture of a block diagram and schematic. You can identify such components as the alternator, the starter, the battery, and the ignition switch, but you do not know the wire sizes needed, or part numbers of any of the components. This type of drawing is extremely helpful for understanding the operation of a system, and of seeing the interrelationship between parts.

3. Schematic Diagrams

This is the type of wiring diagram that is of most importance to us as aircraft maintenance technicians. From a schematic diagram, we can get the information needed to troubleshoot and service the equipment, and since this type of diagram is used in the manufacturer's maintenance manuals, they are FAA-approved data. In the typical diagram of figure 8-3, we have the information we need to service this particular flashing beacon light. In the equipment table, we can find the part number of each of the components, and in the wire table, we have the wire gage and a list of the various electrical connectors.

The symbols used for aircraft schematics vary somewhat from one manufacturer to the next, but the symbols shown in figure 8-4 are typical of the ones you are most likely to encounter.

B. Wire Types

There are many different types of wire used for special applications in aircraft electrical systems, but the majority of the wiring is done with MIL-W-5086 or MIL-W-22759 stranded tinned copper wire, insulated with polyvinyl chloride (PVC), nylon, or Teflon®. The construction of this wire is seen in figure 8-5. The insulation has a voltage rating of 600 volts.

Where large amounts of current must be carried for long distances, MIL-W-7072 aluminum wire is often used. This wire is insulated with either fluorinated ethylene propylene (FEP Fluorocarbon), nylon, or with a fiberglass braid. Aluminum wire smaller than 6-gage is not recommended because it is so easily broken by vibration.

Solid wire may be used for the internal hookup of components where vibration is no problem. The size of solid wire is measured with a wire gage

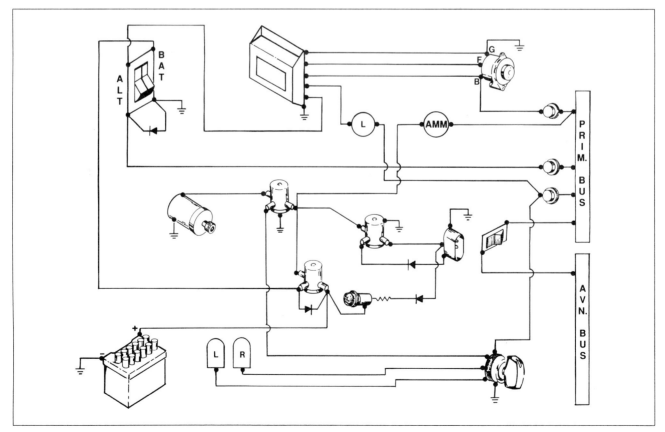

Figure 8-2. A pictorial diagram of an aircraft electrical system.

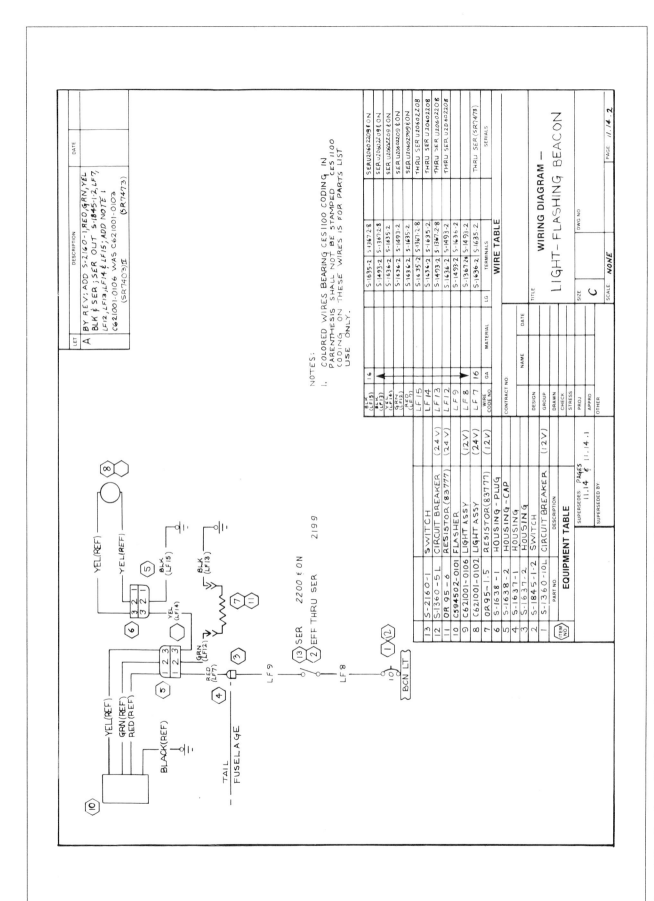

Figure 8-3. A schematic diagram of a portion of an aircraft electrical system.

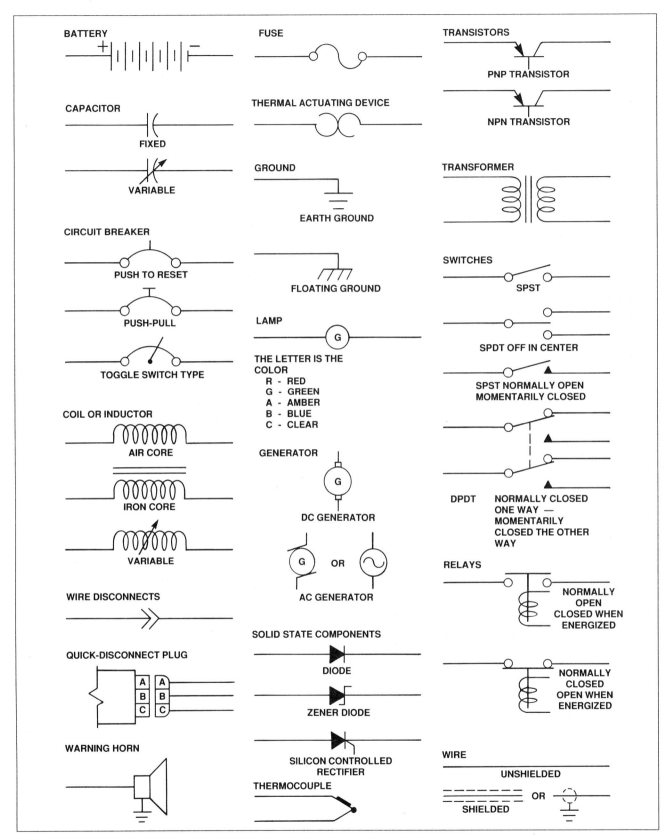

Figure 8-4. Typical symbols used in aircraft electrical system schematic diagrams.

similar to the one shown in figure 8-6. You can see from this tool that the smaller the wire gage number, the larger the wire.

The cross-sectional area of a solid wire is usually expressed in terms of circular mils and is equal to the square of the diameter of the wire in thousandths of an inch.

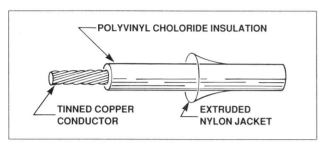

Figure 8-5. MIL-W-5086 copper wire.

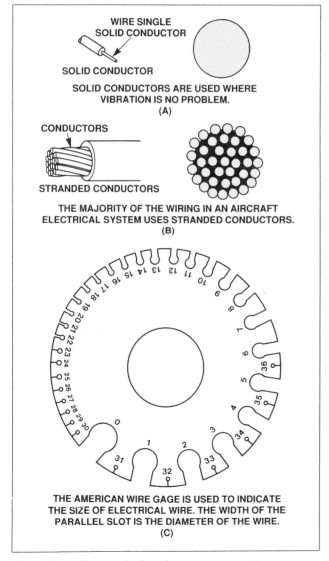

Figure 8-6. Electrical wire size measurements.

For example, a solid 22-gage wire has a diameter of 0.0253″. This is 25.3 thousandths of an inch, or 25.3 mils, and 25.3 squared is 640. A 22-gage wire therefore has a cross-sectional area of 640 circular mils.

Anytime a wire carries electrical current, a magnetic field surrounds the wire, and this field may interfere with some of the aircraft instrumentation. For example, the tiny light that illuminates the card of a magnetic compass is powered with low-voltage direct current, and the field from even this small current can deflect the compass. To minimize this field, a two-conductor twisted wire is used to carry the current to and from this light. By using a twisted wire, the fields cancel each other.

AC or pulsating DC has an especially bad effect on electronic equipment, as its conductors radiate electrical energy much like the antenna of a radio. To prevent radio interference, wires that carry AC or pulsating DC are often shielded. This means that the insulated conductor is encased in a wire braid. The radiated energy from the conductor is received by the braided shielding and is passed to the aircraft's ground where it can cause no interference. A shielded wire may be seen in figure 8-8.

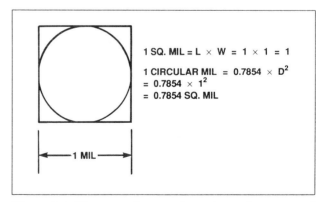

Figure 8-7. The area of a round conductor is measured in circular mils.

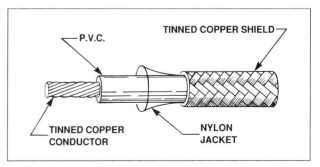

Figure 8-8. Shielded wire is used to prevent interference from radiated electromagnetic fields.

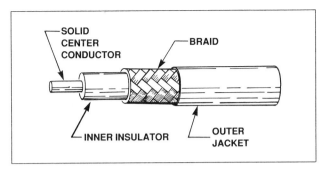

Figure 8-9. A coaxial cable is used to connect an aircraft radio receiver or transmitter to its antenna.

Antennas are connected to most of the radio receivers and transmitters with a special type of shielded wire called a coaxial cable. This type of cable is illustrated in figure 8-9. Coax, as it is commonly called, consists of a central conductor surrounded by an insulator and a second conductor. The spacing and concentricity of the two conductors are critical for the most efficient transfer of energy through the cable.

C. Wire Size Selection

1. Conductor Material and Insulation

The wires installed in an aircraft electrical system must be chosen on the basis of their ability to carry the required current without overheating and to carry it without producing an excessive voltage drop. There are several factors to consider when choosing a wire; the wire material, the required flexibility of the wire, the insulation material, the wire diameter (gage), the length of the wire and the type of installation.

For aircraft, the wire material could be either copper or aluminum. If the conductor is made of copper, the individual strands of wire are typically plated to protect the copper from corrosion. Nickel is the most common coating; however, silver and tin are often used in various situations. The flexibility of a conductor is typically determined by the number of strands which make up the wire and the type of insulation on the wire. The type of insulation around the conductor is also very important. Various insulations have different ratings for heat, abrasion and flexibility. The length and type of installation are factors established by the aircraft manufacturer, and the diameter required can easily be found using the wire chart or schematic diagram.

Whenever possible, choose a replacement wire which has been approved by the manufacturer for that particular installation. If there is no approved data available for this installation, the wire selection should be made with extreme caution and in accordance with the FAA Advisory Circular 43.13-1A (or latest edition).

2. Current-Carrying Capability

In figure 8-10, we have a list of the current-carrying capability of MIL-W-5086 copper wire, and in figure 8-11, we have the same information for MIL-W-7072 aluminum wire. Comparing these charts gives us an indication of the reason it is best to use aluminum wire when it is necessary to carry large amounts of current for long distances. The weight saved by using large aluminum wire in place of copper allows for a lighter aircraft and more payload.

If we need to supply an actuator with 100 amps of current from a 28-volt system, we can use either copper or aluminum wire. Let's find the size we will need if 60 feet of wire is needed between the bus and the actuator. Looking at the chart of figure 8-10, we see that a 6-gage (AN-6) wire will carry 101 amps if it is in free air. If we use aluminum wire, we will have to use a four-gage wire to carry the same amount of current. This observation leads us to an accepted rule of thumb that says whenever we substitute aluminum wire for copper, we should use wire which is two gage numbers larger. Be sure to note here that the larger the number, the smaller the wire. Also, the FAA does not allow aluminum wire smaller (in size, larger in number) than 6-gage to be used on aircraft.

If the wire is routed in a bundle, the heat generated by the flow of current will have no opportunity to escape, so in order to keep the heat down we must derate the wire. If we want to carry our 100 amps in a copper wire and route it in a bundle, we will need a two-gage wire, and if we use an aluminum wire, it will require an advantage (0-gage) wire.

3. Allowable Voltage Drop

When we add any electrical equipment to an aircraft, we must be sure that the current flowing in the wiring does not drop the voltage in excess of that allowed by the FAA. Referring to figure 8-12, we find that in a 28-volt system, we are allowed a one-volt drop if the load is continuous or two volts if the load is intermittent.

The chart in figure 8-13 is familiar to aviation maintenance technicians as that furnished by the FAA to quickly give us an indication of both the current-carrying capability of a wire and the length we can use for the allowable voltage drop.

Let's continue with our example of our intermittent 100-amp flow through 60 feet of wire in a 28-volt system. We must use the chart of figure

Wire size—Specification MIL-W-5086	Single wire in free air—maximum amperes	Wire in conduit or bundled—maximum amperes	Maximum resistance—ohms/1,000 feet (20°C)	Nominal conductor area—circular mills	Finished wire weight—pounds per 1,000 feet
AN-20	11	7.5	10.25	1,119	5.6
AN-18	16	10	6.44	1,779	8.4
AN-16	22	13	4.76	2,409	10.8
AN-14	32	17	2.99	3,830	17.1
AN-12	41	23	1.88	6,088	25.0
AN-10	55	33	1.10	10,443	42.7
AN-8	73	46	.70	16,864	69.2
AN-6	101	60	.436	26,813	102.7
AN-4	135	80	.274	42,613	162.5
AN-2	181	100	.179	66,832	247.6
AN-1	211	125	.146	81,807	
AN-0	245	150	.114	104,118	382
AN-00	283	175	.090	133,665	482
AN-000	328	200	.072	167,332	620
AN-0000	380	225	.057	211,954	770

Figure 8-10. Characteristics of MIL-W-5086 copper wire.

Wire size—Specification MIL-W-7072	Single wire in free air—maximum amperes	Wire in conduit or bundled—maximum amperes	Maximum resistance—ohms/1,000 feet (20°C)	Nominal conductor area—circular mills	Finished wire weight—pounds per 1,000 feet
AL-6	83	50	0.641	28,280	. . . .
AL-4	108	66	.427	42,420	. . . .
AL-2	152	90	.268	67,872	. . . .
AL-0	202	123	.169	107,464	166
AL-00	235	145	.133	138,168	204
AL-000	266	162	.109	168,872	250
AL-0000	303	190	.085	214,928	303

Figure 8-11. Characteristics of MIL-W-7072 aluminum wire.

8-13(B) since it is the wire chart for intermittent loads and this is what our example calls for. First, locate the diagonal line for 100 amps and follow it down until it crosses the horizontal line for 60 feet in the 28-volt column. These two lines intersect between the vertical lines for 6-gage and 2-gage wire. Our current-carrying chart showed us that we could carry 101 amps in a 6-gage wire but we would have an excessive voltage drop. In our example, we are concerned with an intermittent operation, be sure to use the chart in figure 8-13(A) for all circuits with continuous loads.

You will notice in figure 8-13 there are two charts which are very similar. Chart A shows the wire sizes for continuous loads; chart B shows the wire sizes for intermittent loads. Both charts contain one or more heavy diagonal lines. These are the current-carrying curves. When choosing a wire, one must always find the current-carrying curve which represents your circuit. The curves are identified as follows. Curve 1 is for circuits carrying a continuous load in conduit or bundles. Curve 2 is for circuits carrying continuous loads in a single wire in free air. Curve 3 on the intermittent chart is used for all intermittent circuits (two minutes or less) in conduit, a bundle or free air.

When choosing a wire size using these charts you must always be sure that the intersection of current and length falls above the curve which defines your circuit. In other words, if you are choosing a wire for a circuit with a continuous load and the wire is to be placed in a conduit, you cannot use a wire that has an intersection of current and length which falls below curve 1. If your current-length intersection is below curve 1, you must ignore that wire size and choose the wire size which falls at the intersection of curve 1 and your current curve. For example a 14-volt circuit carrying a continuous 70 amps, 12 feet in a conduit must use a 4-gage wire, not a 6-gage wire.

NOMINAL SYSTEM VOLTAGE	ALLOWABLE VOLTAGE DROP — VOLTS	
	CONTINUOUS OPERATION	INTERMITTENT OPERATION
14	0.5	1.0
28	1.0	2.0
115	4.0	8.0
200	7.0	14.0

Figure 8-12. Allowable voltage drops for aircraft electrical systems.

In general, it is always best to use the manufacturers data when selecting a wire if at all possible. If you must choose a wire, always be sure to cross-check your work with figure 8-10 or figure 8-11 to be sure that the wire is actually rated for the current.

Going back to our example, a 4-gage copper wire will carry 100 amps for 60 feet and not exceed the allowable voltage drop for intermittent operation in a 28-volt system. This intersection is well above curve three, so it is okay for intermittent use. By checking the current-carrying capability of 4-gage copper wire, we see that it can carry a maximum of 135 amps in free air, so this is a good choice of wire size.

Let's take an example of a continuous load. If we need to carry 20 amps continuously in a 28-volt system for 40 feet, according to the wire chart, we can use a 10-gage wire. Cross-checking this with figure 14-63, we find that a 10-gage wire is capable of carrying a maximum of 33 amps in a bundle, so this is a good choice.

If we had to carry 20 amps continuously for 40 feet in a 14-volt system, we would have to choose another size wire. In a 14-volt system, we are allowed only a ½-volt drop for continuous operation, and so we will need a wire with less resistance. Referring to the wire chart, we find that for the allowable voltage drop in a 14-volt system we will have to use a six-gage wire.

D. Wire Identification

There is no standard system for wire identification among the manufacturers of general aviation aircraft, but most manufacturers do put some form of identification mark about every 12″ to 15″ along the wires. The identification marking should identify the wire with regard to its type of circuit, location within the circuit, and wire size. Each manufacturer will have a code letter to indicate the type of circuit. Wires in the flight instrumentation circuits are often identified with the letter F. The letter N is sometimes used to identify the circuit ground. For example, the marking F26D-22N tells us that the wire is in the flight instrumentation circuit (F), it is wire number 26 in that circuit (26), it is the fourth segment of this wire from the power source (D), it is a 22-gage wire (22), and it goes to ground (N). The numbers on the wire greatly facilitate our troubleshooting of an electrical system, but always be sure to check the service manual for the correct meaning of the various codes. Even within a given manufacturer, code systems may vary between aircraft.

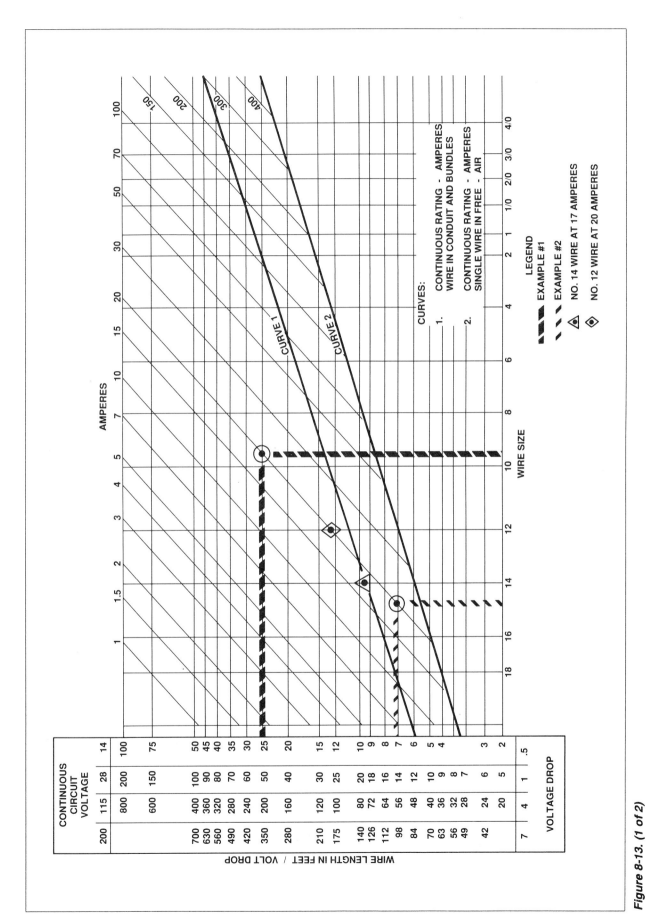

Figure 8-13. (1 of 2)
(A) Electrical wire chart for continuous loads.

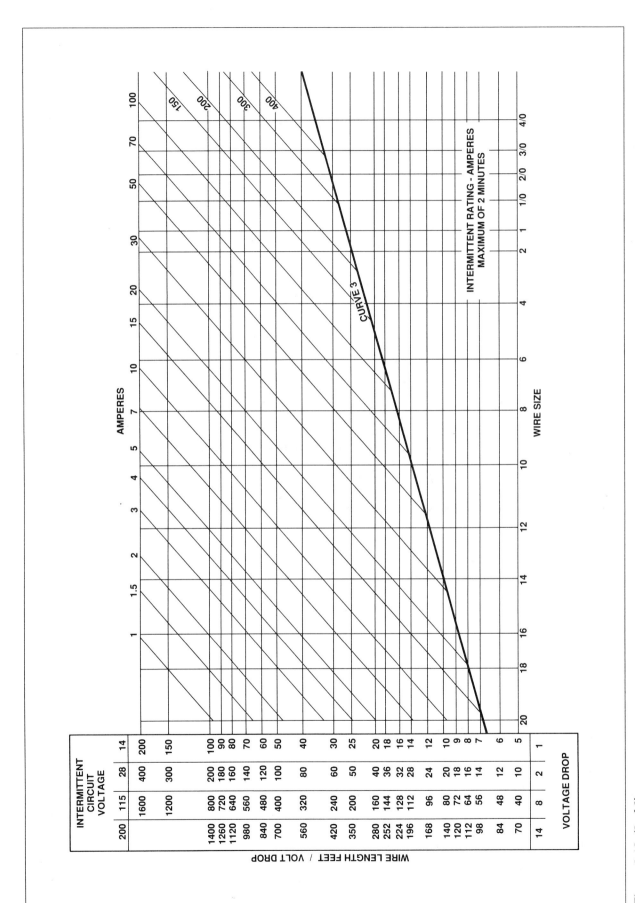

Figure 8-13. (2 of 2)
(B) Electrical wire chart for intermittent loads.

E. Wire Installation and Routing

Aircraft wiring may either be installed in what is known as open wiring, in which it is bundled together and installed with no external protection, or it may be enclosed in a rigid or flexible conduit.

1. Open Wiring

Open wiring is used where there is no great danger of mechanical damage because it is so much easier to install and maintain, and it is lighter in weight.

a. Wire Groups, Bundles and Routing

Wires are grouped and tied together in bundles for the neatest and most efficient routing, but no one bundle should carry wires from circuits that would disable both a main and back-up system if the bundle were damaged. Any wires that are not protected by a circuit breaker, and the wires of the ignition system, are not usually bundled in with other wires.

All of the wire bundles installed in an aircraft should be routed so they will not interfere with any of the controls or moving components. They must be routed where they cannot be damaged by persons entering or leaving the aircraft or by baggage or cargo moving over them or resting on them.

It is better to route the wiring along the overhead or the side walls of the aircraft than in the bottom of the fuselage or the bilges. It is possible for wires routed here to be damaged by fluids that leak into these low areas.

When electrical wires are routed parallel to lines carrying oxygen or any type of liquid, the wiring should be above the fluid line by at least two inches and must not be supported in any way from the line. The wires should also be separated from any control cable by at least 3".

Wire bundles should be secured to the structure with padded clamps, close enough together that the wire bundle will not sag or vibrate between supports. The bundle should be tight enough that it can be deflected only about a ½" with normal hand pressure applied between the supports, as seen in figure 8-14.

The last support should allow enough slack that the connector can be easily connected when the component needs to be replaced.

Generally speaking, the minimum clearances for wires and fluid lines, bleed air ducting and moving parts can be reduced if the proper mechanical protection is provided. Acceptable "secondary" protection typically includes approved continuous sleeving, tape wraps and conduits.

Wires mounted in areas subject to harsh environments, such as extreme temperatures or high moisture, should be protected accordingly. The engine nacelle is typically a high temperature area and often MIL-W-8777 wire is used in this area. This wire is a silver-coated copper wire with silicone insulation. In areas exposed to the environment, such as wheel well areas, a wire with good abrasive resistance and low moisture absorption should be used. The wire should be routed so as to minimize the likelihood of damage from foreign object abrasion.

2. Conduit

Mechanical protection can be provided for the wire by routing the bundles through either a flexible or a rigid conduit. The generally accepted practice for conduit size is to use one whose inside diameter is approximately 25% larger than the diameter of the wire bundle that is being encased.

All conduit, rigid and flexible, should have drain holes at the lowest point in each run (figure 8-15), and these holes should be carefully de-burred to remove any sharp edges that might possibly damage the wiring. The ends of the conduit should also have all of their sharp edges removed to protect the wire insulation from damage.

3. Shielding

Wires carrying alternating current radiate enough energy to interfere with some of the sensitive electronic circuits, and some of the more sensitive control circuits can be disturbed by stray electrical fields that are always present in the aircraft. To protect against the unwanted effect of these fields,

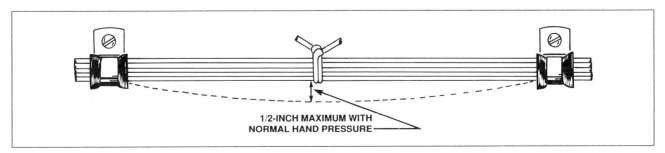

Figure 8-14. Allowable slack for an installed wire bundle.

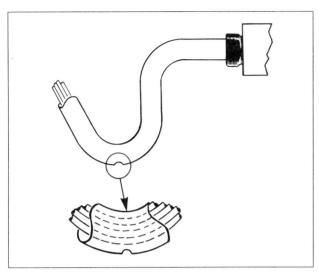

Figure 8-15. A 1/8" diameter drain hole should be drilled into the bottom of all runs of solid conduit at the lowest point in the run.

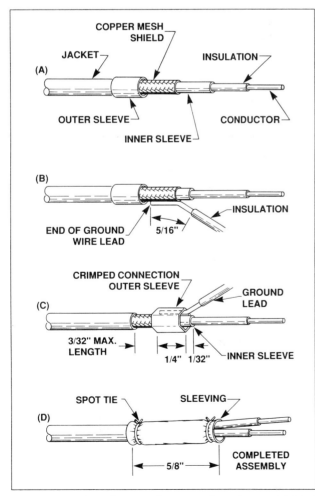

Figure 8-16. Attachment of a ground lead to a shielded wire using a crimped-on sleeve.

wires may be shielded. The insulated wire is covered with a braid of tinned copper or aluminum, and this braid is often covered with an abrasion-resisting nylon outer cover.

The shielding is grounded at only one end to prevent a flow of current within the shield itself. This ground connection should be made by attaching a ground lead to the shield with a crimped-on connector, as seen in figure 8-16. Never solder the ground lead to the shielding, as there is the danger of overheating the insulation and causing it to break down.

4. Bonding and Grounding

Bonding is the practice of connecting various aircraft components to the metal airframe with an appropriate conductor for the purpose of eliminating the build-up of unwanted static charges. Eliminating the build-up of unwanted static charges will reduce radio interference and the possibility of fire. Grounding also connects various aircraft components using appropriate wires; however, grounding is done to provide a return path for electrical components which are not mounted directly to the airframe metal structure. Grounding is also performed on most ground (earth-based) equipment which comes into contact with the aircraft. This is particularly important during fueling. If a static charge causes an electrical spark at the fuel opening, there is an extreme fire hazard.

a. General Procedures

Bonding is essentially the connection of the equipment to ground, normally using a braid or some type of flexible uninsulated wire. In some instances, the bonding connection will also act as the ground connection for an electrical component.

When installing bonding jumpers, be sure that they are as short as practical and have a resistance of no more than about 3 milliohms (0.003 ohms). If the bonding strap must also carry very much ground return current, you must be sure there is no appreciable voltage drop across the bonding connection. If the component that is bonded is shock-mounted (figure 8-17), be sure there is sufficient slack in the bonding braid that it will not be under a strain when the unit flexes to the maximum extent allowed by the shock mounts.

Since the bonding braid carries current, take special care to prevent its flowing through dissimilar metals, which would cause corrosion. Aluminum alloy jumpers are used to connect between an aluminum alloy structure and an aluminum alloy component, and cadmium-plated copper is used to bond stainless steel, cadmium plated steel or brass. If it is impossible to avoid dissimilar metal junctions, be sure to use a bonding jumper that is more susceptible to corrosion than the structure it is bonding. Before connecting a jumper to an anodized aluminum alloy part, remove the oxide coating that protects the metal, as this coating is an insulator. After the connection is made, apply an appropriate protective coating.

b. Testing

In general, most bonding jumpers or ground straps must have each connection made to have 0.003 ohms or less in resistance. This measurement must be taken between the surface being bonded and the bonding jumper as illustrated in figure 8-18.

This test may be performed using an extremely sensitive ohmmeter or a bonding tester, and should be done anytime a connection has been modified, added to or temporally disconnected.

F. Lacing and Tying Wire Bundles

Wire bundles are made up on jig boards in the aircraft factory and are tied together every two to three inches with waxed linen or nylon cord, using two half-hitches secured with a square knot. This procedure is illustrated in figure 8-19.

A convenience item that makes wire bundling fast and neat is a patented nylon strap called a Tyrap®. This small nylon strap is wrapped around the wire, and one end is passed through a slot in the other end and pulled tight. The strap locks itself in place and the end is cut off.

Lacing wire bundles with either a single- or double-lace is approved, and it is faster than tying the bundle with individual spot ties, but it is not as neat and has the disadvantage that if the lacing cord is broken, a good portion of the bundle will be without ties.

Areas of high vibration on many aircraft require special attention to the routing, lacing and tying of wire bundles. As illustrated in figure 8-20, the Boeing 747 areas of high vibration include the wing, engine, engine struts, wheel well, landing gear, empennage and air conditioning bay. The wires routed in these areas will be subject to extreme vibrations and must be installed in accordance with the appropriate service data. In general, wires in this area

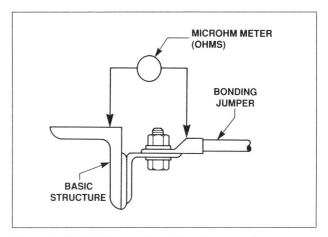

Figure 8-18. Testing a bonding jumper for proper installation.

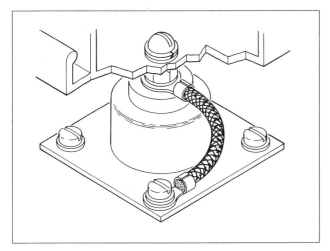

Figure 8-17. A bonding braid connecting a shock-mounted component to the structure must allow full movement of the shock mount.

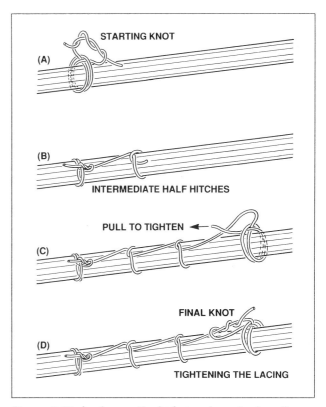

Figure 8-19. Lacing method of securing wire bundles.

should not be bundled with plastic wire wraps (Tyraps) and require special knotting of the string-type lace cord. Additional abrasion protection is also typically provided for wiring installed in high vibration areas.

G. Wire Termination

1. Stripping Wire

In order to attach a wire to its terminal or connection, the protective insulation must be removed. This insulation is typically removed by cutting the insulation and gently pulling it from the end of the wire. This process is known as stripping the wire. The stripping process should be accomplished to expose as little of the conductor as necessary to make the connection and must be done in such a way that the conductor is not damaged beyond limits.

The most common damage to a conductor often occurs when the insulation is being cut. The cutting tool may cut or nick one or more of the conductor strands if improperly used or the tool is out of adjustment. The FAA specifies limits for the allowable nicked or broken strands on any conductor. As can be seen in figure 8-22, a 20-gage copper wire with 19 strands is allowed 2 nicks and no broken strands. In general, the larger the number of strands or the larger the conductor, the greater the acceptable number of broken or nicked strands.

2. Solderless Terminals and Splices

The wiring for a modern aircraft is assembled in the factory on jig boards and tied together in bundles. Groups of wires normally terminate with quick-disconnect plugs, and individual wires have terminals staked, or crimped, onto the wire. The terminal is attached to a screw-type lug on a terminal strip or on the electrical component itself.

a. Copper Wire Terminals

The commercial electrical industry uses many different shapes of terminals, but the one most generally used for aviation applications is the ring-type, either pre-insulated or non-insulated type, made of cadmium-plated copper. The ring-type terminal is the system of choice because the ring terminal is held onto its specific connection with a locknut or lockwasher nut assembly. This type of installation is less likely to fail than slide-on-type terminals. Figure 8-23 shows a typical ring-type terminal.

It is very important to be sure the materials of the splice are compatible to eliminate the possibility of dissimilar metal corrosion. Always be sure to use copper or copper-alloy terminals on copper wire.

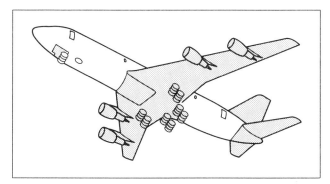

Figure 8-20. *The areas subject to high vibration (indicated by shading), may require special lacing practices.*

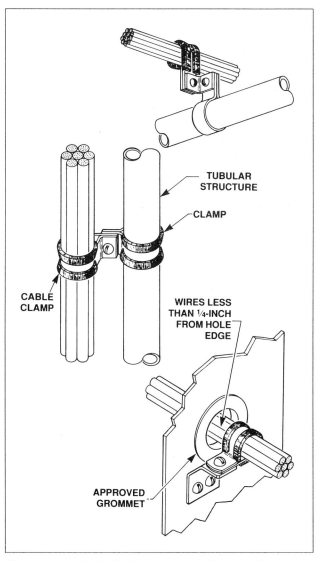

Figure 8-21. *Method of attaching wire bundles to an aircraft structure using cushion clamps.*

WIRE SIZE #	CONDUCTOR MATERIAL	NUMBER OF STRANDS PER CONDUCTOR	TOTAL ALLOWABLE NICKED AND BROKEN STRANDS
24-14	COPPER OR COPPER ALLOY	19	2 NICKED, NONE BROKEN
12-10		37	4 NICKED, NONE BROKEN
8-4		133	6 NICKED, NONE BROKEN
2-1		665-817	6 NICKED, 2 BROKEN
0-00		1,045-1,330	6 NICKED, 3 BROKEN
000		1,665-	6 NICKED, 4 BROKEN
0000		2,109-	6 NICKED, 5 BROKEN
6-000	ALUMINUM	ALL NUMBER OF STRANDS	NONE, NONE

Figure 8-22. Allowable nicked or broken strands as specified by the FAA.

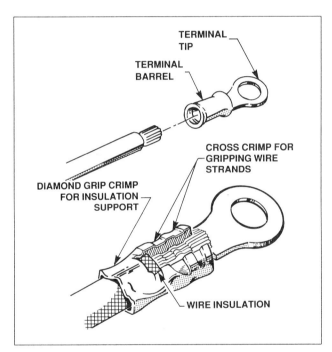

Figure 8-23. Installation of pre-insulated, crimped-on wire terminals.

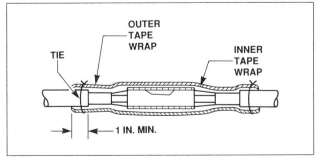

Figure 8-24. Double-tape wrap method of insulating a wire splice.

Splices are often used in situations where a portion of a wire has been damaged or must be lengthened. The FAA recommends against soldered splices, therefore solderless splices are typically used. It is also important to keep splices to a minimum and stagger any splices in a wire bundle.

If a splice or a terminal is installed on a wire it should always be insulated to protect it from shorting to another circuit or ground. On non-insulated terminals, a heat-shrinkable insulation sleeve or double-tape wrap should be installed to protect the exposed wire and terminal. The double-tape wrap method approved for many commercial aircraft applications is shown in figure 8-24. It is also important to use an approved terminal or splice in areas subject to extreme heat, vibration or moisture conditions.

b. Aluminum Wire Terminals

Aluminum wire must be terminated with aluminum terminals or a metal alloy which will not adversely affect the aluminum. Since most of the aluminum wire used in an aircraft is large, the terminals cannot be staked on with hand tools. Most wire manufacturing facilities use pneumatic squeezers that produce enough pressure to ensure a proper connection.

In cases where it becomes necessary to splice copper to aluminum wire, be sure to use a splice connector compatible with both metals. Since the aluminum wire will most likely be a larger diameter than the copper wire, a build-up sleeve must be slid over the copper conductor before installing the crimp splice. Extra care should always be taken to protect this type of splice from moisture.

c. Pre-Insulated Terminals

Pre-insulated terminals have become commonplace in the aircraft industry due to their ease of installation and proven effectiveness. On pre-insulated terminals and splices, a color-coded insulator typically made of nylon covers the barrel of the terminal and is crimped when the terminal is crimped onto the wire. A second crimp on the insulation below the terminal barrel causes the terminal to grip the insulation of the wire. This not only provides good electrical protection but helps prevent any vibration stresses from being concentrated where the barrel grips the wire.

The color of the terminal insulation identifies the wire sizes the terminal will accommodate. Red terminals are used on wire gages 22 through 18, blue terminals are used on 16- and 14-gage wire, and yellow terminals are used on 12- and 10-gage wires.

To install a terminal, strip the insulation from the end of the wire so the end of the strands will just stick through the barrel. Be sure to use a stripper with the proper size notch so none of the strands will be nicked. Slip the stripped end of the wire into the barrel of the terminal and put the terminal in a crimping tool until the barrel rests against the stop in the tool, and squeeze the handles of the tool together.

d. Crimping Tools

There are a number of types of crimping tools available, but the best ones have a ratchet mechanism that will not allow them to open until they have crimped the terminal to the proper size. These tools should be periodically calibrated against a standard so you will be assured that the terminal is properly installed. If a terminal is properly crimped on the wire, the wire will break before the terminal slips off. For limited applications, the non-ratchet-type tool may be used; however, the technician should be sure to close the tool completely for a proper crimp. In production-type applications, many factories use a pneumatically operated machine.

3. Emergency Splicing Repairs

In general, splices in electrical wire should be kept to a minimum and should be avoided completely in areas of high vibration. Splices in bundles shall be staggered and there shall be no more than one splice in any one segment of wire. Splices shall not be used to salvage scrap wire and cannot be installed within 12" of a termination device. In certain instances, these rules can be amended if the splice installation has been authorized by an ap-

Figure 8-25. Crimping tools for pre-insulated wire terminals.

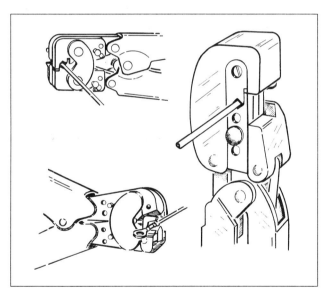

Figure 8-26. Heavy-duty crimping tool for installing pre-insulated wire terminals.

proved engineer. In some cases, emergency splices have been approved for use in commercial aircraft with the understanding that the wire would be repaired with a permanent installation within a limited number of flight hours.

4. Junction Boxes and Terminal Strips

Most of the wires that are not terminated in AN- or MS-type, quick-disconnect plugs are connected to terminal strips housed in junction boxes. The

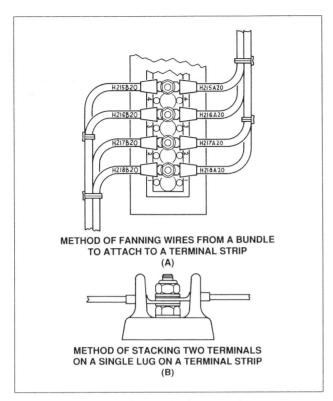

Figure 8-27. Terminals attached to a stud.

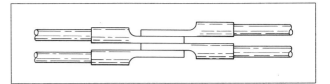

Figure 8-28. Method of stacking four terminals for attachment to a single lug on a terminal strip.

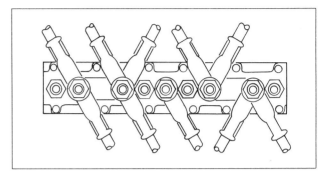

Figure 8-29. Bus straps may be used to join two lugs on a terminal strip.

typical terminal strips are made of a plastic or paper-base phenolic compound, which has high mechanical strength as well as good electrical insulation properties. Studs from size six up to about ¼" in diameter are used, with the smaller sizes suitable only for control circuits having very low current and for areas where the wires can be protected from mechanical vibration and strain. It is standard practice to use no stud smaller than a number ten for power circuits.

Terminal strips normally have barriers between the adjacent studs to keep the wires properly separated. Ideally, we should have only two wires attached to any stud, as we have in figure 8-27, but in no case should we have more than four terminals stacked on any single stud. Notice that the ring of the terminal is not in line with the center of the wire but is offset to one side, and so when stacking terminals on a stud, stack them as we show in figure 8-28 so that all of the wires will lie flat.

If it is necessary to attach more than four wires to a stud, you should use a cadmium-plated copper bus strap to join two of the studs, and divide the wires between them. Bus straps (figure 8-29) are available for terminal strips, so you may make up a busbar with as many studs connected as is necessary.

The terminal strips in an aircraft electrical system are mounted in junction boxes made of either aluminum alloy or fiberglass. These boxes are mounted where they will be least in the way and where there is a minimum chance of their being damaged. All boxes must have drain holes to prevent any water accumulating in them that could cause a short circuit of the wiring. It is a good policy to mount the boxes either vertically or on an overhead portion of the compartment, to minimize the possibility of washers and nuts dropping between terminals and causing a short.

It is extremely important when working in any junction box that you do not wear any kind of metal or electrically conductive ring. Some technicians recommend the removal of any wristwatch. If you should short between a hot terminal and ground with a ring or watch, the jewelry will possibly weld to the stud and give you a serious burn. Of course, the same short is very likely to damage the electrical circuit.

5. AN/MS Connectors

Most of the electrical components in an aircraft are designed so they may be serviced with a minimum amount of time needed for their removal and installation. The electrical wiring is usually connected through quick-disconnect plugs. As we can see in figure 8-30, there are many different types of plugs, but they are all somewhat similar. The individual wires are fastened to pins or sockets inside the plugs and are clamped tight to prevent mechanical strain on the cable being transmitted into the connectors themselves.

The shells of these connectors may be fitted with various types of inserts that may carry either pins or sockets. It is customary, to help eliminate possible short circuits to ground, that the end of the connector that carries the power use socket connectors, and pins in the connector that receives the power.

In figure 8-31 we have the explanation of the code used to describe an MS conductor.

The MS number is the basic configuration of the connector:

MS3100 — wall receptacle
MS3101 — cable receptacle
MS3102 — box receptacle
MS3106 — straight plug
MS3107 — quick-disconnect plug
MS3108 — angle plug

The letter following the configuration tells the class of connector it is.

A — General purpose, solid aluminum alloy shell
B — General purpose, split aluminum alloy shell
C — Pressurized, solid aluminum alloy shell
D — Environmental-resistant, solid aluminum alloy shell
K — Fire- and flame-proof, solid steel shell

Figure 8-30. Types of MS quick-disconnect plugs and receptacles for aircraft wiring.

The size of the connector is indicated with a code number—the higher the number, the larger the connector. The insert arrangement is a code number to identify the number and size of the connectors and their physical arrangement. You must use a chart to identify each arrangement by the code number.

The contact style may be either an "S" or "P" to indicate a socket or pin (female or male) arrangement. The final letter in the identification is one of the last letters in the alphabet: W, X, Y or Z. These letters indicate the rotation of the insert in the connector. If a component has two or more connectors of the same type, it is possible to connect the wrong plug to a receptacle. To prevent this, the inserts may be rotated in their relationship to the index slot so that only the correct plug may be inserted into the receptacle.

There are two ways the wires may be connected to the contacts in a connector. Some of the newer plugs are designed for high-speed production, and a tapered pin is crimped onto the wire end with an automatic machine. This type of connection is illustrated in figure 8-32. The pin is slipped into a tapered hole in the pin or socket in the connector. A special tool is used to remove or replace the pin if it is ever necessary.

Many of the MS connectors have the wires soldered into solder pots in the end of the pin or socket. When soldering a wire into a connector, leave about 1/32" of stranded wire between the top of the solder pot and the end of the wire insulation. When the solder in the connection has cooled, slip a piece of close-fitting insulation tubing over the end of the connector to insulate it. This procedure may be seen in figure 8-33.

6. Coaxial Cable

The connection between a piece of electronic equipment and its antenna is often made with coaxial cable, which is a special form of shielded wire. Coaxial cable has an inner conductor, a layer of insulation and a braided outer conductor, all of this encased in a vinyl jacket. There are a number of sizes and types of coaxial cable used for electronic installation, and each type must be terminated in the way specified by the manufacturer of the connectors. BNC connectors are perhaps the most widely used type, and in figure 8-34, we have the method used for their installation.

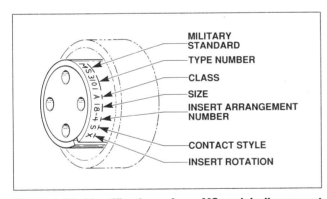

Figure 8-31. Identification of an MS quick-disconnect plug.

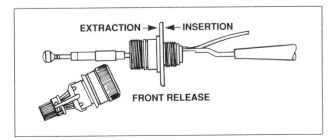

Figure 8-32. Attachment of wires into a quick-disconnect plug, using tapered pins crimped onto the wires.

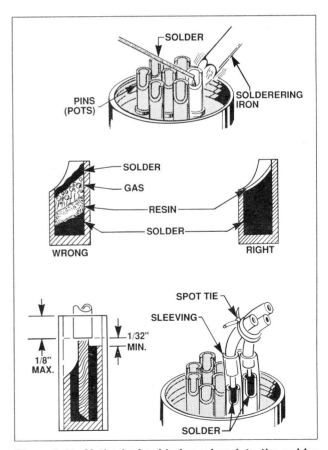

Figure 8-33. Method of soldering wires into the solder pots in an MS connector.

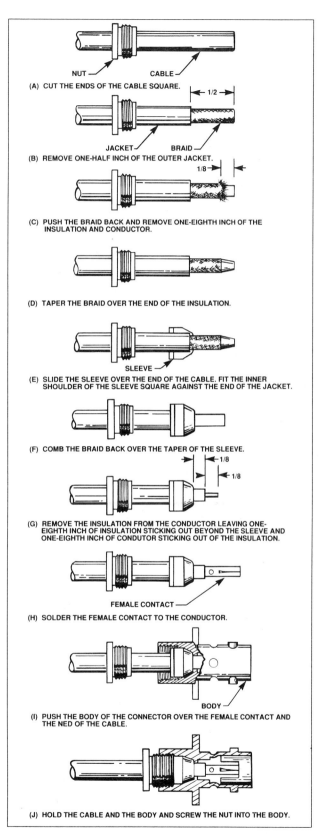

Figure 8-34. Installation of a BNC connector on a coaxial cable.

II. ELECTRICAL SYSTEM COMPONENTS

A. Switches

The purpose of a switch is to interrupt the flow of current to the component it controls. Switches are rated with regard to the voltage they can withstand, and the current they can carry continuously. The continuous current is not typically the limiting factor with a switch, since damage to a switch occurs as the contacts begin to open, and current continues to flow across them.

Inductive loads create the high degree of "rush-in" current and require derating of the switch current to prevent switch failure. An inductive load, such as a motor, will force current to flow across the contacts as their resistance increases when they start to open. The resistance of the filament of a lamp (an inductive circuit) is very low when it is cold. Therefore, when the switch is first closed in a lamp circuit, the initial rush-in of current is extremely high, as much as 15 times as high as the current the lamp uses for normal operation.

Because the switch must be able to carry its rated current continuously and at the same time be able to operate the type of load in which it is installed, a derating factor has been devised that allows us to match the switch to the type of load. For example, a typical aircraft toggle switch is rated at 24 volts for 35 amps. If this switch is to be installed in a lamp circuit, it should not be used for more than 4.4 amps. Be sure to consult the derating factor table in AC 43.13-1A when choosing a switch for any circuit.

As much as is possible, switches should be installed in such a way that upward or forward movement of the control will turn the switch on. If the switch is used to control circuits in the right and left engines, the operating control should point to the engine that is affected.

Certain circuits must be operated only in an emergency. For these circuits, the switch is normally enclosed in a cover that must be lifted before the switch can be actuated. These switches are said to be guarded. If the circuit is one that could endanger a system if it were operated inadvertently, the cover may be safety-wired shut with a very light safety wire that can be broken if it becomes necessary to actuate the switch.

1. Toggle and Rocker Switches

Toggle switches have been one of the more popular type switches for aircraft electrical systems, but

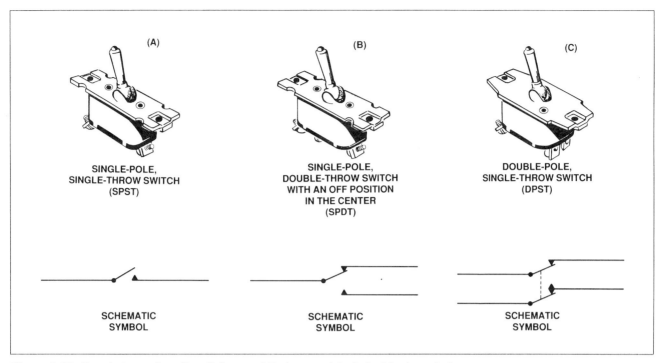

Figure 8-35. Toggle-type aircraft switches and their associated electrical symbol.

in the past few years, this type of switch with its bat-like operating handle has been replaced by rocker-type switches that allow for easier operation and make for a more attractive panel. Rocker switches are also safer in the event that the pilot or passenger should accidentally strike a switch during flights in rough air, a crash or hard landing.

Electrically, both toggle and rocker switches operate the same and are available with a number of different contact arrangements. These switches are named according to the number of circuits they control and the number of sets of poles they have. An assortment of toggle switches may be seen in figure 8-35. A single-pole, single-throw (SPST) switch has two contacts and controls only one circuit. It can only be either on or off. A single-pole double-throw (SPDT) switch selects two conditions for a single circuit. A flap motor switch would, for example, have two positions for sending current through the flap motor, driving the motor in the direction either to raise the flaps or to lower them.

A double-pole, single-throw (DPST) switch controls two circuits but with only an open and closed position. The advantage of a double-pole switch is that both circuits are controlled together, but they can be fused separately. If one circuit fails it would not effect the other. An example of this type of switch might be found on the aircraft to control the navigation and panel lights. Both circuits could be turned on with one switch, and yet the circuits are completely independent. Double-pole, double-throw (DPDT) switches control two circuits in two conditions. Double-throw switches may be a two or three position switch. A two position switch controls the circuit(s) in two conditions only. A three position switch would also have a center position for turning the circuit off.

2. Rotary Switches

When it is necessary to select several conditions for a circuit, a rotary switch may be used. These switches are made up of wafers with contacts arranged radially around the central shaft and contact arm. This construction is seen in figure 8-36. Any number of wafers can be stacked onto the shaft to control as many circuits as are needed.

3. Precision (Micro) Switches

Precision switches, generally known by the trade name Microswitch® because of the popularity of this particular manufacturer, are used in many applications in an aircraft. Most commonly, these switches are actuated when some mechanical device reaches a particular position. These switches require only a slight movement (generally less than $1/16"$) of the operating plunger to cause the internal spring to snap the contacts open or closed. Figure 8-37 shows the internal mechanism of this type of switch. When precision switches are used to limit the movement of a mechanism, they are typically referred to as limit switches. For example, the electric motor of a flap actuator would be turned off by a limit switch when the flaps reach their up or down limit. The up-and-locked and down-and-locked lights are controlled through a precision switch when the landing gear is firmly secured in the appropriate positions.

B. Relays and Solenoids

One of the big advantages of an electrical system is the ease with which components may be remotely

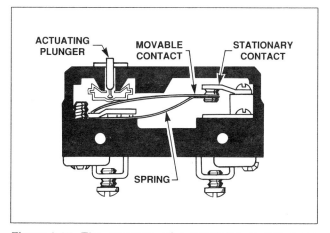

Figure 8-37. The contacts of a precision switch snap open with only a small movement of the actuating plunger.

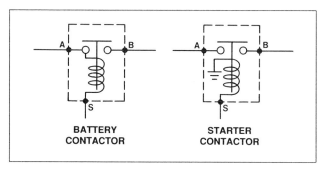

Figure 8-38. Internal connections of two types of solenoids.

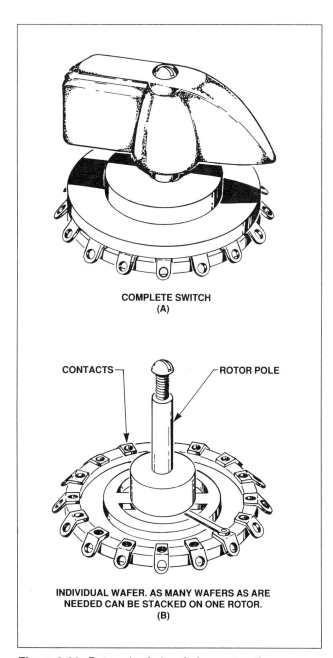

Figure 8-36. Rotary (wafer) switch construction.

controlled. By using a solenoid or relay, a very small switch can be used to control a high current, such as that needed to operate an aircraft engine starter. Relays and solenoids are similar in their application, differing only in how they operate.

Normally, a relay has a fixed soft iron core around which an electromagnetic coil is wound. Movable contacts are closed by the magnetic pull exerted by the core when the coil is energized, and are opened by a spring when the coil is de-energized. A solenoid has a movable core that is pulled into the center of an electromagnetic coil when the coil is energized. Because of the movable core, solenoids respond quicker and are stronger than relays.

Both relays and solenoids are used for electrical controls. Solenoids are typically used for high current applications and also find important use as mechanical control devices; for example, to move locking pins into and out of mechanically actuated devices. To eliminate the confusion between relays and solenoids, many aircraft manufacturers often refer to magnetically activated switches as contactors.

Two very commonly used solenoid-type switches are the battery contactor and the starter contactor used on most light aircraft. The basic difference between these two devices is the operation of the coil. One end of the coil in the battery contactor is internally connected to the main terminal that connects to the battery. The other end of the coil comes out of the contactor case through an insulated terminal and goes to the battery master switch, through which it is grounded when the master switch is turned on. These are seen in figure 8-38.

One end of the control coil in the starter solenoid is grounded inside of the housing and the other end comes out through an insulated terminal and goes to the ignition switch. When the ignition switch is in the START position, a small control current flows through the starter contactor coil and closes it, allowing the high current to flow to the starter. It should be noted that starter contactors are often intended for intermittent duty cycles. The electromagnetic coil may overheat if activated for long periods of time.

C. Current Limiting Devices

Excessive current will flow in an electrical circuit when it becomes shorted. This high current flow can generate enough heat to damage the wiring, and can even cause an in-flight fire. To prevent damage from electrical shorts, all circuits in an aircraft must be protected. This can be accomplished using either a fuse or a circuit breaker.

1. Fuses

A fuse is simply a piece of low-melting-point alloy encased in a glass tube with metal contacts on each end (figure 8-39). It is placed in the circuit, usually at the busbar, so all of the current flowing to a circuit must pass through it. If too much current flows, the link will melt and open the circuit. To restore the circuit, the blown fuse must be replaced with a new one.

There are two types of fuses used in aircraft circuits: the regular glass tubular fuse and the slow-blow fuse. The regular fuse has a simple narrow strip of low melting point material that will melt as soon as an excess of current flows through it. The slow-blow fuse has a larger fusible element that is held under tension by a small coil spring inside the glass tube. This fuse will pass a momentary surge of high current such as you have when the switch in a lighting circuit is closed, but it will soften under a sustained current flow in excess of its rating, and the spring will pull the link in two, opening the circuit.

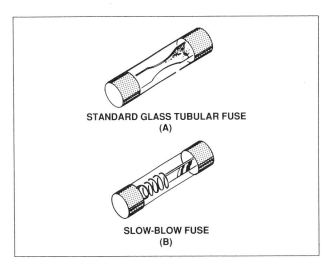

Figure 8-39. Aircraft fuses.

According to the flight regulations, aircraft that are equipped with fuses are required to carry at least 50% of any one type of fuse installed on the aircraft. If one fuse of a particular type is installed, one spare of the same rating must be carried on the aircraft at all times.

Fuses which are accessible by maintenance personnel only are found on some aircraft. These fuses, commonly called current limiters, are often used to isolate a complete distribution bus in the event of a short to that bus. The pilot would simply continue the flight without use of the isolated bus and the have problem corrected upon landing.

2. Circuit Breakers

Circuit breakers are used rather than fuses because it is so much easier to restore a circuit in flight by simply resetting the circuit breaker than it is by removing and replacing a fuse. There are two basic types of circuit breakers used in aircraft electrical systems, those that operate on heat and those that are opened by the pull of a magnetic field. Most aircraft breakers, however, work on the principle of heat. When more current flows than the circuit breaker is rated for, a bimetallic strip inside the housing warps out of shape and snaps the contacts open, disconnecting the circuit.

There are three basic configurations of circuit breakers used in aircraft electrical systems: the push-to-reset, the push-pull type and the toggle type. All three types are illustrated in figure 8-40. The button of the push-to-reset type of circuit breaker is normally in, and it pops out only when the circuit has been overloaded. This type of circuit protection device cannot be used as a switch, as there is no way to grip the button to pull it out.

169

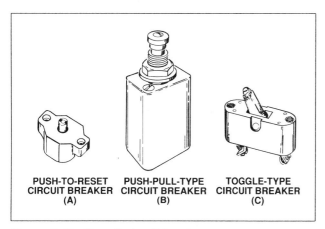

Figure 8-40. Aircraft circuit breakers.

The push-pull-type circuit breaker has a small lip that can allow the breaker to be pulled to open the circuit. Normally these circuit breaker buttons are in, but if a breaker is overloaded, the button pops out and is easily identified. To aid in this identification, many of these circuit breakers have a white band around the button that is visible when the breaker has popped. This type of circuit breaker should be used to switch the circuit only for maintenance purposes.

The toggle-type circuit breaker is normally used as a control switch as well as a circuit breaker. When the toggle is up, the circuit is closed, but if the circuit is overloaded, the breaker will pop partially down. To restore the circuit, move the toggle all of the way down and then back up.

All aircraft circuit breakers must be of the trip-free type, which simply means that the breaker contacts will remain open as long as a short circuit exists, regardless of the position of the actuating control. This prevents the circuit breaker from being closed and manually held closed if a fault exists that could generate enough heat to cause a fire.

Anytime a circuit breaker opens a circuit, allow it time to cool and then reset it. If it was popped because of a transient condition in the electrical system, it will remain closed and the electrical system should operate normally, but if it was opened by an actual fault, it will pop open again and you should leave it open and find the cause of the excessive current.

On many large commercial aircraft, electronic circuit breakers are being used to control overloaded power distribution circuits. The current flowing through a conductor is measured using inductive pickups called current transformers. This signal is monitored by the systems control unit computer and if an over-current condition exists, the computer opens the circuit breaker.

III. AIRCRAFT LIGHTING SYSTEMS

As aircraft have grown in complexity, and the airspace has become more crowded, the lighting systems required for safe flight and passenger comfort have increased in number and sophistication. Lighting systems illuminate everything from cargo compartments to the pilots instrument panel. Exterior

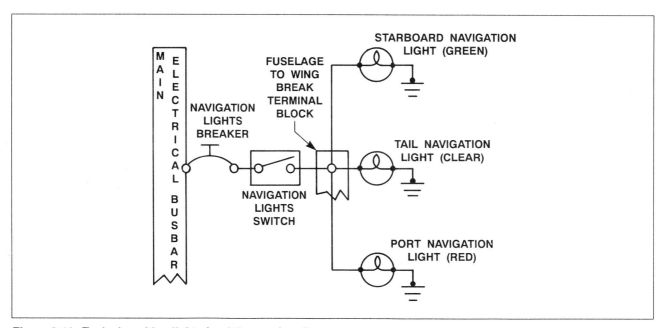

Figure 8-41. Typical position light circuit for an aircraft.

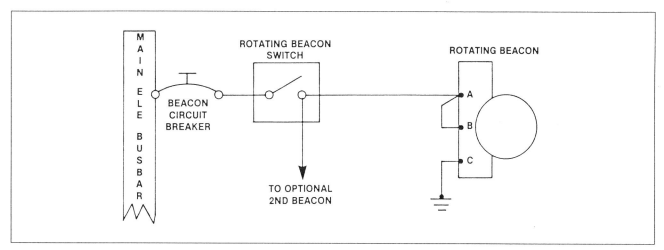

Figure 8-42. Motor-driven rotating beacon.

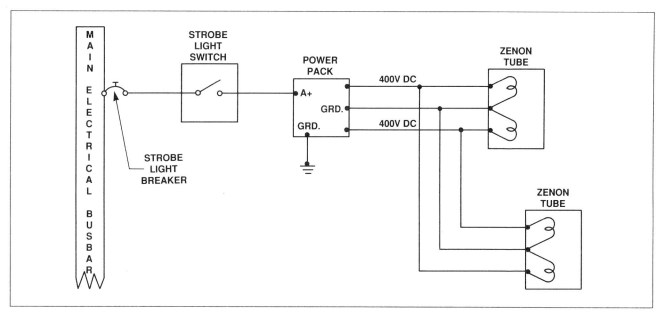

Figure 8-43. Strobe-type anti-collision light circuit.

lights are required to ensure safe operations during night flights. The aircraft technician must become familiar with aircraft lighting circuits in order to maintain these systems properly.

A. Exterior Lights

There are a variety of exterior lighting systems. These include position, landing, taxi, anti-collision and wing inspection lights.

1. Position Lights

Position lights are used to indicate the position of an aircraft during night operations. If pilots can identify the position of another aircraft from its position lights, they may safely navigate around that aircraft, hence, position lights are often referred to as navigation lights. One or more position lights must be located on each wing tip, and the tail of the aircraft. The right wing tip must have a green colored light, the left a red light, and the tail must have a white light. These lights are required on any aircraft certified for night flight. The actual light bulb is covered with a clear glass, the color is achieved by placing a red or green lens over the bulb. Most navigation lights operate in one mode only; however, some older model aircraft have dimmer or flasher circuits incorporated with position lights. A typical position light circuit is shown in figure 8-41.

2. Anti-Collision Lights

Anti-collision lights are found in two basic styles and the difference typically comes as a function

171

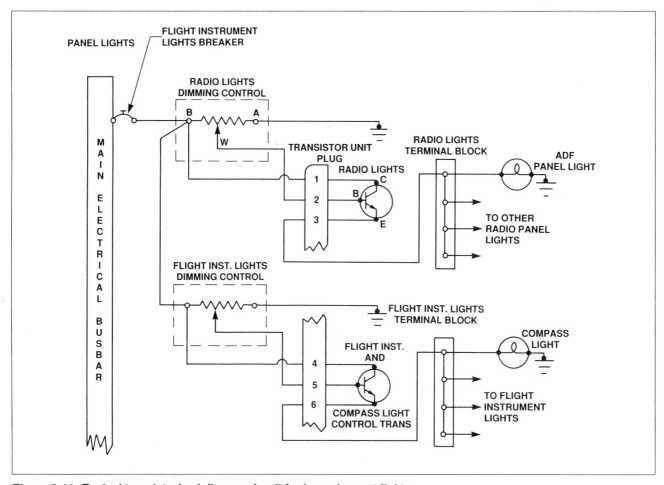

Figure 8-44. Typical transistorized dimmer circuit for incandescent lights.

of age. Older aircraft were originally equipped with rotating beacons, either on top of the vertical stabilizer or, on the top or bottom of the fuselage. The rotating beacon system typically contained a stationary light bulb and a rotating reflector covered by a red glass lens as shown in figure 8-42.

Newer systems utilize solid state electronics to create a flashing- or strobe-type anti-collision light. The strobe-type flashing anti-collision light has an extremely bright flash produced by a xenon tube which requires approximately 400 volts. The high voltage is produced by the strobe power supply which uses a capacitor charging system to achieve the high voltage. Typically, two white wing tip strobe lamps and one red tail strobe are used on modern aircraft. Figure 8-43 shows the power supply and wiring arrangement for an aircraft strobe.

3. Landing and Taxi Lights

In order to provide night visibility for landing and taxiing, two or more sealed beam lamps are mounted on the aircraft facing forward. These lights can be mounted in the leading edge of the wing, in the cowling or on landing gear struts. Some aircraft employ a retractable light which extends from the wing during use. Since the airplane is in a nose-high attitude during landing, lights must point forward and slightly down. Taxi lights point mostly straight forward of the airplane.

Since both landing and taxi lights are relatively high power circuits, they are often controlled, or switched, by a solenoid. The pilot activates the appropriate cockpit switch; this engages the solenoid and turns on the lamp.

4. Wing Inspection Lights

Many aircraft are equipped with anti-icing or deicing equipment. The purpose of this equipment is to remove, or prevent the build-up of, ice on critical surfaces such as the leading edge of the wing. Wing inspection lights are typically flush mounted in the fuselage or engine nacelle and are directed towards the leading edge of the wing. If the pilot suspects the formation of ice, the wing can be illuminated and the proper inspection made.

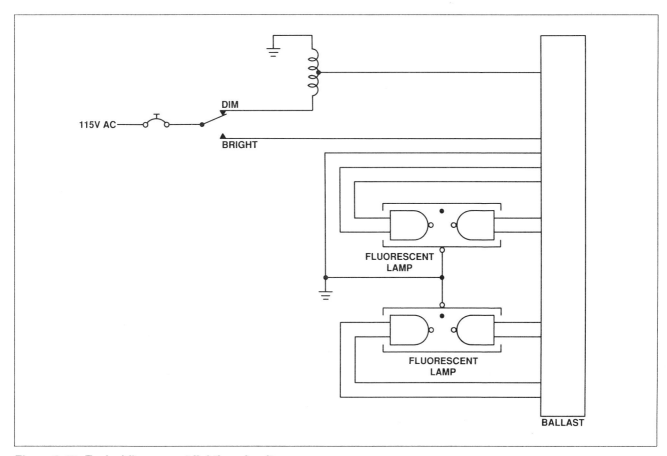

Figure 8-45. Typical fluorescent lighting circuit.

B. Interior Lights

There are a variety of interior lights found on modern aircraft, including instrument lights, overhead lights, step lights and reading lights to mention a few. In general, these lights can be divided into two basic categories: incandescent and fluorescent. Incandescent lights use a small coil of wire, called a filament, which glows in a white light when current flows through it.

Fluorescent lights are made of a gas-filled glass tube which glows when a high AC voltage is applied to electrodes at each end. Applying current releases the electrodes and they emit free electrons. The free electrons strike atoms of mercury vapor in the tube and this produces an ultraviolet light. The invisible ultraviolet light strikes the phosphorous coating on the inside of the tube and it glows in a white light. The conversion of one kind of light to another is known as fluorescence. Fluorescent lamps are much more efficient than incandescent lamps, however they require the use of transformers and AC voltage. Therefore, fluorescent lamps are found only on large commercial aircraft.

Both incandescent and fluorescent lamps can operate in a bright or dimmed position. Incandescent lamps are often dimmed using a solid-state circuit to control the current to the lamps. As illustrated in figure 8-44, a potentiometer is used to control the input signal to the transistor, thus controlling the current to the light. As can be seen by figure 8-45, the fluorescent tube is in the dim position when only a single voltage source is applied to the ballast transformer. In the bright position, an additional voltage is applied and the ballast and more current is sent to the fluorescent tube.

Another type of interior lighting system has recently been introduced to aircraft instrument panels. The Electro Luminescent (EL) panel contains a fluorescent paste sandwiched between two layers of plastic. The paste glows when an AC voltage is applied to the panel. The light glows through the unpainted areas of the plastic, typically displaying the needed letters and/or numbers. Since Electro Luminescent panels operate only with alternating current, most light aircraft with EL systems use a static inverter specifically designed for the panel.

C. Maintenance and Inspection of Lighting Systems

Most lighting circuits are relatively low maintenance items. Periodic inspections of the wire for chaffing and hardware security, corrosion of components, and general condition of the circuit should be performed during routine inspections. Lamp replacement is generally the most needed repair for lighting systems, and one must always be careful to install the correct bulb. Several variations of a given lamp may fit the same socket. Be sure to install a bulb with the correct voltage and amperage requirements.

When dealing with any high-intensity flashing lamp or strobe system, be careful to avoid electrical shock. The system operates on a high voltage and requires a few minutes to discharge itself if the lamp is defective. Always allow a strobe system to sit in the OFF position for approximately five minutes prior to maintenance.

IV. SMALL SINGLE-ENGINE AIRCRAFT ELECTRICAL SYSTEMS

The electrical systems for most of the small single-engine aircraft are relatively simple. The power is distributed to the various loads through the aircraft busbar, or bus. The bus receives power from the battery or aircraft alternator. Aircraft systems have a master switch to disconnect the battery from the electrical system when the generator is not in use, and there is also a switch to open the generator or alternator field and shut off its output in case of an electrical system malfunction.

Most of the aircraft service manuals have all of the electrical circuits broken down with only one basic circuit per page. This makes troubleshooting much easier by isolating the individual circuit. The term "schematic" is often used to refer to the electrical diagrams found in such manuals. The intent of these diagrams is to show the aircraft technician the relationship between all the electrical components and their associated wiring of a particular system. The electrical diagrams do not show the physical location of components; they do however sometimes contain code numbers which help technicians locate a component on the aircraft.

A. Battery Circuit

The electrical systems for most small aircraft use lead-acid batteries rated at either 12 or 24 volts, and, almost without exception, they all use a negative-ground, single-wire system. This means that the negative terminal of the battery is connected directly to the aircraft structure, and this places the entire metal aircraft structure at a negative potential. All of the electrical components need only a single wire from the positive voltage source to supply them, and their negative terminals should be connected to the metal structure to provide a return (negative) path for the current. The exception to this system occurs when the aircraft is built using composite materials in place of aluminum. The composite materials are non-conductive. On composite aircraft, or portions of the aircraft made of composites, a two-wire system must be used. One wire is needed for the positive connection and one for the negative connection.

In figure 8-46, we have a typical battery circuit for a light aircraft. The positive terminal of the battery connects to the battery contactor, which is a normally-open heavy-duty relay. One end of the control coil of the contactor connects internally to the main terminal, and the other end is sent to ground through the battery master switch on the instrument panel. A clipping diode is installed across the coil of the relay to eliminate spikes of voltage that are induced when the master switch is opened and the magnetic field from the contactor coil collapses. The voltage spikes may damage sen-

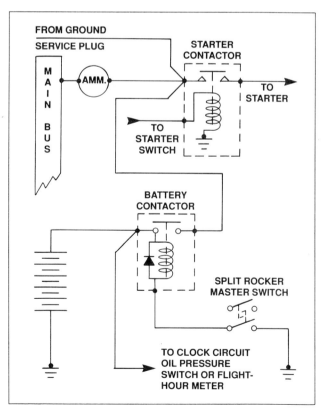

Figure 8-46. Battery circuit for a small single-engine aircraft.

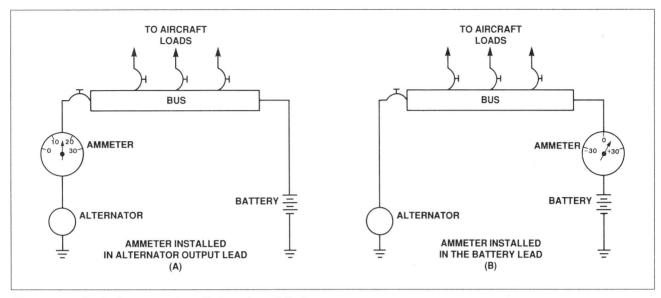

Figure 8-47. Typical ammeter installations for a DC alternator.

sitive electronic components if they are not controlled using the clipping diode.

Most aircraft have some circuits that require power all of the time, whether the master switch is on or not. These low-current components, such as the clock and the flight-hour meter circuit, are connected to the main terminal of the battery contactor and supply current to these components anytime a battery is installed in the aircraft. These circuits often employ an in-line fuse or one mounted adjacent to the battery to protect the circuit.

The battery and battery contactor on a light aircraft may be located on the engine-side of the firewall, or they may be back in the rear baggage compartment, depending upon the weight-and-balance characteristics of the aircraft. To cut down on the amount of large wire that is needed, the starter contactor, which is always on the engine-side of the firewall, normally serves as a handy junction point for the heavy wire from the battery contactor and the ground service circuit, as well as for the wire that supplies the aircraft electrical loads through the ammeter.

B. Generator Circuit

An aircraft generator or alternator is the primary source of electrical energy. It must have sufficient capacity to supply all of the electrical loads and to keep the battery fully charged.

It is possible for an aircraft to have electrical demands that exceed the rating of the generator. If this occurs for two minutes or less, no action need be taken. If the overload continues, the aircraft must have placards in plain sight of the flight crew that tell them exactly what equipment can be used at any given time.

Most of the smaller electrical systems use a current limiter in the generator control to limit the current the generator can put out. Aircraft which use a DC alternator typically have a circuit breaker as their current limiter. An ammeter is typically installed in either the alternator output lead or battery positive lead. An ammeter installed in the battery lead should not show discharge except during the momentary operation of high-current devices such as landing gear motors or flaps. If a discharge is shown on the ammeter, the generator/alternator is incapable of supplying enough current to meet demands. If the ammeter is installed in the alternator output lead, the meter will indicate a positive value whenever the alternator is working correctly. Figure 8-47 illustrates two different connections for the ammeter in a DC alternator circuit.

Some installations, instead of having an ammeter, use a load-meter which is similar to an ammeter but is marked in percentage rather than in amps. The flight crew should never allow the electrical load to exceed 100% on the meter. The shunt for the load-meter is matched to the generator output so it will indicate 100% when the generator rated current is flowing.

In an installation where there is no way to monitor the load and where it is important that the battery be kept charged in flight, you should be sure that the total continuous load connected to the generator does not exceed 80% of the generator's rated output.

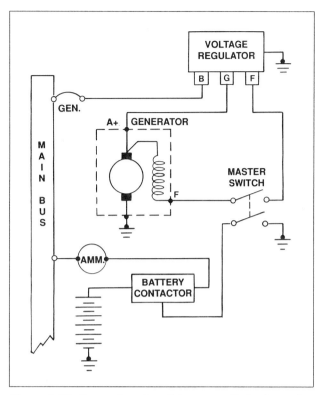

Figure 8-48. Generator circuit for a small single-engine aircraft.

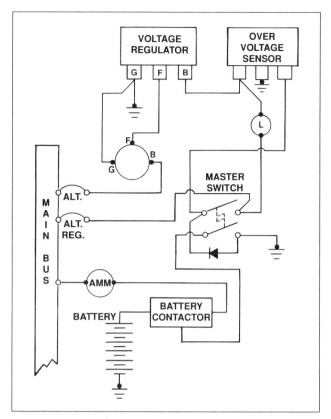

Figure 8-49. Alternator circuit for a small single-engine aircraft.

In figure 8-48, we have a typical light-aircraft generator circuit. The generator output goes to the voltage regulator and through the reverse current relay inside the regulator box to the main bus. It must pass through a circuit breaker of adequate capacity to protect the generator in case the current limiter should malfunction. The field current flows from the field terminal of the generator, through one side of the master switch to the F-terminal of the voltage regulator, and to ground inside the regulator.

You will notice that the generator and the battery are in parallel with their common terminal at the main bus. The ammeter measures only the current flowing into or out of the battery and only indirectly tells what the generator is doing.

C. Alternator Circuit

The advantages of more current for less weight has made the alternator the modern choice for production of DC electrical power on most aircraft. The exception to this is the use of starter generators found on turbine-powered aircraft. The external circuit of a DC alternator is similar to that used by a generator, with a few exceptions. In figure 8-49, we have a typical DC alternator circuit for a light aircraft.

The G-terminal of the alternator is connected to ground at the G-terminal of the voltage regulator. The B-terminal of the alternator is its positive output terminal, and it connects to the aircraft main power bus through a circuit breaker that protects the alternator from exceeding its current limits. The F-terminal of the alternator connects directly to the F-terminal of the voltage regulator, and this circuit continues out the B-terminal of the regulator, through the over-voltage sensor to the alternator side of the master switch and then to the main bus through the alternator regulator circuit breaker.

Some alternator systems incorporate the over-voltage sensor, an under-voltage sensor, and the voltage regulator into one unit. This unit, often referred to as an Alternator Control Unit (ACU), is typically a solid-state unit with no moving parts.

The electrical system master switch is an interlocking double-pole single-throw switch. The battery side of the switch can be turned on and off independent of the alternator, but the alternator side of the switch cannot be turned on without also turning on the battery. The alternator can be turned off, however, without affecting the battery side of the switch.

An over-voltage relay in the field circuit senses the output voltage, and if it becomes excessive, it will open the field circuit, shutting off the alternator output. It also turns on a warning light on the instrument panel informing the pilot that the generator is off the line because it has produced an excessively high voltage.

A solid-state diode is placed across the master switch from the field connection of the alternator side to the ground terminal of the battery side. This allows any spikes of voltage that are induced into the system when the field switch is opened to pass harmlessly to ground rather than getting to the main power bus.

D. External Power Circuit

Because of the heavy drain the starter puts on the battery, many aircraft are equipped with external power receptacles where a battery cart, or external power supply, may be plugged in to furnish power for engine starting. This circuit is seen in figure 8-50.

Power is brought from the battery cart through a standard three-terminal plug. Two of the pins in the aircraft receptacle are larger than the third, and they are also longer. When the cart is plugged in, a solid contact is made with the two larger plugs. The external power relay in the aircraft remains open and no current can flow from the external source until the plug is forced all the way into the receptacle and the smaller pin makes contact. This small pin then supplies power through the reverse-polarity diode to the external power relay which closes, connecting the external power source to the aircraft bus.

The reverse-polarity diode is used in the circuit to prevent an external power source with incorrect polarity from being connected to the aircraft's bus. The diode simply blocks current from flowing to the external power relay if the applied power is connected backwards or offering reverse polarity.

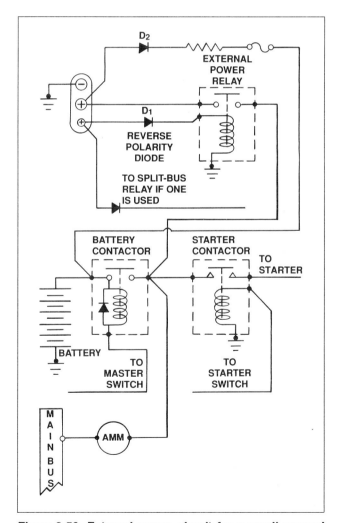

Figure 8-50. External power circuit for a small general aviation aircraft.

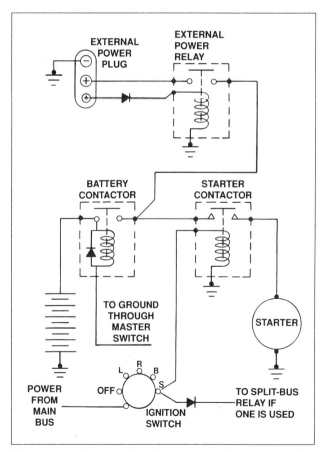

Figure 8-51. Starter circuit for a small single-engine aircraft.

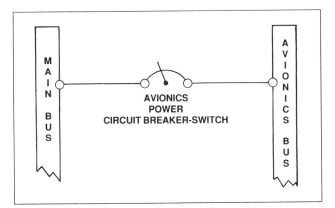

Figure 8-53. Avionics power source using a circuit-breaker switch.

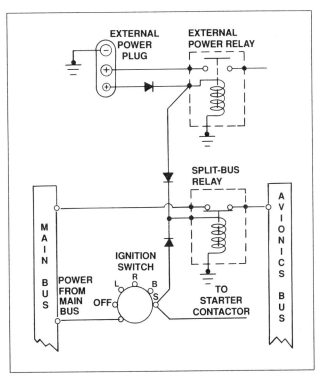

Figure 8-52. Avionics power source using a split-bus relay.

It is quite possible that the aircraft battery can be so completely discharged that the battery contactor cannot get enough current to close. This would prevent the external power source from charging the battery. A circuit consisting of a diode, a current-limiting resistor, and a fuse is connected between the positive terminal of the external power plug and the battery side of the battery contactor. With this arrangement, enough current can flow from the external power source to energize the battery contactor coil so that it can close and allow the battery to be charged. The diode D_2 is in the circuit to prevent the positive pin in the external power receptacle from being "hot" when no external power plug is connected.

Today, with many of the smaller aircraft having 12-volt systems while others have 24-volt systems, it is extremely important that you have the correct voltage when you use an external power source. Diode D_1 prevents the external power relay closing if you have a power source with the wrong polarity, but there is normally no protection against the improper voltage—so be careful.

E. Starter Circuit

In most aircraft the starter is activated, through a solenoid, by the ignition switch or start button found on the instrument panel. This circuit may be seen in figure 8-51. A spring-loaded START position on the switch sends current from the main bus to the coil of the starter contactor. When the starter contactor closes, the high current required for the starter motor flows from either the battery, or the external power source to the starter.

F. Avionics Power Circuit

The vast majority of the avionics equipment used today has solid-state components, which can be damaged by voltage spikes such as those produced when a magnetic field collapses and sends its induced voltage into the system. To prevent this type of damage to the avionics equipment, the radios often receive power from a separate bus that may be isolated from the main bus when the engine is being started, or when the external power source is connected.

The avionics bus may be connected to the main bus with a split-bus relay. This is a normally-closed relay that is opened by current from the ignition switch when it is in the START position, and from the control pin of the external power receptacle. Diodes are used in the circuit to both of the sources of current to isolate one when the other is being used. This system is illustrated in figure 8-52.

The avionics bus of some aircraft is isolated from the main bus by a circuit breaker switch rather than with a relay. This system is shown in figure 8-53. When an avionics power switch is installed, it is the responsibility of the pilot to be sure this switch is turned off before the master switch is either turned on or off, and any time the engine is started or external power is connected to the aircraft.

This arrangement has a couple of very definite advantages. The circuit breaker-type switch can be used as a master switch for all of the avionics equipment so their individual switches will not have to be used. And in the case of a fault in any of the avionics equipment, this circuit breaker will open to protect the avionics wiring.

G. Landing Gear Circuit

In figure 8-54, we have the landing gear circuit for a typical general-aviation airplane. The landing gear is hydraulically operated by a reversible DC electric motor. It turns in one direction to lower the gear and in the opposite direction to raise it.

Let's follow the system as the gear is operated. Consider the airplane to be in the air with the landing gear down and locked, but with the landing gear selector switch in the GEAR UP position. We see this in figure 8-54(A). The down-limit switches in each of the three landing gears are in the DOWN position and the up-limit switches are all in their NOT UP position. Current flows through the NOT UP side of the up-limit switches through the FLIGHT side of the squat switch and through the hydraulic pressure switch to the gear up relay.

The squat switch in the circuit is mounted on one of the struts such that it is in one position when the weight of the airplane is on the landing gear, and in the other position when there is no weight on the wheels. The purpose of the squat switch in this circuit is to prevent the gear from being retracted while the aircraft is still on the ground. The hydraulic pressure switch is used to turn off the pump motor in the event hydraulic pressure exceeds its limits. This switch is closed when the hydraulic pressure is low, but opens when the hydraulic pressure rises to a preset value.

From the pressure switch, the current goes through the coil of the LANDING GEAR UP relay and to ground through the landing gear selector switch. This flow of current creates a magnetic field in the LANDING GEAR UP relay and closes it, so that current can flow from the main bus through the relay contacts and the windings that turn the motor in the direction to raise the landing gear. As soon as the landing gear is fully up, the up-lock switches in each gear move to the UP position and current is shut off to the landing gear motor relay and the motor stops. The down-limit switches have moved to the NOT DOWN position, and the three down-and-locked lights go out. If the throttle is closed when the landing gear is not down and locked, the warning horn will sound.

The landing gear may be lowered in flight by moving the landing gear selector switch to the GEAR DOWN position. Follow this in figure 8-54(B). Current flows through the NOT DOWN side of the down-limit switches, through the coil of the LANDING GEAR DOWN relay, and to ground through the landing gear selector switch. The motor turns in the direction needed to produce hydraulic pressure to lower the landing gear.

When the landing gear is down and locked (figure 8-54(C)) current flows through the DOWN sides of the down-limit switches, and the green GEAR DOWN AND LOCKED lights come on. In the daytime, current from these lights goes directly to ground through the closed contacts of the light dimming relay, but at night when the navigation lights are on, current from this light circuit energizes the relay, and current from the indicator lights must go to ground through the resistor. This makes the lights illuminate dimly so they will not be distracting at night.

If any of the limit switches are in a NOT UP or a NOT DOWN position, a red UNSAFE light will illuminate. But when the landing gear selector switch is in the LANDING GEAR DOWN position and all three gears are down and locked, the light will be out. If the selector switch is in the LANDING GEAR UP position and all three gears are in the up position, the light will also be out.

If the airplane is on the ground with the squat switch in the GROUND position, and the landing gear selector switch is moved to the LANDING GEAR UP position, the warning horn will sound but the landing gear pump motor will not run.

H. Alternating Current Supply

Small aircraft for which DC is the primary source of electrical power have little use for AC except for certain instruments that require 26-volt, 400-Hz AC and some lighting circuits known as Electro Luminescent Panels. Most aircraft needing this type of power are equipped with solid-state, or static, inverters. These units consist of a solid-state sine-wave oscillator followed by a transformer that produces the required power. A block diagram of this system is seen in figure 8-55. Some instruments that use 26-volt AC have built-in inverters and can be operated with a DC input voltage.

Before solid-state electronics became such an important factor in aircraft instrumentation, most DC-powered aircraft got their AC for instruments from small single-phase rotary inverters (figure 8-56) which had a small DC motor driving an AC generator. These units may still be found in some of the older aircraft, but because they are noisy, subject to mechanical problems, and are electrically inefficient, the rotary inverters in most cases are being replaced by solid-state inverters.

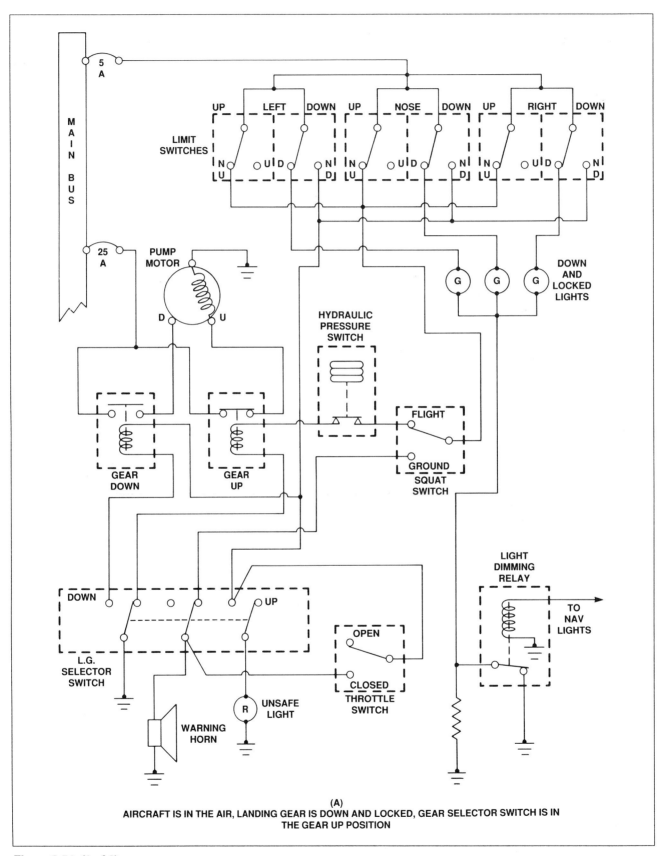

Figure 8-54. (1 of 3)
(A) Landing gear circuit.

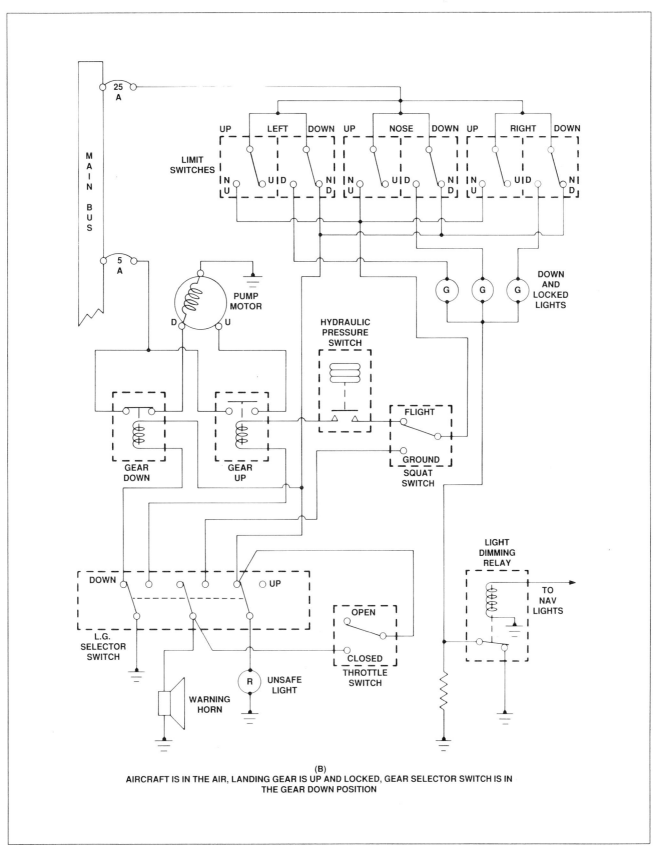

Figure 8-54. (2 of 3)
(B) Landing gear circuit.

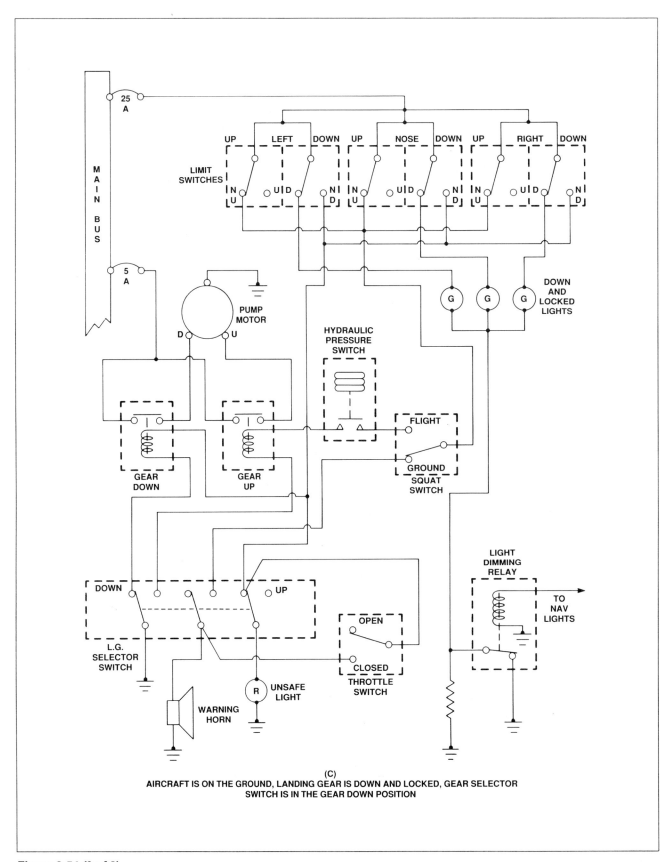

Figure 8-54 (3 of 3).
(C) Landing gear circuit (3 of 3).

V. SMALL MULTI-ENGINE AIRCRAFT

The popularity of the light twin-engine aircraft in the general aviation fleet has made this a type of electrical system that we need to become increasingly familiar with. The vast majority of light twin-engine aircraft use two generators or alternators and have their voltage controlled so that when they are connected together on the power bus, the regulators work together to keep the output voltages of the two sources the same. This is called paralleling.

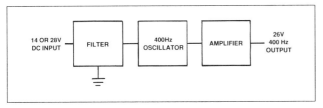

Figure 8-55. A solid-state inverter uses an oscillator to produce 400-Hz AC, which is amplified so that it can drive AC instruments.

Figure 8-56. A rotary inverter, used to produce AC from DC.

A. Paralleling with Vibrator-Type Voltage Regulators

Some of the lower-output electrical systems use vibrator-type voltage regulators with provisions for paralleling the two generators. These generator controls work in the same way as the three-unit controls for a single-engine installation, except that the voltage regulator relay has an extra coil that is connected through either a paralleling switch or relay to a similar coil in the voltage regulator of the other engine.

In figure 8-57, we have the basic principle of the paralleling circuit for a twin-engine system using vibrator-type regulators. With the paralleling switch open, each generator acts in exactly the same way it would act in a single-engine system. The generator with the highest voltage will produce the most current. Let's assume that the paralleling switch is closed, and the left generator is producing a higher voltage than the one on the right engine. Current will flow from the left-engine regulator to the regulator on the right engine through both paral-

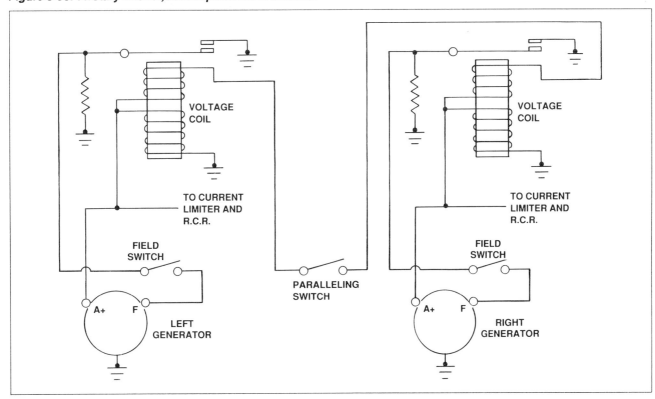

Figure 8-57. Paralleling circuit for a light twin-engine aircraft using vibrator-type voltage regulators.

leling coils. The flow through the left coil produces a magnetic field that assists the voltage coil, and the contacts open sooner, inserting the field resistor more often and lowering the output voltage of the left generator. This same current flows through the paralleling coil in the right voltage regulator, except that it flows in the direction that will cause its magnetic field to oppose the field from the voltage coil. This allows more field current to flow through the right generator and increase its voltage.

The only current that flows in the paralleling coils is that caused by the difference in the output voltages of the two generators, and this small current will produce just enough magnetic field difference to keep the generators putting out the same voltage and thus sharing the load equally. This theory of operation applies to vibrating-type voltage regulators, however, transistorized regulators can provide the same paralleling functions using solid-state devices.

B. Paralleling with Carbon-Pile Voltage Regulators

Light twin-engine aircraft with generators of greater output than can be controlled with a vibrator-type voltage regulator use carbon-pile voltage regulators. As we have explained for the single-engine installation, the field current is controlled by varying the pull of an electromagnet on the voltage regulator armature. The higher the output voltage, the more current through the electromagnet and the more pull on the armature. With a carbon pile regulator this reduces the pressure on the carbon pile, increasing its resistance and decreasing the field current.

The only difference between a carbon-pile voltage regulator used on a single-engine installation and that used for a twin-engine system is a paralleling coil. This coil (figure 8-58) is connected in the same way as the paralleling coil of a vibrator-type regulator; the basic difference being that the current that flows through it is produced by the voltage drop across paralleling resistors in the ground leads to the armatures of both generators. All of the generator output flows through these resistors. If the left generator puts out more current than the right, the voltage drop across the left paralleling resistor will be higher than that across the right and electrons will flow through the paralleling coils of both voltage regulators. The magnetic field of the left paralleling coil will assist the left voltage coil and loosen the carbon pile. This will decrease the field current in the left generator and lower its voltage. At the same time, the field of the right paralleling coil will oppose that of its voltage coil and the spring will compress the carbon pile and increase the output voltage of the right generator.

C. Paralleling Twin-Engine Alternator Systems

Most modern light twin-engine aircraft contain DC alternator systems for electrical power generation. These alternators are typically controlled through

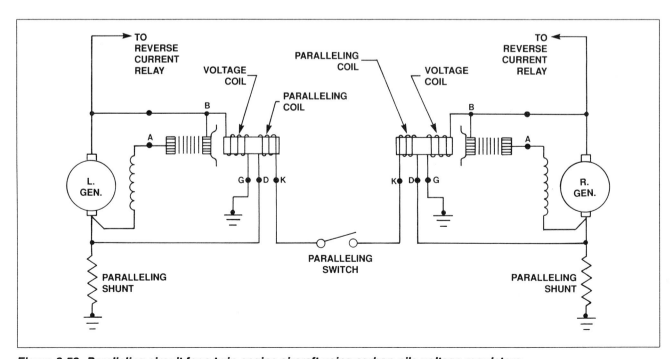

Figure 8-58. Paralleling circuit for a twin-engine aircraft using carbon-pile voltage regulators.

two relatively complex solid-state alternator control units (ACU). The alternator control units receive input from both alternators in order to provide paralleling. The ACU also provides voltage and current regulation as well as over-and-under voltage sensing and protection. In the event of a fault condition the ACU would automatically isolate the defective alternator and alert the flight crew.

VI. LARGE MULTI-ENGINE AIRCRAFT

There are two basic types of large aircraft power distribution systems: the split-bus and the parallel system. The split-bus system is typically found on twin engine aircraft such as the Boeing 757 and 767, the McDonnell Douglas MD-80 and the Airbus A320. The parallel system is typically used on aircraft containing three or four engines such as the DC-10 and Boeing 727. Most four-engine aircraft employ a modified split-bus system.

Figure 8-59 shows a simplified version of a split-bus power distribution system. This schematic shows that the AC generator power from the right engine is connected to the right distribution bus and isolated from the left bus by the bus tie breakers (BTB). The left AC generator supplies power only to the left bus. In the event of a generator failure, the failed generator is isolated by the generator breaker (GB) and BTB 1 and 2 close to connect the isolated bus. On some aircraft, the Auxiliary Power Unit (APU) generator could be started during flight and used to carry the load of the failed generator. In that case, the left and right busses would once again be isolated.

A parallel power distribution system is shown in figure 8-60. With this system all three generators are connected to a common bus and share the electrical loads equally. In the event of a generator failure, the failed unit would be isolated from the bus by its generator breaker and the flight would continue with two generators supplying the electrical power.

A modified split-bus system found on some four-engine aircraft connect the two right-side generators in a parallel configuration. The two left-side generators are also paralleled. The right- and left-side busses are kept isolated by a split-system breaker. In the event of a generator failure, the associated paralleled generator will carry the entire load for that bus. If both generators on the same side should fail, the split-system breaker would close and send power to the inoperative bus from the working generators.

The Boeing 727 is one of the most popular jet airliners, and its electrical system is typical for this type of aircraft. A block diagram of this system is seen in figure 8-61. The B727 uses a parallel power distribution system where each of the three

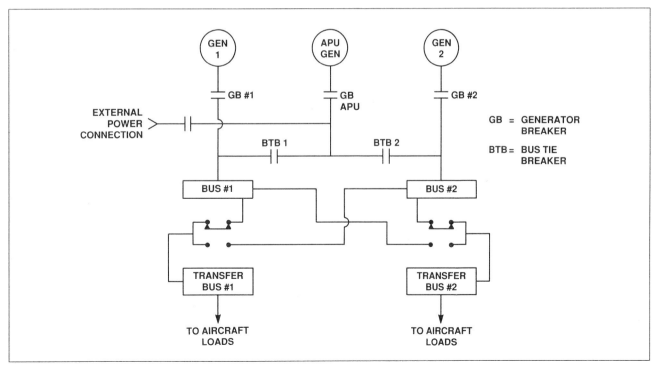

Figure 8-59. A typical split-bus power distribution system for a twin-engine commercial airliner.

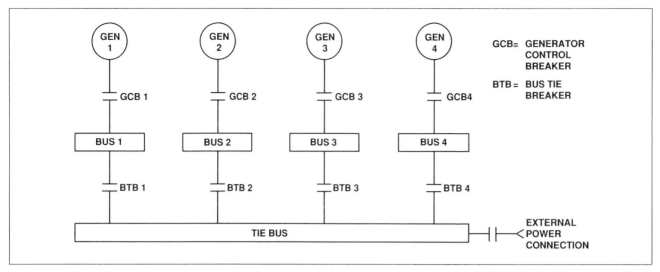

Figure 8-60. A typical parallel power distribution system for a three-engine commercial airliner.

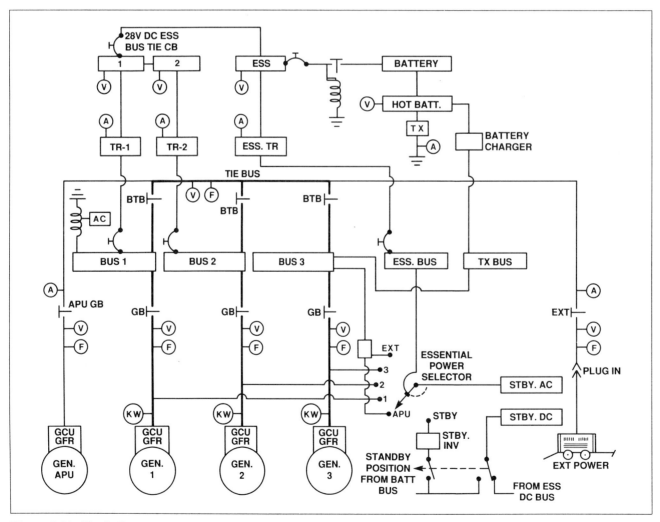

Figure 8-61. Block diagram of the electrical system of a Boeing 727 jet transport aircraft.

engines drives a three-phase alternator through a hydraulic constant-speed-drive (CSD) unit. Each alternator produces 115 volts at 400 Hz. Single phase transformers reduce some of this power to 28 volts AC to operate most of the lights, and there are three transformer-rectifier (T-R) units that convert the 115 volts AC to 28 volts DC for emergency lights and system control circuits. The 28-volt DC is also used for charging the battery.

A gas turbine-powered auxiliary power unit (APU) located in the wheel well drives a three-phase, 115-volt alternator identical to the ones driven by the engines. The APU is used to supply all of the needed electrical power when the aircraft is on the ground. The control panel for the APU is shown in figure 8-62. Unlike some aircraft, the 727's APU cannot be operated when the aircraft is in the air.

Each engine-drive alternator feeds its own bus through a generator breaker, and all three buses can be tied to a common sync, or tie, bus through bus tie breakers (BTBs) when the alternators are synchronized. In this way, the alternators can divide the electrical load among themselves.

While most of the electrical power used by this aircraft comes directly from the AC alternators, the aircraft is equipped with a battery to start the APU and to provide power for certain essential lighting, avionics and instrument equipment. In normal operation, the battery is disconnected from the DC loads. DC power is supplied by transformer-rectifier units connected to the main AC buses. The battery is kept fully charged by a battery-charger circuit that draws its power from the AC transfer bus.

Equipment that is essential for flight gets its power from an AC and a DC essential bus, and the flight engineer has a selector on the upper panel that allows the engineer to supply the essential buses from any of the three main AC buses, the external power supply, or the APU. In the event that all three AC generators fail, the engineer can select the STANDBY position of the selector switch, and

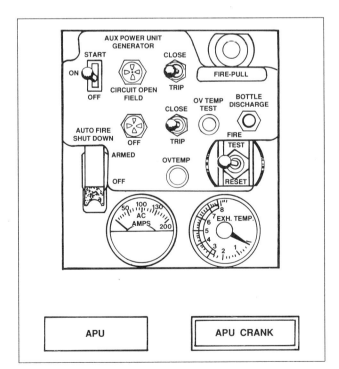

Figure 8-62. Control panel for the auxiliary power unit of a Boeing 727. The APU can be started from this panel, and the output of its alternator monitored.

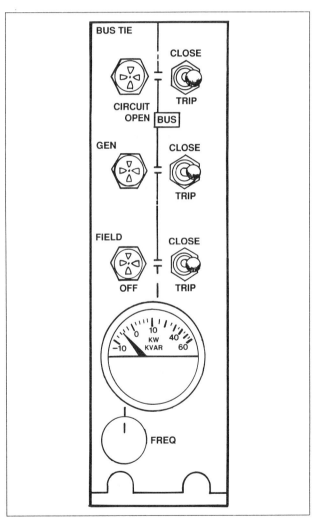

Figure 8-63. Generator control panel for a Boeing 727. The switches control the bus tie breaker, the generator breaker, and the field relay. The KW-KVAR meter indicates the power the generator is putting out, and the frequency control varies the output of the CSD unit to control the generator frequency.

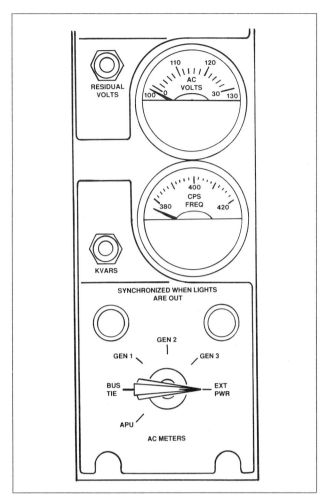

Figure 8-64. The AC electrical panel of a Boeing 727.

the standby AC and DC buses are supplied directly from the battery bus with a standby inverter providing the AC.

A. AC Alternator Drive

Each of the alternators is driven through a hydraulic constant-speed drive unit. The flight engineer has controls that can disconnect the CSD in case of an alternator malfunction. The CSD has its own lubrication system, and the flight engineer has a low oil pressure warning light and an oil temperature gage for the unit. The temperature gage can either indicate the temperature of the oil entering the CSD unit, or it can show the amount of temperature rise as the oil passes through the drive unit to give an indication of the load that is being carried by the CSD.

The paralleling process of two or more alternators requires that the flight engineer connect the generators to the tie bus. If there are no other power sources connected to the tie bus, the first alternator can be connected without complications by simply closing its BTB. But, before the second or third alternator can be placed on the tie bus, they must be synchronized with regard to voltage output, phase and frequency. The phase rotation is determined when the alternator is installed and it will not change in flight. The voltage is controlled by the generator control unit to hold the output to 115 volts, plus or minus five volts, and the frequency is maintained by the CSD. There is a control on the flight engineer's panel (figure 8-63) that allows the frequency to be adjusted for paralleling. The CSDs maintain the frequency at 400 Hz, plus or minus 8 Hz. It is the flight engineer's job to determine that all generators are producing equal values before they are connected to the tie bus. If a generator is paralleled improperly, damage may occur to the generator or the CSD.

B. Alternator Instrumentation and Controls

A frequency meter and an AC voltmeter (figure 8-64) may be selected to monitor the voltage and frequency of the APU, the external power source, or any of the three alternators as well as that of the tie bus. A push-button beside the voltmeter allows it to read residual voltage, which is the voltage produced by the selected alternator when no field current is flowing. If there is residual voltage, you know the alternator is turning, but no voltage tells you the CSD has been disconnected or the residual has been lost.

Two synchronizing lights on the panel illuminate when the selected alternator is not synchronized with the power on the tie bus, and it must not be connected until it is synchronized. The alternator frequency should be adjusted until the lights blink at their slowest rate and the bus tie breaker closes when both lights are out.

The generator control unit used with each of the engine-driven alternators and with the APU contain all of the circuitry needed to maintain a constant voltage as the load current varies, and the control unit will open the bus tie breaker, the generator breaker, or the generator field relay if a fault occurs in either the system or the alternator. The control unit will also alert the flight engineer as to the defect, in order that corrective action can be taken.

Each of the three main alternator control panels have a control for the bus tie breaker and a light that is on when the breaker is open, a generator breaker switch, and the generator field switch. There is a light that is on when the generator breaker is open, and one when the field is off. The generator

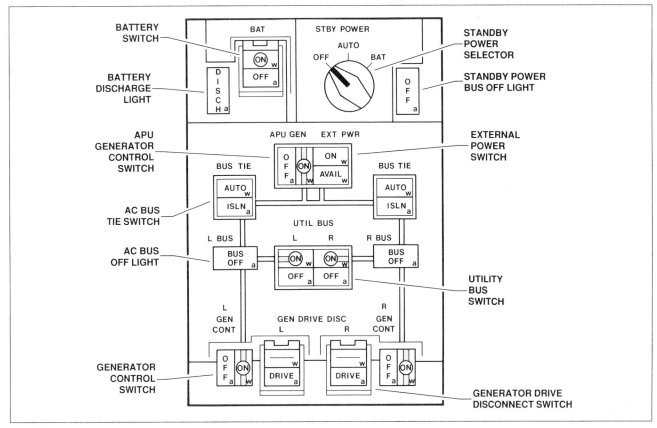

Figure 8-65. The generator control panel for a B-757.

field switch connects voltage to the generator field through the voltage regulator. The generator breaker connects the generator output leads to the correct AC power distribution bus. The bus tie breaker connects the number 1, 2 or 3 AC bus to the tie bus.

Each alternator is equipped with a KW-KVAR meter that normally indicates the amount of electrical power being produced by the alternator in kilowatts. When a button on the instrument panel is pressed, the indicator shifts circuits and indicates the reactive power in kilo-volt-amps. This indication is to inform the flight crew of the real power of the generator. Real power is the measure of both the resistive and reactive loads.

C. Automated AC Power Systems

Many of the latest generation commercial aircraft use automated systems for controlling the various AC power distribution functions previously mentioned. The use of automated systems has made it possible to reduce flight crew work loads enough that most modern commercial airliners require only two flight crewmembers. The flight engineer's position has been replaced by automated electronic sys-

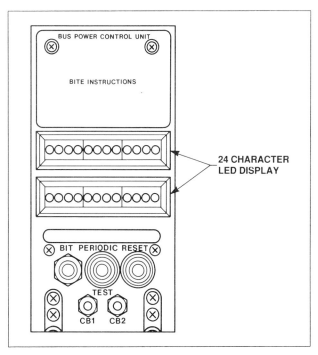

Figure 8-66. The generator control unit from a B-757. This unit contains a 24-character LED display for visual display of system faults.

tems. The generator control panel shown in figure 8-65 illustrates the simplicity of a modern power distribution system.

The automated systems not only are designed to lighten flight crew loads, they also incorporate built-in test circuitry to ease maintenance troubleshooting. The Built-In Test Equipment (BITE) circuitry is typically contained in the control units located in the aircraft's electrical equipment bay. The control units, known as Line Replaceable Units (LRU) are easily removable to facilitate maintenance. The BITE system employs an LED display which can be viewed by maintenance personnel during system troubleshooting. The technician then refers to a code book to determine the defective component and necessary repairs. Figure 8-66 shows the generator control unit and the associated BITE displays.

It should be noted that the complex electronic circuitry which makes the automated systems possible are very sensitive to stray electrical currents. The current produced from the static electricity which is commonly produced by a technician's movements around the aircraft could be harmful to the LRU. It is therefore very important to ground yourself using an approved wrist strap before touching any electrical component which is labeled as Electro Static Discharge Sensitive (ESDS). Be sure to look for a symbol as shown in figure 8-67 which identifies this type of component.

Figure 8-67. The commonly used Electro Static Discharge Sensitive (ESDS) symbols.

Chapter IX
Types of Electrical Circuits

The principles of electronics used to be so simple that all we needed to understand was the way to change the voltage and current of high-frequency alternating current electricity. But because of the technological developments, we now have two major types of electronic systems: analog and digital, and both are used in modern avionics.

The example in figure 9-1 shows a light bulb being turned on with an analog circuit and a digital circuit. To turn on the light with the analog circuit, the light must go from off through dim, get brighter, and then be on full intensity. The reverse is true for turning the light off. To turn on the digital circuit, the switch is closed and the light illuminates with full intensity. With the digital circuit, there is no dim or half-on position. Only two conditions exist in the digital circuit: the light is full on or full off. It can be seen from the graph of the current flow for each circuit that both circuits have different wave forms or current flow patterns. It can also be said that both types of circuits have definite advantages when used in the correct situation.

In analog electronics, the voltage and current values of the circuit can take on an infinite number of possibilities. The ratiometer-type fuel quantity indicator is a good example of an analog system. The float in the tank rides on the surface of the fuel, and as it moves up and down, it varies the resistance of the tank unit. This resistance change causes the current through the meter to vary and move the pointer of the indicator over the dial in such a way that its position indicates the quantity of fuel in the tank. The movement of the float is smooth and stepless, and the changes are said to be made in a line, or in a linear fashion. There are an infinite number of positions that the float can obtain. The gage can indicate 20 gallons of fuel or 20.5 gallons of fuel. In fact, the gage is actually capable of indicating 20.534 gallons of fuel. Of course, it is impossible for the pilot to distinguish

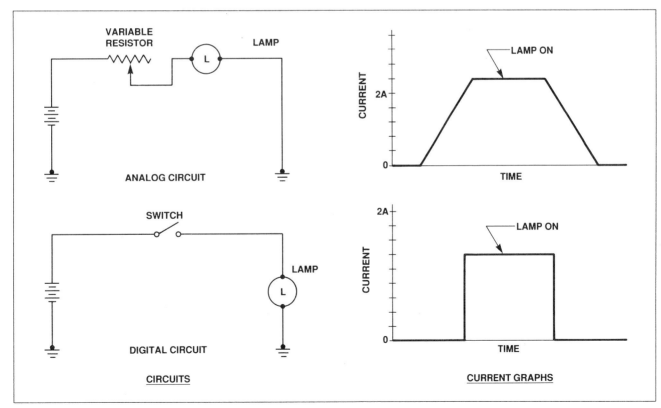

Figure 9-1. Modern electrical systems are divided into two types of circuits. Analog circuits vary the voltage or current, while digital circuits can only allow current to flow or not allow it to flow.

191

any changes of such small value. The analog system, in this case, is too variable for its own needs.

On the other hand, a digital fuel quantity indicator would have only a finite number of indications which could show up on the gage. For example, the digital unit may not be able to indicate 20.534 gallons; it may be limited to 0.5 gallon increments. A reading of 20.5 gallons could be shown and very easily read by the pilot. As shown in figure 9-2, the 20.5 gallon reading is easier to read on a digital system than on an analog system. The analog is capable of showing smaller changes, but the pilot cannot analyze the information easily. In this case a digital readout to the pilot is more practical.

Digital electronics must not be confused with digital displays. A clock that uses a motor to drive a series of drums with numbers around its periphery gives its display in numbers, or it has, as we say, a digital display, but it is still an analog device. Digital electronics do not deal with varying amounts of current and voltage, they employ discreet circuit conditions; that is, simply, on and off. An electrical switch is a good example of a digital device. It has only two electrical conditions; it can be only on or off. When the switch is on, current can flow. When it is off, current cannot flow. Information is handled by a digital circuit by controlling the number and sequence of the bits of information, that is, by the number and sequence of times a circuit is on and the number and sequence of times it is off. Digital electronics have found so many applications because of the development of the integrated circuit and microprocessor, which are digital devices that incorporate tremendous numbers of devices on a tiny chip of silicon.

Digital devices are precise and fast in their operation, but they are usually quite limited in their power handling capability, so for this reason, many devices have a combination of digital and analog circuits, using each type of circuit for the particular function for which it is best suited.

I. ANALOG ELECTRONICS

There are three basic types of circuits used for analog (linear) electronics. They are the rectifier, the amplifier, and the oscillator. We will examine each of them briefly. A more detailed study of these circuits will be found in a subsequent chapter.

A. Rectifiers

The simplest of these three, the rectifier, is used in both analog and digital circuits to change AC into DC for operating the circuit. A rectifier, or detector as it is sometimes called, uses a diode to act as a check valve to allow electron flow in one direction, but not in the other.

1. Half-Wave Rectifier

The simplest form of a rectifier is the half-wave rectifier which uses one diode. This circuit is illustrated in figure 9-3. It is inefficient in its use of the AC it converts into DC, as its output voltage is only one-half of the available AC. The output of a half-wave rectifier is pulsating DC whose frequency is the same as that of the input AC.

2. Full-Wave Rectifier

A bridge-type rectifier makes more efficient use of the AC. It uses four diodes arranged as we see in figure 9-4. In this circuit, the pulsating DC output voltage is the same as the available AC, and the frequency of the pulsations is twice the frequency of the input AC. This rectifier utilizes the entire AC wave (full-wave rectification) and creates a more useful direct current.

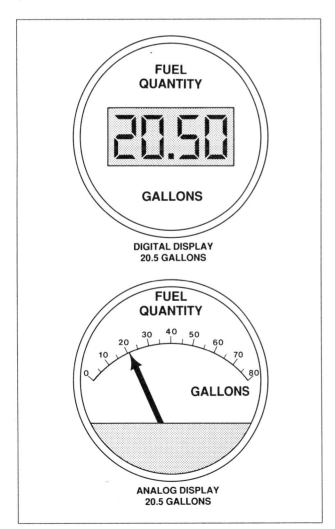

Figure 9-2. Digital and analog data displays.

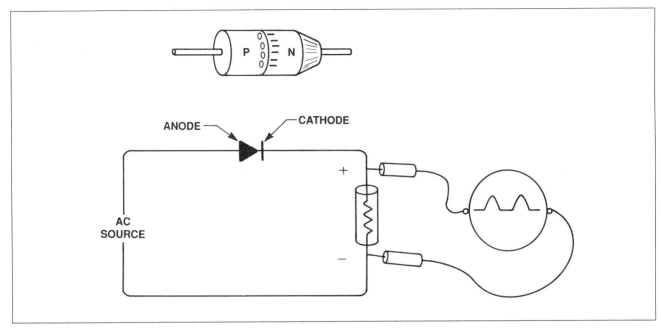

Figure 9-3. A semiconductor diode half-wave rectifier with its output waveform.

When the bottom of the transformer secondary winding is negative, electrons leave it and go through diode 1, the load resistor, and diode 2 to the top of the transformer. During the next half cycle, the electrons flow from the top of the transformer through diode 3 and through the load resistor in the same direction as before, then through diode 4 back to the bottom of the transformer.

B. Amplifiers

One of the fundamental circuits of analog electronics is the amplifier. Its function is to change the amplitude of the input signal. Most amplifiers increase the amplitude, but a few specialized circuits decrease the amplitude. The phase of the output may be the same as that of the input, but some amplifiers invert the signal.

A transistor is an electron control device that varies the current flow between the emitter and the collector by varying the current through the base. The circuit is considered an amplifier because the large signal in the collector circuit will duplicate exactly the waveform of the weaker base signal except the collector signal will be of a much higher, or amplified, value.

In figure 9-5, we have an NPN transistor with one voltage source between the emitter and the base and another source between the collector and the base. Anytime base current flows, there will be a much larger flow between the emitter and the collector. And base current will flow any time

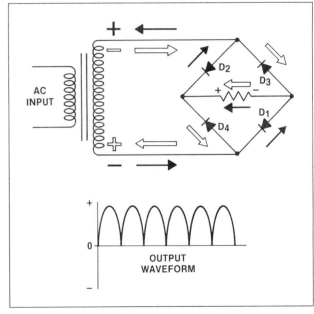

Figure 9-4. A four-diode, bridge-type, full-wave rectifier circuit.

the base of an NPN transistor is positive with respect to its emitter. If a switch (S) is placed in the base circuit, collector current (IC) will flow when the switch is closed and stop flowing when the switch is open. The amount of current may be varied by controlling the voltage of the source EEB. The greater the voltage, the stronger the forward bias of the emitter-base junction and the more base current will flow, and therefore the more collector

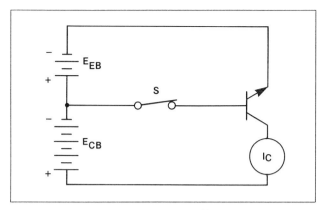

Figure 9-5. An NPN transistor amplifier using two sources of voltage.

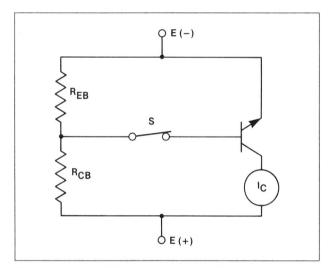

Figure 9-6. An NPN transistor amplifier using a single power source.

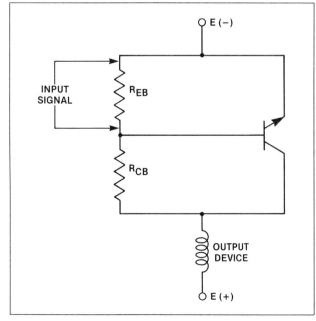

Figure 9-7. An NPN transistor amplifier showing the input signal.

current. Collector current is also limited by the resistance and the voltage of the collector/emitter circuit. It should also be noted that the collector circuit receives no power from the base circuit; it is only controlled by the base.

The two voltage sources are usually replaced with a single source and a couple of resistors (figure 9-6). The voltage drop across resistor REB provides the forward bias for the emitter-base junction, and the resistor RCB gives the reverse bias needed between the emitter and the collector.

We can continue the evolution of a practical amplifier by looking at figure 9-7, where we are using an input resistor REB across which we impress the input signal. The input signal varies the current through resistor REB to control the emitter-base bias which varies the amount of collector current flow through the output device.

This is a workable amplifier circuit, but the amount of current we can use to drive an output device is dependent upon the current-carrying capability of the transistor, so a more practical circuit uses this transistor to control a larger transistor, and all of the load current flows through the output transistor.

In figure 9-8(A), we have added the output transistor Q_2 to the circuit. When the input signal drives the base of transistor Q_1 positive, it conducts, and draws base current from Q_2. This makes Q_2 conduct, and the collector current of Q_2 drives the output device which may be an item such as a motor, a solenoid or a speaker. The greater the forward bias of Q_1 the greater will be its collector current, and its collector current is the base current for transistor Q_2.

There are many different types of amplifiers used in practical electronic circuits, but as aviation maintenance technicians we are primarily concerned with what they do, rather than the inner workings of the amplifier itself, so we will use the symbol we see in figure 9-8(B) to represent our amplifier. The abbreviation $-V_{CC}$ is used to represent the negative terminal of the power supply and the triangle represents the amplifier. Many functional diagrams of analog electronic circuits omit the power supply and its connections, but they are assumed to be there, and in tracing the functions of a circuit, we are primarily concerned with only the control and output portions of the circuit.

C. Oscillators

Rectifiers convert AC into DC, amplifiers change the value of either the current, voltage, impedance or phase, and oscillators make alternating current

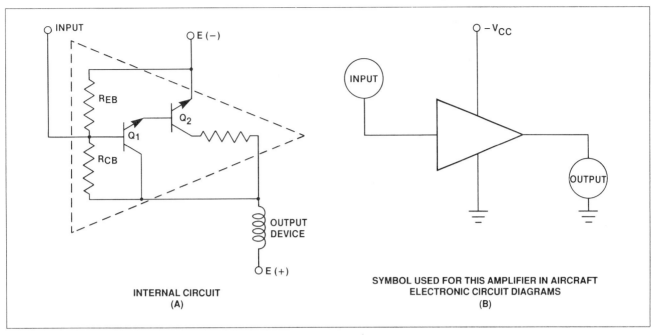

Figure 9-8. A two-transistor amplifier in aircraft electronic circuit diagrams.

from direct current. AC produced by a rotary generator normally has a nice smooth sine waveform, but electronic oscillators are able to produce almost any waveform. The most generally used waveforms are the sine wave, the square wave and the sawtooth wave. These waveforms are illustrated in figure 9-8. Each has a specific function in electronic circuits and each may be generated to have the exact frequency, amplitude and waveform needed.

II. DIGITAL ELECTRONICS

An electrical or electronic system can do one of two functions, or both. It can handle information, or it can perform work. Electricity used in a radio, for example, handles information, while that used by an electric motor performs work. There can be a very fine line between the two, but we can think of a control circuit as handling information and the output device as performing the work. The control devices must normally sense all of the input parameters and determine what action is needed by the output device.

A system of logic has been developed in the past couple of decades that allows control circuits to use only two electrical conditions: OFF-ON or LOW-HIGH, to monitor almost any type of variable and to manipulate the information to arrive at an output that is appropriate to the input condition. This system, which is based on Boolean algebra, uses the binary number system, and it is able to solve complex computations.

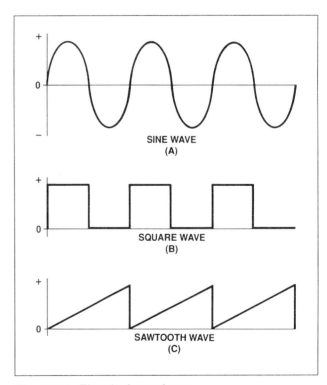

Figure 9-9. Electrical waveforms.

The base-two (binary) system utilizes only the digits zero and one. When the circuit is ON, a one is represented, when it is OFF, a zero is indicated. By converting the ON or OFF messages to represent numbers found in the decimal system, a computer system can be designed that will perform complex tasks based on the simple choice of ON or OFF.

To build a binary number system that will correspond to the decimal system, let us begin with one switch. When this switch is in the OFF position, a zero is indicated. When it is in the ON position, it represents one. Because these are the only possibilities for a single switch, additional switches must be added to represent larger quantities. A second switch would represent the quantity 2. If the first switch were OFF and the second switch were ON, the quantity 2 would be indicated. If both the first and second switch were ON, you would add 1 + 2 and the quantity 3 would be indicated. This procedure of adding switches would continue with each switch representing a larger value. The table found in figure 9-10 will help you understand the conversion of decimal to binary numbers.

The basic requirement to handle information in this manner is that an extremely large number of simple manipulations must be made, and it was the development of the integrated circuit chip containing hundreds or thousands of logic gates, which are arrangements of switches, that made this type of electronic control of information practical.

A. Digital Building Blocks

Logic gates are the fundamental building blocks of digital electronics, and they are essentially arrangements of semiconductor switches that produce a specific output for a given combination of input conditions. We can analyze these gates by a truth table that tells us the condition at the output terminals for each possible combination at the inputs.

When we discuss logic gates, we will consider two conditions: One and Zero. Typically a logic One is a condition of positive voltage or a high voltage, while a logic Zero is a negative voltage, a low voltage, or zero volts.

1. AND Gate

This is one of the most fundamental gates and it can have a logical One at its output, C only when input A *and* input B are both logical One. If either input becomes a Zero, the output of the AND gate will also become Zero. You can easily see that this is true by looking at the truth table illustrated in figure 9-11. It should be noted that the output signal of a logic gate is always represented on the right side of the symbol, the inputs of the gate are on the left.

2. OR Gate

The OR gate is designed to give a One output any time either input A *or* input B is one. The output will only go to Zero if both inputs are zero as seen by the truth table.

3. Amplifier or Buffer

This device has only one input and one output, and it may be used to isolate one portion of a circuit from another. Its output is always the same as its input.

BINARY NUMBER							DECIMAL NUMBER
64	32	16	8	4	2	1	
0	0	0	0	0	0	1	1
0	0	0	0	0	1	0	2
0	0	0	0	0	1	1	3
0	0	0	0	1	0	0	4
0	0	0	0	1	0	1	5
0	0	1	1	0	1	1	27
0	1	1	0	0	0	0	48
1	0	1	1	1	0	0	92
1	1	1	0	1	0	1	117

Figure 9-10. Binary conversion chart.

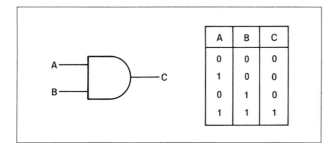

Figure 9-11. AND gate and its truth table.

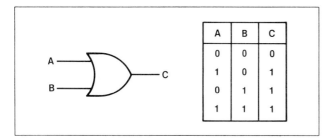

Figure 9-12. OR gate and its truth table.

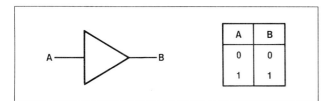

Figure 9-13. Amplifier, or buffer, and its truth table.

4. NOT Gate

This device also has only one input and one output. But its output is always the opposite of its input. When the input is logic One, its output will *not* be logic One, it will be logic Zero. Its symbol is the same as that of the amplifier except it is followed by a small circle.

5. NAND Gate

This is a NOT AND gate, and its symbol is the same as that used for the AND gate followed by a circle. The output of a NAND gate will be logic One when input A *and* input B are *not* logic One.

6. NOR Gate

The NOR gate is the same as a NOT OR gate and there will be no signal, a logic Zero, at the output anytime there is a logic One on any of the inputs.

7. EXCLUSIVE OR Gate

This gate has a logic One on its output when one and only one of its inputs has a logic One. If both inputs become One or both become Zero, the output will become Zero.

Each logic gate is made up of one or more transistors, resistors and/or diodes. These components are each formed onto the silicon chips or integrated circuits which contain thousands of logic gates. Modern integrated circuits are built into various gate combinations to perform specific functions. The various functions performed are a function of the gate types and arrangements. These various combinations are known as the software of the chip. The chips containing various software patterns can be assembled into computers, autopilots, radios and a multitude of avionics equipment.

III. MICROCOMPUTERS

Computers were, first, mechanical analog devices. These were later developed into electrical analog computers that could add, subtract, multiply and divide electrical quantities whose voltages were carefully made into an analogy of the numbers being handled. But the advancements in digital circuits made analog computers obsolete, and they have been replaced almost entirely with digital units.

The first logic gate was fabricated using a chip of silicon during the early 1960s. Very quickly, the number of gates per chip was increased and the term Integrated Circuit (IC) became part of nearly everyone's vocabulary. The arithmetic logic unit (ALU) developed in these early years could manipulate numbers just as the microprocessors of today. However, the microcomputer chip of today was not possible until the ALU could be combined with many other functions, on a single chip. This meant that chips had to be constructed that would contain at least 1,000 transistors on a single chip. This technology was developed during the 1970s and today's chips contain input-output (I/O) circuits, read-only memory (ROM), random access memory (RAM), and clock circuits in addition to the ALU. The 80286 microprocessor, which is far from "state-of-the-art" technology has the equivalent of 125,000 transistors on a single chip.

Figure 9-18 illustrates how the microprocessor chip serves as the central device in a microcomputer. Some microprocessor chips contain more elements

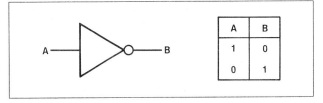

Figure 9-14. NOT gate, or inverter, and its truth table.

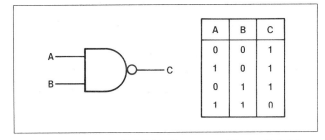

Figure 9-15. NAND (NOT AND) gate and its truth table.

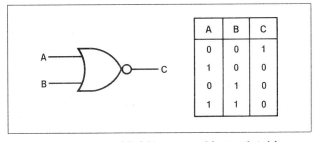

Figure 9-16. NOR (NOT OR) gate and its truth table.

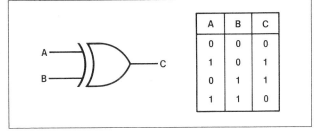

Figure 9-17. EXCLUSIVE OR gate and its truth table.

than those listed, some contain less. The elements listed are described below:

Arithmetic Logic Unit (ALU)—The arithmetic logic unit performs arithmetic and logic operations on data determined by signals from various control lines. The ALU can perform binary addition, subtraction, multiplication and division, and also perform a variety of logical comparisons.

Controller-Sequencer—This section manages (controls) and supervises (sequences) the activities of the microprocessor. This is accomplished by producing a variety of control signals to various input pins to carry out an instruction.

Data Register—The data register provides temporary storage for data received from the data bus. This may also be called the instruction register.

Address Register—The address register is a temporary storage device to hold the location (address) of a data word in memory or I/O which is being used in a current operation.

Accumulators—The ALU is capable of only a few operations. The microprocessor gains additional power using storage registers like those described above. The ALU can operate on data from only one or two of the registers at a time. Some registers, which are more versatile than the others, are known as accumulators. During an arithmetic and logic operation the accumulator may perform a dual function. Before an operation it holds an operand. After, it holds the resulting sum, difference, or logical answer.

Instruction Decoder—The instruction decoder is a circuit that analyzes the contents of the data register to determine what operation is to be performed. Once the operation code is recognized, the controller-sequencer can supply the appropriate preprogrammed signals for operation.

Clock—The precise timing of events in the microprocessor is controlled by a free-running oscillator. The clock input may be driven by a crystal, an LC-tuned circuit, or an external clock source.

Volatile Memory—Data stored in volatile memory units will be lost when the power to the computer is turned off. Memory of this type is usually stored on IC circuits containing latch or flip-flop circuits. Volatile memories are used to hold data input for processing.

Nonvolatile Memory—Data stored in nonvolatile memory will not be affected when the computer is turned off. This type of memory may be contained on a variety of magnetic tapes or disks, or on special memory chips. Nonvolatile

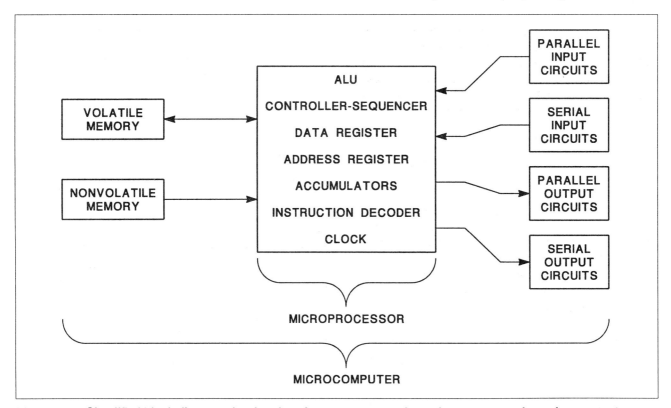

Figure 9-18. Simplified block diagram showing the microprocessor as the major component in a microcomputer.

memories are used basically for program data storage.

Input-Output (I/O)—To become fully functional, the microprocessor must access memory and peripheral devices. This is accomplished by placing data on either an address bus or data bus, depending upon the function of the operation. The I/O data travels through logical circuits called ports.

Serial Ports—These will transmit data one binary digit at a time. After one bit of information is sent, the next binary digit follows. This process is repeated until all of the information has been transmitted.

Parallel Ports—The data is transmitted using two wires and is called parallel because each circuit is wired in parallel with respect to the next. This is a continuous type of transmission and is capable of transmitting large amounts of serial data.

In later chapters, we will see how microprocessors are being used in a variety of aircraft systems. As the technology has progressed to make these units smaller, and more affordable, they have been designed in to many systems. This is a fast moving technology and it is nearly impossible to keep up with the current applications of these versatile little pieces of silicon.

Chapter X

Basic Semiconductor Circuits

There are only three basic types of circuits in most of our electronic equipment. Rectifiers change AC into DC. Amplifiers change the amplitude of either the voltage or the current, and oscillators produce AC from DC. We have introduced each of these, by function, in an earlier chapter and will now take a careful look at their construction.

I. THE RECTIFIER

There have been a number of different types of rectifiers used to change AC into DC. All of these devices are simply electron check-valves that allow electrons to pass in one direction, but block them when they attempt to flow in the opposite direction. The number of diodes used, and the manner in which they are connected can provide a variety of results.

A. Half-Wave Rectifier

A half-wave rectifier circuit uses a single diode in series with the voltage source, which is the secondary of the transformer in figure 10-1, and the load. Electrons can flow only during the half-cycle when the cathode, represented by the bar across the arrowhead, is negative. The output waveform of this type of rectifier is one-half of the AC wave, and because of this, it is a very inefficient type of circuit.

B. Full-Wave Rectifier

In order to change both halves of the AC cycle into DC, we can use a transformer with a center-tapped secondary winding and two diodes.

To simplify tracing the flow through the solid-state diodes, we will follow the flow of conventional current, which is from the positive side of the source, through the load to the negative side. Conventional current, whose direction is opposite that of electron flow, follows the direction of the arrowheads in the diode symbols. Follow the flow in figure 10-2. During the half-cycle when the top of the secondary winding is positive, current flows through diode D_1, and passes through the load from the top to the bottom. This causes the top of the load to be positive. After leaving the load, the current flows into the secondary winding at the center tap which is negative during this half-cycle. During the next half-cycle, the bottom of the secondary winding will be positive and the center tap will be negative with respect to the bottom. Current will flow through diode D_2 and through the load resistor in the same direction it passed during the first half-cycle. The output waveform is pulsating direct current, whose frequency is twice that of the pulsating DC produced by a half-wave rectifier.

C. Bridge-Type, Full-Wave Rectifier

The two-diode, full-wave rectifier requires a transformer that will give the desired output voltage across just one-half of the secondary winding. This is inefficient, and this inefficiency may be overcome by the bridge-type, full-wave rectifier, in which we can use the entire secondary winding. To do this, we must use four diodes instead of two.

We can trace the flow through the load in figure 10-3. During the half-cycle when the top of the transformer secondary is positive, the current flows through diode D_1 and through the load resistor from the right to the left, and then down through diode D_2 and back to the bottom of the secondary winding, which is negative. During the next half-cycle, the polarity of the secondary has reversed, and current flows through diode D_3, through the load in the same direction as before, and up through diode

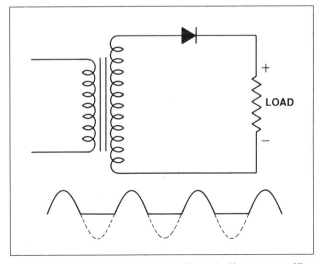

Figure 10-1. A semiconductor diode half-wave rectifier with its output waveform.

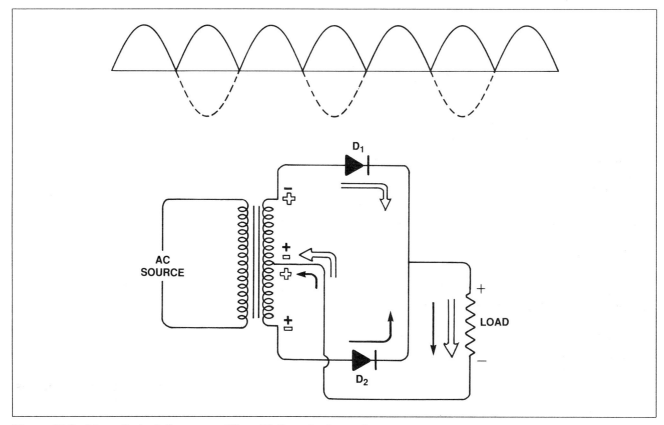

Figure 10-2. A two-diode, full-wave rectifier with its output waveform.

D4 to the top side of the transformer, which at this time is negative. The output waveform is similar to that produced by the two-diode, full-wave rectifier, but the voltage is quite a bit higher because with this circuit the entire secondary winding is used.

D. Full-Wave, Three-Phase Rectifier

The most familiar 6-diode, three-phase, full-wave rectifier is the one found in the DC alternator, which is rapidly replacing the generator as a device for producing direct current electricity for our modern aircraft. This alternator uses a three-phase stator and six silicon diodes arranged as those in figure 10-4.

Let's examine the current flow through the load resistor for one complete cycle of all three phases. Remember that we are tracing conventional current which is opposite the electron flow, but which follows the direction indicated by the arrowheads in the diode symbols. In that portion of the cycle when the output end of phase A is positive, current leaves it and flows through diode D_1 and down through

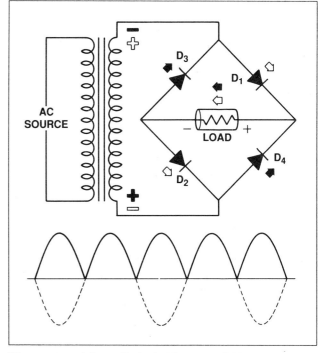

Figure 10-3. A four-diode, bridge-type, full-wave rectifier with its output waveform.

the load, making its top end positive. After leaving the load, the current flows through diode D_2 and coil C, whose output lead is negative. Now, as the alternator field rotates, it causes the output end of the coil B to become positive and the output of coil A to be negative. Current flows out of B, through diode D_3, the load, diode D_4, and back through coil A. Continued rotation causes the output of coil C to be positive and B to be negative, so the current leaves C, passes through diode D_5, the load and diode D_6 and back into coil B.

The output waveform of this rectifier gives us a very steady direct current as the current from the three phases overlaps and there is never a time when the current drops to zero.

E. Filters

The output waveform for any of the single-phase rectifiers we have just discussed has been pulsating DC, which drops to zero, then rises to a peak. In the half-wave rectifier of figure 10-1, the voltage drops to zero and remains here during

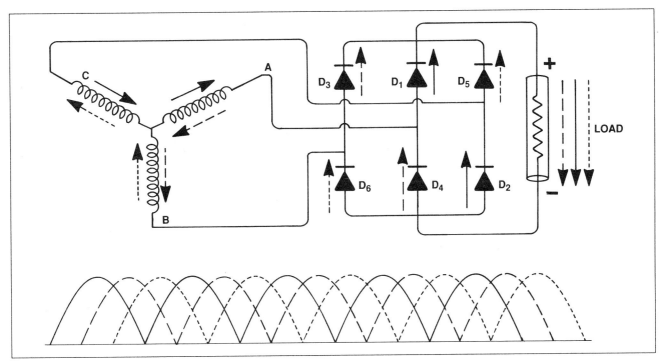

Figure 10-4. *Three-phase, 6-diode, full-wave rectifier and its output waveform.*

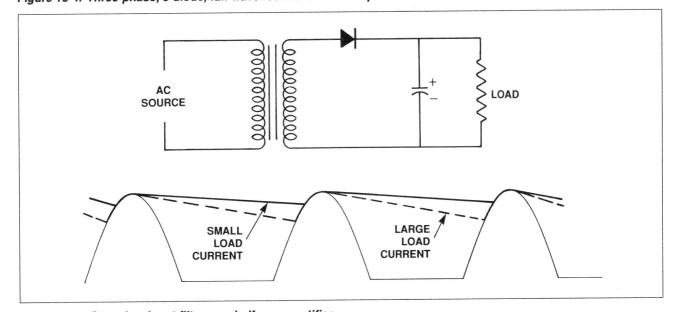

Figure 10-5. *Capacitor-input filter on a half-wave rectifier.*

one-half of the cycle; then it rises to the peak, and back to zero. This type of voltage would not be satisfactory for very many applications, and so some form of filter must be used to smooth out the DC so it will remain at a fairly constant value—at least, so it will not drop all of the way to zero.

1. Capacitor-Input Filters

If we install a large capacity electrolytic capacitor across the load, as we see in figure 10-5, the capacitor will charge as the voltage rises, and then as the voltage drops, it will discharge and prevent the voltage dropping to zero. The voltage will follow the curve shown in figure 10-5. A large flow of load current will cause the voltage to drop more than a small current.

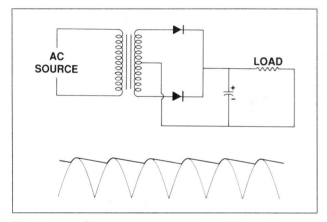

Figure 10-6. Capacitor-input filter on a full-wave rectifier.

In the full-wave rectifier of figure 10-6, the action of the capacitor input filter is the same as it was for the half-wave circuit, except that the frequency of the output ripple is twice as high as it was with the half-wave circuit.

2. Choke-Input Filter

If a choke (inductor) is placed in series with the load, the effect of the inductance will oppose any change in the load current. During the part of the cycle in which the current is increasing, the magnetic field in the inductor will induce a current that opposes the rise. When the source current starts to drop off, the inductor will produce a current that opposes the drop-off. The resulting waveform is a low alternating voltage whose frequency is twice that of the source.

3. Pi Filter

There are a number of arrangements of capacitors and inductors for filtering the ripple from rectified AC, but one of the most popular is that of a capacitor input filter whose circuit diagram resembles the Greek letter pi (π). In figure 10-8, we see a pi filter connected to a full-wave rectifier. The two capacitors are high-capacity electrolytics which have a minimum reactance to the ripple frequency of the rectified AC, but which will not allow any flow of DC to ground. The inductor has a high opposition to the AC caused by the ripple, but offers very little opposition to the flow of the DC. With this ar-

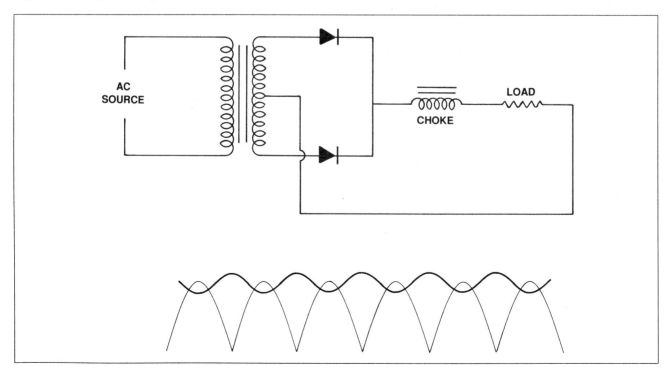

Figure 10-7. Choke-input filter on a full-wave rectifier.

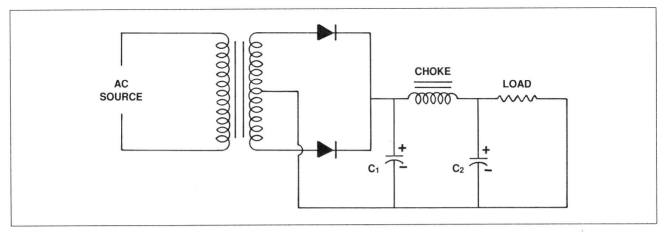

Figure 10-8. Pi filter on a full-wave rectifier.

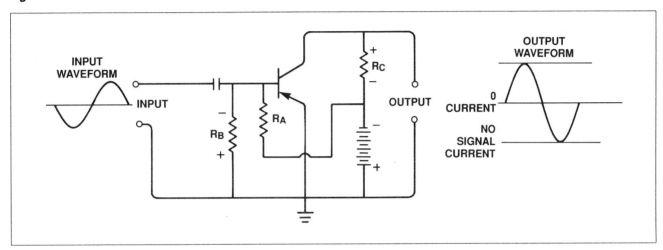

Figure 10-9. A small voltage applied to the input of a transistor amplifier will cause a large change in the output current, and the output voltage will be 180° out.

rangement, the ripple frequency AC is passed to ground and leaves an almost pure DC output.

II. AMPLIFIERS

An amplifier uses, as an input, a very small change in either voltage or current to produce a large change in the output voltage or current. A transistor is a low-impedance device, which means that in a transistor amplifier, we will use a small change in current in the emitter-base circuit to cause a large change in the current in the emitter-collector circuit.

A. Transistor Amplifiers
1. Common Emitter Amplifier

In the common-emitter amplifier, the transistor is installed in the circuit of figure 10-9 in such a way that the forward bias for the emitter base junction is provided by the voltage drop across resistor R_B, and the reverse bias for the collector base junction is provided by the voltage drop across resistors R_A and R_C. When there is no signal voltage across R_B, there will be a steady current flowing through the transistor and resistor R_C, whose voltage drop is opposite in polarity to that of the battery. The result is that the steady output current is negative, but not as negative as that of the battery.

When a negative signal is put on the input between the base and the emitter, across R_B, the forward bias across the emitter base is increased. There will be more current flow through the transistor and R_C, and the voltage drop across R_C will increase, causing the output to become less negative; or, if you would like to think of it that way, it would go positive.

During the half-cycle of the input signal when it is positive, the forward bias of the emitter base junction decreases and the transistor conducts less, decreasing the voltage drop across the resistor R_C so the output will go more negative.

205

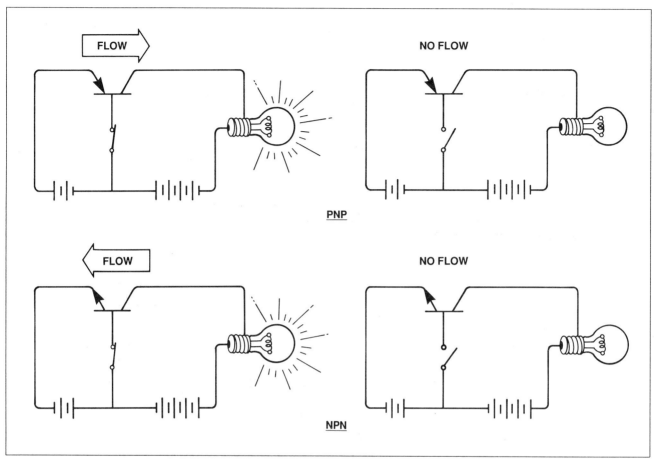

Figure 10-10. Current will flow in the emitter-collector circuit of a transistor amplifier when current flows in the emitter-base circuit.

In a common-emitter transistor amplifier, the emitter collector current is much greater than the emitter base current, and the output current is similar to but opposite in phase to the input current.

When we look at this basic transistor amplifier, we see a very important principle. When the base of a PNP transistor is negative with respect to its emitter, the transistor will conduct. When it is not negative, it will not conduct. An NPN transistor works in the same way, except its base must be positive for it to conduct.

2. Common Base Amplifier

If a transistor is connected into an amplifier circuit with the base common to both the input and output, a condition similar to that shown in figure 10-11 results. The battery E_{eb} provides a forward bias for the emitter-base junction and battery E_{cb} reverse-biases the collector-base junction. Resistor R_e is in series with battery E_{eb}, and it is across this combination that the input signal is applied.

Consider the portion of an AC input signal when the voltage is going positive; the forward bias will

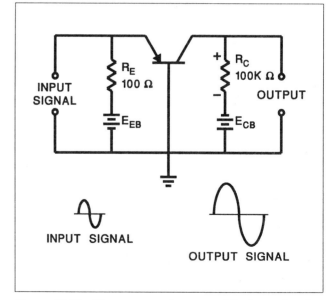

Figure 10-11. The common-base amplifier has a low-impedance input and a high-impedance output, and does not invert the phase of the amplified signal.

be increased, and the transistor will conduct more electrons in its collector-emitter circuit. This increase in collector current increases the voltage drop across R_c which, because of its polarity, will cause the output voltage to become less negative (−) or, in essence, positive. It is clear that this form of amplifier does not invert the phase of the signal as the common emitter amplifier did.

The current flow through the emitter is the sum of the current that flows through the collector and the base, so there can be no current gain in a common base amplifier. In fact, the current amplification will always be less than one. The voltage gain, however, is another matter. The impedance of the output circuit is high and that of the input circuit is low, to the current that flows through R_c will generate a much larger voltage drop than the voltage change across R_e.

3. Common Collector Amplifier

It is sometimes desirable to use a transistor circuit to transform a high impedance input into a low impedance output. This may be done using a common-collector amplifier circuit like the one shown in figure 10-12.

In this circuit, battery E_{eb} provides the forward bias for the emitter-base junction, and the input signal is applied across this battery and resistor R_b. A negative-going signal will increase the forward bias, increasing the collector-emitter current and making the top resistor R_e more negative with respect to ground; and so the output voltage follows the input. Base resistor R_b has a high resistance and therefore only a small base current is required to control a rather large emitter current. The low resistance of emitter resistor R_e provides a rather small voltage change. All of this means that the common collector amplifier has a voltage amplification of less than one, a rather high current amplification, and it does not invert the phase of the amplified signal. It provides an impedance step-down.

B. Alphabet Classification of Amplifiers

In our discussion above, we have classified amplifier circuits according to which lead is common to both the input and output terminals. This type of classification will tell us about input and output impedances, and the gain we can expect in current and voltage. Amplifiers may also be classified according to function or operating level. When the amplifier is classified according to operating level, an alphabet system is utilized.

1. Class A Amplifiers

A class A amplifier provides a fixed amount of base current which flows at all times. This means that even when the amplifier is not processing a signal, there will be a small collector current, known as the quiescent current. The class A amplifier is constantly using power, and the maximum useful power

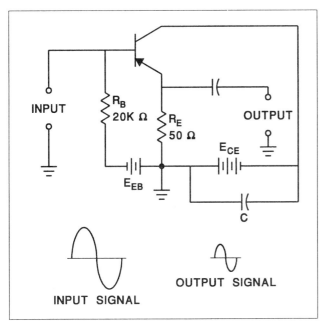

Figure 10-12. The common collector amplifier has a high-input impedance, a low-output impedance, and does not invert the phase of the signal being amplified.

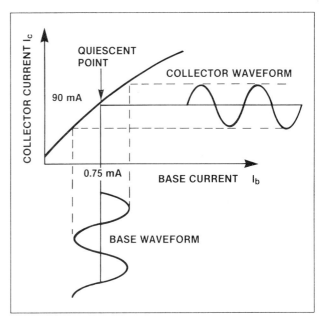

Figure 10-13. Class A amplifier operation.

207

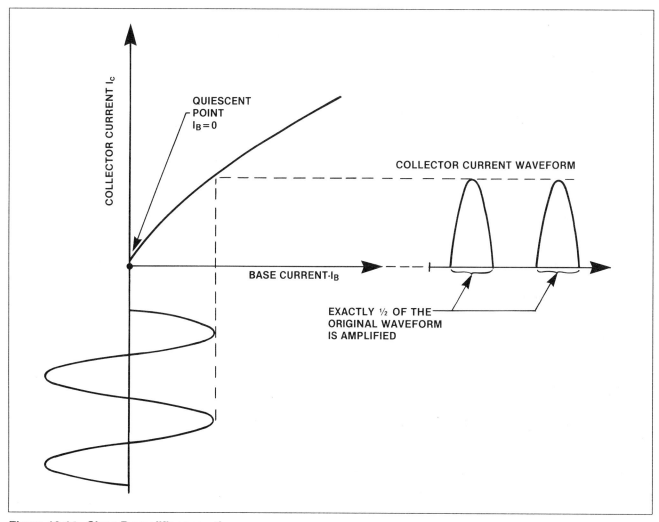

Figure 10-14. Class B amplifier operation.

that comes out of the amplifier is rarely more than 10 or 20% of the power consumed. Class A amplifiers have excellent fidelity and a minimum of distortion.

2. Class B Amplifiers

Class B amplifiers are made to be more efficient than class A. The class B amplifier almost always uses two transistors which are biased so that each one amplifies one-half of the signal. The outputs of the two transistors are then combined to produce the amplified output. These are commonly called push-pull amplifiers.

3. Class C Amplifiers

The final amplifier in radio transmitters are often class C amplifiers. This class of amplifier is intended for efficient power amplification where a powerful signal is needed at a single frequency. In a class C amplifier, the collector current is zero for most of the input sine wave cycle. The output is a series of short, rounded, current pulses which repeat at the desired frequency.

III. OSCILLATORS

There are many different types of oscillators used in electronic circuits, but in our introduction to analog electronics we will mention only two of the types most frequently used. The Hartley oscillator can be used to produce a sine wave output, and the multivibrator oscillator produces a square wave.

For electronic oscillation to occur, two conditions must be met; there must be amplification, and there must be feedback of the correct phase from the output circuit back into the input.

1. Tank Circuits

In order to fully understand oscillation, let's review one of the basic principles of AC electricity, that of resonance. A coil and a capacitor connected in

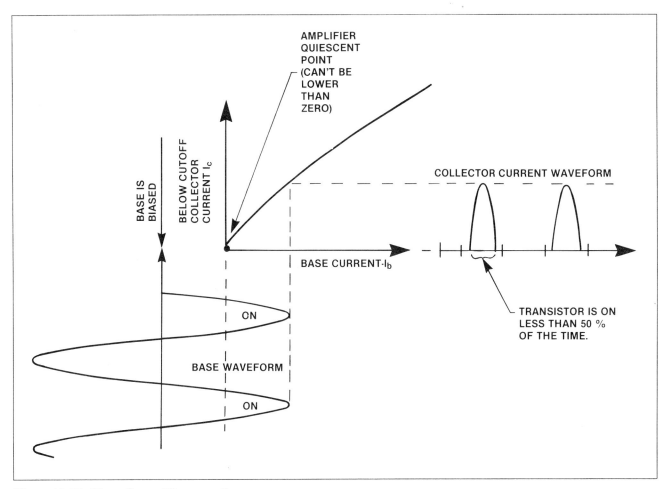

Figure 10-15. Class C amplifier operation.

parallel give us what is known as a tank circuit. If we place the switch in figure 10-16 in position A, current will flow until the capacitor charges to the voltage of the battery and current will cease to flow. If the switch is placed in position B, the capacitor will remain charged, and keep the electrical energy stored in the form of electrostatic fields within the capacitor. When the switch is placed in position C, the capacitor will discharge through the coil, and the energy in the electrostatic field will be transformed and stored in an electromagnetic field around the windings of the coil. When the capacitor is completely discharged, the current will stop flowing, and the electromagnetic field will collapse, pushing electrons back into the capacitor on the plates opposite the ones originally holding them. By the time all of the energy has dissipated from the coil, the capacitor will be fully charged, and it begins to discharge through the coil again. If there were absolutely no resistance in the circuit, and the values of the capacitor and the coil were correctly chosen, the charging and discharging, or oscillation, would continue indefinitely. But, un-

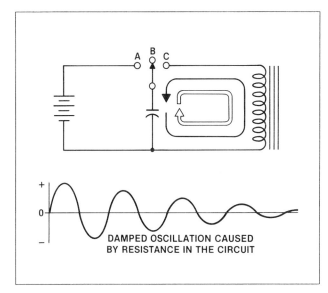

Figure 10-16. Electronic tank circuit. Unless energy is added to an oscillator circuit to replace that which is lost in resistance, the oscillations will die out.

fortunately, there is always some resistance, and it dampens the oscillations, eventually causing them to stop altogether.

2. Hartley Oscillators

In order to sustain oscillation, the exact amount of energy which is lost must be restored. This is done by the feedback circuit such as the one we see in the Hartley Oscillator illustrated in figure 10-17.

When switch S is closed, electrons begin to flow in the circuit, biasing the transistor through the voltage divider R_A and R_B, and flowing through L_1, the lower half of the center-tapped inductor. As the current rises, it induces a voltage into the upper half of the coil and charges capacitor C_2 in such a direction that it increases the forward bias of the transistor, further increasing the

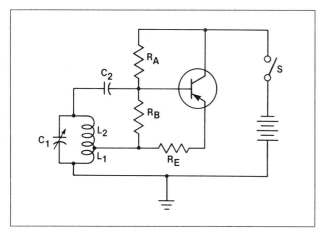

Figure 10-17. Hartley oscillator. The output frequency is determined by the capacity and the inductance.

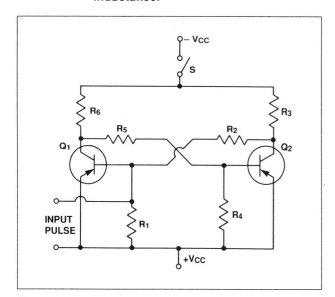

Figure 10-18. Bistable multivibrator oscillator. In digital circuits, this is called a flip-flop.

electron flow. By the time the circuit is saturated, that is, when there is no more rise in the current, no more voltage will be induced into L_2 from L_1, and all of the energy stored in the tank circuit will be in electrostatic fields in capacitor C_1. Now that there is no more force to push the electrons into the capacitor, it will discharge, and the energy lost will be made up of that stored in capacitor C_2 as it discharges. As C_2 discharges, the forward bias of the transistor decreases, and it begins to conduct less, decreasing the current through L_1.

This changing current induces a voltage into L_2 so it will charge C_2 and decrease the forward bias, and eventually reverse-bias the transistor. This drives it all of the way to cutoff. When all flow is stopped, C_1 and C_2 will be fully charged and the procedure will repeat itself.

The frequency of oscillation of this Hartley Oscillator is determined by the amount of inductance and capacitance in the tank circuit. Increasing the product of L and C will decrease the frequency of the AC the oscillator produces. The frequency in hertz may be found by the formula:

$$fr = \frac{1}{2\pi \sqrt{LC}}$$

In this formula, 2π is a constant (6.28); L is the inductance of the coil in henries; and C is the capacitance of the capacitor in farads.

3. Multivibrator Oscillators

A multivibrator oscillator is one type of electronic circuit that finds use in both analog and digital electronics. In analog circuits, an astable, or free-running, multivibrator is used to produce square wave AC and DC. In digital circuits the bistable multivibrator is used as a flip-flop, which is a handy switching circuit used to store information.

Since the bistable multivibrator is the most widely used, we will look at its operation. It is essentially two amplifiers with the output of one connected to the input of the other. When one turns on, it turns the other off. This flip-flops back and forth with the circuit having two stable conditions. We see in figure 10-19 that either transistor Q_1 or Q_2 will conduct, but both cannot conduct at the same time. A positive pulse at the input will cause Q_1 to conduct and a negative pulse will force Q_2 into conduction.

Let's assume that power is supplied to the circuit and a negative pulse is applied to the input of Q_1 across resistor R_1. The base of Q_1 is driven negative with respect to its emitter and since this is a PNP transistor, it is forward-biased and begins to conduct. The voltage drop across R₆ keeps the base of Q_2 positive enough to hold it in shut off, and since there is not enough voltage drop across R_3 to reverse-bias Q_1, it will continue to conduct. Now, if a positive pulse is applied to the input, Q_1 will shut off and since there is no longer a voltage drop across R_6, Q_2 will be forward-biased into conduction. The voltage drop across R_3 reverse-biases Q_1 and holds it cut off until a negative pulse is applied to the input.

IV. VOLTAGE CONTROL CIRCUITS

A. Full-Wave Voltage Doubler

A circuit consisting of three capacitors and two diodes can be used to provide direct current with a voltage almost twice the peak voltage of the input AC.

In figure 10-19(A), during the half of the cycle when the top of the power source is positive, current flows through diode D_1 to charge capacitor C_1 to full line voltage. During the next half-cycle, when the bottom terminal of the power source is positive, current flows through D_2 and charges capacitor C_2 to the full line voltage. Capacitors C_1 and C_2 are in series across capacitor C_3, which will charge to twice the voltage of the input. Actually, the DC output voltage will depend to a great extent on the amount of load current that flows. With low-output current, the DC voltage will be approximately twice the peak value of the source AC.

B. Voltage Regulator

The zener diode, as you will remember, is a special form of diode with a specifically controlled reverse voltage. Below this voltage, the diode acts as a check valve, allowing flow when it is forward biased, but preventing flow when it is reverse biased. When the reverse-bias voltage exceeds the zener voltage of the diode, it will break down and conduct in its reverse direction.

This characteristic of the zener diode and the fact that the zener voltage is always exactly the same value make the zener diode a good voltage regulator.

In figure 10-20, we have a 15-volt zener diode in series with a bleeder resistor across the filtered output of a 24-volt DC power supply, with the anode of the zener diode connected to the negative terminal. This is the reverse bias direction. The zener diode will break down at 15 volts and allow enough current to flow through the bleeder resistor to maintain a nine volt drop across it.

The load is connected across the zener diode, and since it will maintain a 15-volt drop across it, there will always be exactly 15 volts across the load. When the load current increases, the current through the zener diode will decrease enough to maintain the nine volt drop across the bleeder resistor, and when the load current decreases, rather than allowing the voltage to rise, the current through the zener diode will increase.

The current through the zener diode will vary so that the total current through the load and the zener diode will be the correct amount to maintain the nine-volt drop across the bleeder resistor.

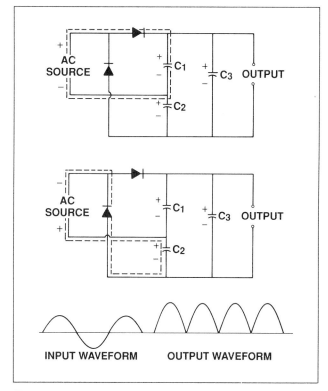

Figure 10-19.
(A) Capacitor C_1 of this full-wave voltage doubler is charging to full line voltage.
(B) During this half-cycle, capacitor C_2 charges to full line voltage, and since C_1 and C_2 are in series across C_3 it will charge to double the line voltage.

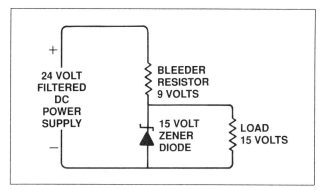

Figure 10-20. A zener diode voltage regulator circuit.

Chapter XI

Radio Transmitters and Receivers

I. AVIATION FREQUENCY SPECTRUM

The spectrum of radio frequencies has been divided up by international treaties into various bands and ranges, and some of them have been designated for aeronautical communications. Figure 11-1 shows the frequency divisions with their names and the aeronautical facilities in each band.

The vast majority of aeronautical communications for civilian aircraft is done in the very high frequency (VHF) band between 108.0 and 135.95 MHz, and it is with equipment that operates within this band that we as aviation maintenance technicians are most concerned.

VHF communications is, because of the nature of its waves, restricted to line-of-sight distances. This means that you can receive a signal from a transmitter only if there are no obstructions between the two antennas. And because of the curvature of the earth, the maximum range of line-of-sight communications is restricted. Maximum reception range can be extended by raising the receiver and/or transmitter antenna off of the earth's surface. For example, an airplane flying 10,000 feet above the ground would have a maximum reception distance of about 140 miles. Of course, there are many things that could decrease this distance. The same aircraft flying at 1,000 feet may be restricted to approximately 30 miles of reception.

II. RADIO WAVES

A. Composition

In the chapter on basic electricity, we saw that lines of magnetic force surround any conductor carrying current. This field is illustrated in figure 11-2. If the conductor is carrying alternating current, the magnetic field starts at the conductor and expands until it reaches its maximum when the current flow is the greatest. Then it collapses back into the conductor, reverses, and expands again as the current flow builds up in the opposite direction.

When a voltage source is connected to a capacitor (figure 11-3), an electrostatic field is established between the two plates, and if the source is AC, the electrostatic field will build up, collapse, and reverse its polarity each half cycle of the AC.

Every conductor has some inductance, some capacitance, and some resistance. For each frequency, there is a particular length of conductor that will be resonant. That is the length where the capacitive reactance is exactly equal to the inductive reactance. Electrical energy travels 186,000

FREQUENCY BAND	UTILIZATION
LOW FREQUENCY BELOW 300 kHz	LF/MF COMMUNICATIONS 200 - 416 kHz ADF RANGES 90 - 1800 kHz LORAN 1750 - 1950 kHz
MEDIUM FREQUENCY 300 - 3000 kHz	COMMERCIAL AM BROADCAST 535 - 1605 kHz
HIGH FREQUENCY 3 - 30 MHz	HF COMMUNICATIONS 2 - 25 MHz
VERY HIGH FREQUENCY 30 - 3000 MHz	MARKER BEACONS 75 MHz VHF NAVIGATION AND COMMUNICATIONS 108.0 - 135.95 MHz
ULTRA HIGH FREQUENCY 300 - 3000 MHz	TRANSPONDER AND DME 960 - 1215 MHz
SUPER HIGH FREQUENCY 3000 MHz - UP	RADAR 220 - 5200 MHz

Figure 11-1. Frequency utilization spectrum for aviation communication and navigation.

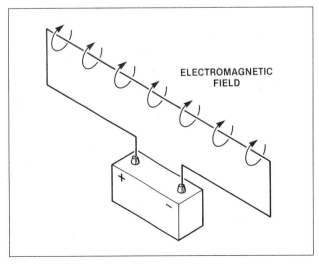

Figure 11-2. Electromagnetic field surrounding a wire carrying electrical current.

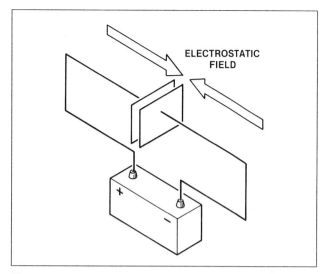

Figure 11-3. Electrostatic field across a capacitor that is connected to a source of electrical voltage.

miles per second, or 300,000,000 meters per second, and the wavelength of AC varies with its frequency. For example, AC, whose frequency is one hundred megahertz (100,000,000 cycles per second), has a wavelength of 9.82 feet, and the higher the frequency, the shorter the wavelength.

If a piece of wire whose length is exactly one-half of the wavelength is suspended in the air between towers and is excited, or fed, in its center by an AC generator, both electromagnetic and electrostatic fields will surround the wire. This is seen in figure 11-4. At the instant, one end of the wire is negative the other end will be positive, and at this time the electromagnetic field will be expanded to its maximum. As the polarity of the wave begins to change, the magnetic field begins to collapse and the electrostatic field begins to pull in from the ends of the wire.

At frequencies above about 10 kHz, the electrostatic field does not have enough time to completely pull in before it starts to build back up in the opposite direction, and an independent field forms which is repelled as the new electrostatic field builds up. This newly formed field is then radiated out into space as radio waves. These waves are in phase with the two fields that cause them, but perpendicular to both fields.

The relationship between the direction of the three fields may be visualized by holding the fingers of the right hand as we see in figure 11-5. If the forefinger represents the direction of the electromagnetic field, and the thumb, the direction of the electrostatic field, the middle finger will point in

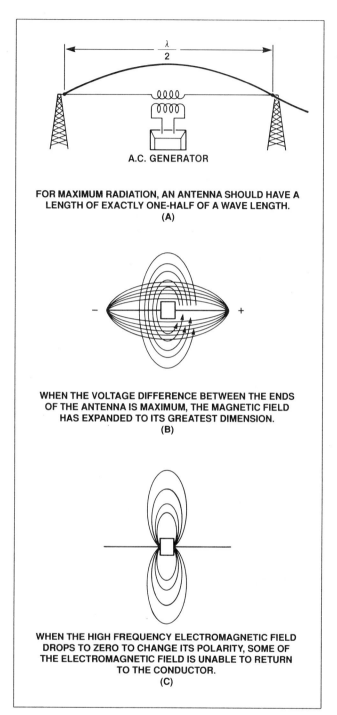

Figure 11-4. Half-wave antenna.

the direction of the radiated wave which is called the direction of propagation.

B. Propagation and Reception

For a radio wave to be of practical use, it must carry information. The first type of radio transmissions were made using the continuous wave (CW) technique. This type of transmission is called continuous wave since the AC waveform does not change its frequency or amplitude, it is simply turned on and off at given intervals. In communication by CW, information is transmitted by short spurts of radio frequency energy, with some spurts shorter than the others. Figure 11-6 illustrates the waveform of a CW signal. The shorter spurts are called dots and the longer ones dashes, and by arranging them according to the Morse code, they carry information.

Radio communication can be made more effective by carrying voice on a modulated carrier wave, MCW. In MCW transmission a carrier wave, having a frequency appropriate to the type of communication and assigned by the Federal Communications Commission, is generated by an extremely accurate crystal-controlled oscillator. This carrier is then amplified and fed into an antenna that is resonant to this frequency, and a maximum amount of the energy is radiated out into space.

The carrier wave by itself is of little use, but we can modulate, or change it, according to the information we want to transmit. For communication radios, the information we wish to transmit are the voices of the operators. In some cases, such as navigation radios, the information we wish to transmit is a signal to indicate the location of a particular navigation radio transmitter. The two most generally used methods of modulation are amplitude modulation (AM) and frequency modulation (FM). In AM, the voltage of the carrier wave is changed, and with FM, the frequency of the carrier is changed.

1. Amplitude Modulation

When amplitude modulation is used, the signal we wish to transmit is amplified and is used to change the amplitude of the carrier. After leaving the amplifier, the audio signal is a direct current with a varying voltage which can be used to change the AC voltage of the carrier. The amplitude modulated carrier has a waveform on both its positive and its negative sides that is the exact shape of the audio frequency information we want to transmit. Study figure 11-7 to see how the AM signal is constructed, transmitted, and detected.

The modulated carrier wave is now radiated into space from the transmitting antenna. The receiving antenna acts like an inductor and receives an induced current from the electromagnetic waves which left the transmitter. As the carrier wave cuts across the antenna, it generates in it an AC voltage that, while extremely weak, is an exact copy of the transmitted wave.

Amplifiers in the receiver build up the voltage of this received signal, and then it is detected, or rectified, which removes one-half of the modulated carrier. This is further treated by demodulating it, which is the process of removing the carrier frequency energy. We now have a varying DC voltage,

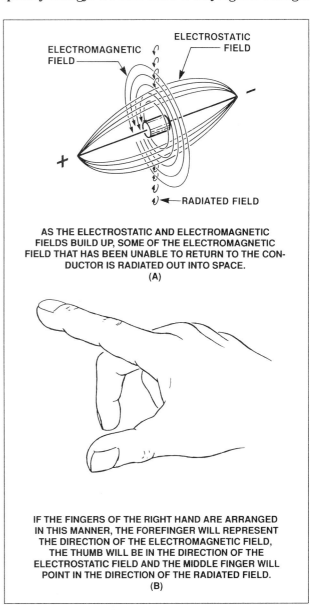

Figure 11-5. The production of a radiated field.

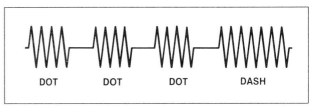

Figure 11-6. Information is carried by CW communication by transmitting a series of short and long spurts of radio frequency energy.

215

which, when amplified and sent to the speaker, becomes a varying amplitude alternating current that is an exact copy of the original audio signal.

2. Frequency Modulation

One of the primary limitations of amplitude modulation (AM) is its susceptibility to static interference. Static interference is caused by the magnetic fields created during rapid changes in electron movement. For example, static interference can be created by arcing of an electrical connection, spark plug ignition and atmospheric lightening. The magnetic fields set up by this interference radiate from their sources and are picked up by AM radio receivers. Since the magnetic fields have a changing amplitude, the radio amplifies the interference and sends the signal to the speaker or headphones. To improve the reception of intelligible information, a system which does not rely on the variation of amplitude was invented. This system is called frequency modulation (FM).

In an aircraft frequency modulated system, AC is generated in the very high frequency (VHF) band for the carrier. The carrier wave changes its frequency to transmit the information desired. The frequency changes from its center, or resting, frequency by an amount that is proportional to the amplitude of the information being transmitted. As

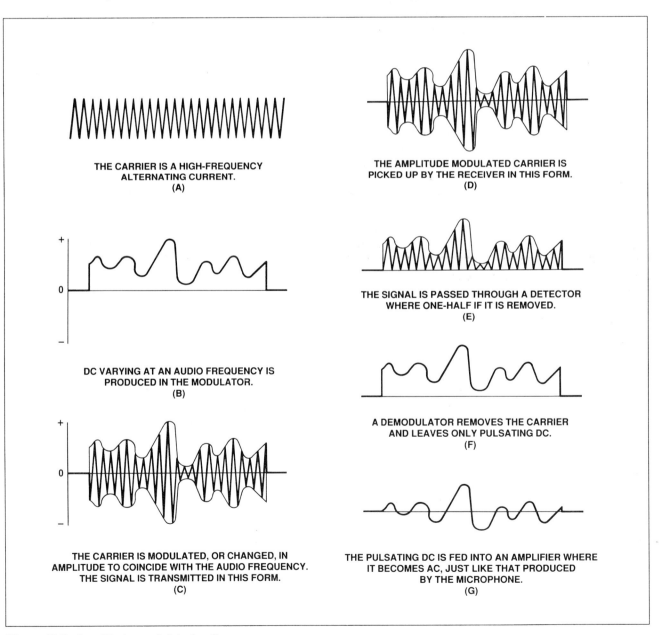

Figure 11-7. Amplitude modulated radio waves.

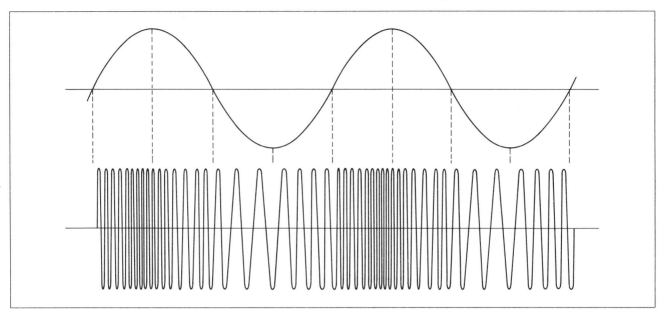

Figure 11-8. Example of a frequency modulated wave.

shown in figure 11-8, if the transmitter has a center frequency of 112.0 MHz and it is modulated with an audio frequency of, say, 1,000 Hz, the carrier will vary from its center frequency by an amount that is proportional to the amplitude of the 1,000-Hz signal. The carrier will vary one thousand times each second (1,000 Hz). This varying, or swinging, of the carrier from one side of the center to the other, generates sidebands similar to those in amplitude modulation. But these sidebands, instead of having to be added to the carrier power, are simple changes in frequency, while the carrier amplitude remains constant. This may be seen in figure 11-8. Modulation of an FM signal requires much less power because it is the oscillator that is modulated rather than the power amplifier as is done with AM.

3. Wave Propagation

Radio wave propagation can be identified as one of three types. They are (1) ground wave, (2) sky wave, and (3) space wave. These classes result from the effects of the earth and the atmosphere have on the radio wave. The type of propagation that takes place is dependent upon the frequency of the radio wave.

Ground wave, also known as surface wave, travels along the surface of the earth. This wave extends from the surface to a height which can range from a few feet to thousands of feet above the earth. This wave is sort of pulled along the curvature of the earth because of its contact with the surface. The frequency range of normal AM broadcast radios

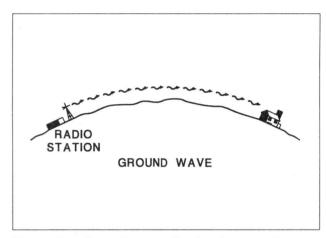

Figure 11-9. Ground wave radio signal propagation.

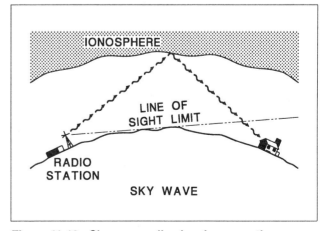

Figure 11-10. Sky wave radio signal propagation.

217

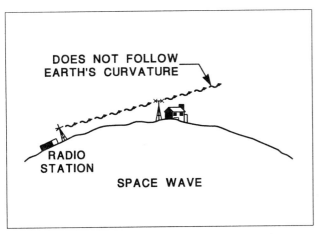

Figure 11-11. Space wave radio signal propagation.

and the aircraft ADF receiver are characterized by this type of wave.

Sky wave propagation is the result of radio waves being reflected back to earth from the layers of the ionosphere. Radio waves traveling in this manner are reflected back to the earth from a very high altitude and are able to travel great distances. This type of wave propagation is used for high frequency (HF) communications over great distances.

The ionosphere is made up of charged atoms (ions) whose energy levels are subject to a variety of factors. Even the change in solar energy from day to night will affect the makeup of the ionosphere. Because of these changing conditions the reflection of radio waves off the ionosphere will vary. At any time, day or night, there is a "window" of usable frequencies created by the reflecting properties of the ionosphere. At night this will usually be in the lower range of HF channels, and during the day it will be the higher range of HF channels.

Space waves are those types of radio signals we would expect to receive if we were on the moon. This type of wave must travel in a straight line from the earth station to the receiver. The wave is not held to the earth as in a ground wave. This is known as "line-of-sight". VHF aircraft radios rely on this type of propagation. This line of sight propagation limits the travel of VHF signals. Signals will not bend to the curvature of the earth and will not "go through" mountains. There are, however, some circumstances that may permit reception of a station beyond the line-of-sight range.

Density, temperature, and humidity of the atmosphere change with altitude, and may cause a "bending" of these VHF radio waves. These atmospheric conditions cause a change in the speed of the radio wave and result in the bending effect.

Bending can cause a 10 to 15% increase in the apparent line-of-sight distance. It is also possible for the atmospheric conditions to bend the signal away from the earth and result in a reduced line-of-sight distance.

III. RADIO RECEIVERS

A. Amplitude Modulated

When a modulated carrier wave cuts across the antenna of a receiver, it generates a voltage that is an exact copy of the transmitted signal, but it must be separated from all unwanted signals, energy must be added to it, and the audio signal removed before it can actually be used. By studying the simplest form of radio receiver, using a diode as the detector (figure 11-12), it will help us understand the principle of radio reception.

Not only the signal we want, but every other signal that is radiated into space is picked up in the antenna and sent to ground through the tank circuit. The capacitor and inductor of the tank circuit are a specific value so the circuit is resonant to only the frequency we want, this is the tuner section of the receiver. Frequencies below that desired go to ground through the inductor, and those above the resonant frequency pass to ground through the capacitor. The resonant frequency of the tuner section follows the lower impedance path to ground through the diode and the earphones.

The diode rectifies the signal and allows only one-half of the modulated carrier to get through. Then the capacitor offers an easy path to ground for the carrier, but the changes in amplitude of the modulation envelope are at an audio frequency, and so they are opposed by the capacitor but easily pass to ground through the coils of the electromagnets in the earphones.

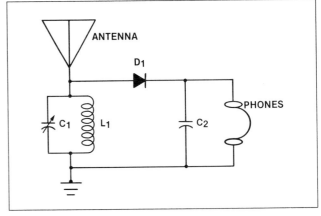

Figure 11-12. A simple diode detector radio receiver circuit.

The earphones are made up of a coil of wire wrapped around a small permanent magnet (figure 11-13), and this coil carries the demodulated signal, which is varying DC. The magnet is near a thin spring-steel diaphragm and holds it partially deflected toward the magnet. When the current through the coil increases in such a direction that aids the electromagnet, the combined field pulls the diaphragm over, but when the field of the electromagnet opposes that of the permanent magnet, the diaphragm relaxes. The diaphragm moves back and forth in step with the audio voltage which has been taken from the carrier, and its movement produces sound waves which affect the ear in exactly the same way as the original sound at the transmitter.

The simple crystal receiver allows us to hear a very strong radio signal, but it has very little output power, as there has been no energy added to it, nor does it give much choice of the stations it can receive.

The super-heterodyne receiver uses one of the more popular circuits in current use, as most of its stages of amplification are selective, and it concentrates its energy in the amplification of only the desired signal.

We can follow the super-heterodyne circuit by using the block diagram of figure 11-14. A pre-tuned amplifier selects a rather narrow band of frequencies from those picked up by the antenna in a manner similar to that used by the simple crystal receiver. This signal is amplified by using it to control a flow of electrons from the power supply. The amplified, modulated carrier is now directed into the mixer which is sometimes called the converter. Also fed into this mixer is the output of the local oscillator. The frequency of the local oscillator is controlled by the pre-amplifier, so it will always oscillate at a frequency a specific number of megahertz away from that of the radio frequency energy being amplified by the preamplifier. In the block diagram we have here, the preamplifier is tuned to 121.5 MHz, and the local oscillator is automatically tuned so it will produce a frequency of 110.7 MHz. Both of these signals are fed into the mixer where the process of heterodyning takes place. This is a procedure in which two frequencies are mixed to produce a desired third frequency.

Actually, when two frequencies are fed into this stage of the receiver, four frequencies will appear at its output: the two original frequencies, the sum of the two, and the difference between the two. If we input 121.5 and 110.7 MHz signals, we will have in the output, 10.8, 110.7, 121.5, and 232.2 MHz. All four of these frequencies are fed into the first stage of the intermediate frequency (IF) amplifier. The IF amplifier is a highly selective circuit, and it amplifies only the lower of these frequencies, the 10.8 MHz signal. All of the other frequencies pass to ground in this stage.

When another frequency is selected on this receiver, for example, 122.9 MHz, the local oscillator will automatically produce a frequency of 112.1 MHz. Now, the first IF amplifier will receive 10.8, 112.1, 122.9, and 235.0 MHz signals. All of these frequencies besides the 10.8 MHz are passed to ground, and only the 10.8 MHz signal will be amplified and sent through the rest of the circuits.

This intermediate frequency is modulated and is exactly like the original received signal except for its frequency. The reason for using this intermediate frequency is to get the maximum efficiency from the equipment. The preamplifier must have broad band characteristics, since it must amplify all of the frequencies in the range the equipment is designed to receive. It cannot have optimum ef-

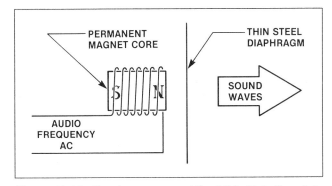

Figure 11-13. *Earphones convert the AC fed into them into pulsations of the thin steel diaphragm. This produces sound waves just like the ones which produced the signal in the microphone.*

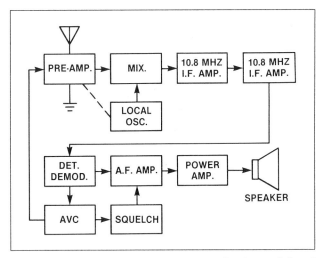

Figure 11-14. *Block diagram of an amplitude modulated super-heterodyne radio receiver.*

ficiency over the entire range, but by converting all of these frequencies into one, into 10.8 MHz, the IF amplifiers have only one frequency to amplify, and so they can be made highly efficient.

The second stage of intermediate frequency amplification further amplifies the signal and then delivers it to the detector-demodulator stage where the modulated IF signal is detected, or rectified, changing it into varying voltage DC, but it still has the 10.8 MHz component. The demodulator removes this and delivers a changing DC with no component of either the original carrier or the intermediate frequency to the audio frequency amplifier. The AF amplifier changes the varying DC into AC, builds up its voltage and delivers it to the power amplifier where sufficient current is controlled to drive the speakers with the desired volume.

The automatic volume control, or AVC, circuit maintains the audio output at a rather constant value. If the input signal is strong and causes the output to get too loud, a signal is sent back to the preamplifier to cut down its voltage output, or gain, so it will produce the desired audio output.

Aircraft communications receivers have so much gain that they amplify all of the internal noise in the receiver, and when they are tuned to a channel with no signal coming in, there will be a hissing noise heard that is distracting; but when a signal comes in and is amplified, it will override this hissing noise. To minimize this distraction, a squelch circuit is installed. The same voltage that drives the automatic volume control is applied to the audio amplifier, so when there is no signal being received, the audio will be low enough that no hissing will be heard. When a signal is received, however, the squelch will allow the audio frequency amplifier to amplify it to the proper volume.

B. Frequency Modulated

An FM receiver, such as the one whose block diagram we have in figure 11-15, uses a super-heterodyne circuit, but there are two stages that differ from those in an AM super-heterodyne receiver.

The signal is received by the antenna and taken into the tuned RF amplifier. The local oscillator is tuned with the RF amplifier so the proper intermediate frequency will be generated. This frequency is amplified through two stages of IF amplification and then it is passed to a limiter. The signal, up to this point, varies in its amplitude because of noise and static, but in the limiter stages it is clipped in both polarities; so the signal fed to the discriminator varies in frequency but not in amplitude.

The discriminator changes the frequency variations into amplitude variations, and the signal sent to the audio frequency amplifier is the same as it would receive in an AM receiver.

IV. RADIO TRANSMITTERS

Regardless of whether we are talking about a broadcast transmitter or the communications transmitter in an aircraft, the principles are the same. We must have an oscillator to generate the AC of the frequency assigned for this particular transmission; and we must have a modulation system that changes the sound energy in our voice into electrical signals to modulate, Remember, modulation is means by which we change the carrier wave so it can carry the intelligence. There must be the necessary transmission lines to conduct the modulated carrier wave to the antenna with the minimum loss of energy.

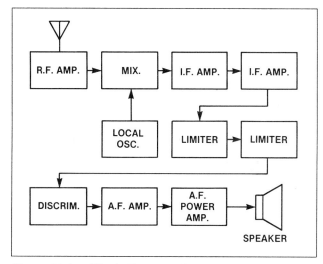

Figure 11-15. Block diagram of a frequency modulated super-heterodyne radio receiver.

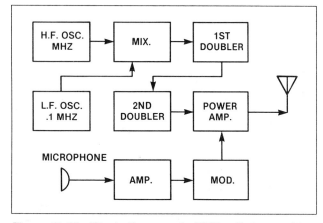

Figure 11-16. Block diagram of a VHF radio transmitter.

And finally, there must be an antenna that can efficiently radiate this energy into space.

In figure 11-16, we have a block diagram of a VHF transmitter. The two oscillators are crystal-controlled and produce a stable frequency for the carrier. The high-frequency oscillator has its frequency selected by the megahertz selector and the low-frequency oscillator by the kilohertz selector.

The output of both oscillators is fed into a mixer and then filtered to get the desired frequency. In the VHF range, the carrier frequencies are higher than it is possible for a crystal to vibrate, so the output of the mixer is fed into a circuit which doubles the frequency as it comes from the mixer, and then it goes into another doubler. The frequency that reaches the power amplifier is therefore four times that which came from the oscillator.

The sound waves that enter the microphone are converted into electrical signals and amplified, then taken into a modulator which works with the power amplifier to modulate, or superimpose, the audio signal on the carrier.

V. ANTENNA

A. Principles of Operation

We saw in figure 11-5 that the signal in an antenna consists of both electrostatic and electromagnetic fields, and if the signal is fed into a wire whose length is exactly one-half of the wave length of the carrier, the wire will act as a resonant circuit. Electrons will flow to one end of the antenna and pile up there; then during the next half cycle, they will all travel to the opposite end where they will again pile up. When all of them are at one end, there will be no antenna current flow, and the electromagnetic fields will be zero. The electrostatic field, however, will be maximum because of the voltage difference between the two ends. As the electrons start back to the other end, the electrostatic field decreases and the electromagnetic field builds up until the ends are at the same voltage and the current is maximum. As long as the signal is supplied to the antenna, it will oscillate, and since it is oscillating at ratio frequencies, the field expands so far during each half cycle that a part of it cannot get back into the antenna before the next cycle forces it out and part of the energy is radiated out into space.

B. Length

As we can see, the greatest amount of energy will be radiated into space when the antenna length is exactly one-half of the wave length. If the antenna is used for more than one frequency, there must be some means of adjusting its length, or else the antenna must be a compromise and it will not operate at its peak efficiency. Since an antenna acts as a resonant circuit, and resonance may be controlled with either capacitance or inductance, adding either of these in series with the antenna will change its resonant frequency and effectively change its electrical length. If a capacitor is added in series with the feed line, the antenna will be electrically shortened, and if an inductor is added, it will be electrically lengthened.

The formula for finding the required length needed for a half-wave antenna is:

$$\text{Length (feet)} = 468/F(\text{MHz})$$

The constant 468 comes from the fact that radio waves travel at 300 million meters per second, and therefore one wave of a 1-MHz signal is 300m long,

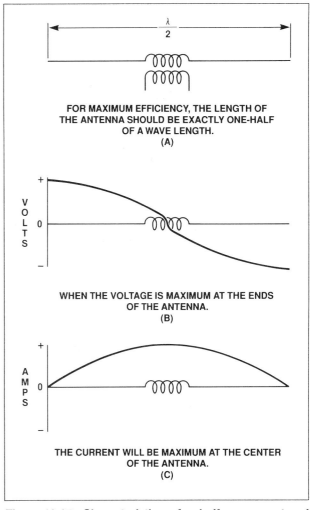

Figure 11-17. Characteristics of a half-wave, centered radio antenna.

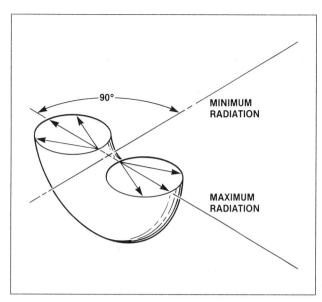

Figure 11-18. Radiation pattern from a horizontally polarized radio antenna.

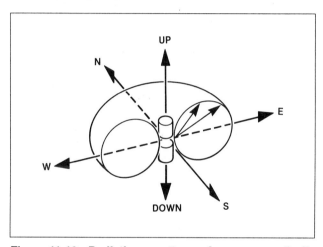

Figure 11-19. Radiation pattern from a vertically polarized radio antenna.

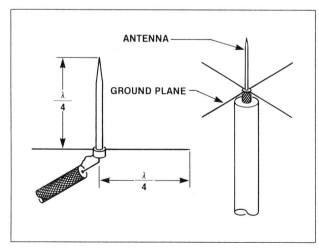

Figure 11-20. Quarter-wave antenna using a ground plane as a reflector.

and one-half of a wave has a length of 150m, or 492 ft. The dielectric effect of the air at the end of the half-wave antenna effectively lengthens it by about five percent, so the antenna should be computed at 95% of 492 or 468 ft. This is the correct length for one megahertz, and to find the length for any other frequency, divide 468 by the frequency in megahertz.

C. Polarization and Field Pattern

If the antenna is horizontal to the earth (figure 11-18), the energy will be radiated out from the antenna in a doughnut-shaped pattern, and there will be no lengthwise radiation from it. The maximum signal will be radiated perpendicular to the conductor.

Maximum reception will be obtained from a horizontally oriented transmitter antenna if the antenna for the receiver is also horizontal. But if a transmitter antenna is mounted vertically (figure 11-19), it will still put out a field pattern resembling a doughnut. This time, however, it will be as though the doughnut were laid on the ground and it will radiate its energy equally in all directions.

D. Types

Antennas used on aircraft come in many shapes, sizes, and types of construction. We will discuss here, only the basic forms.

1. Hertz

The half-wave antenna we have been discussing is called a hertz antenna and may be fed either in its center or from its end. If it is fed in the center where the current is maximum, it is called a current-fed antenna, but if it is fed from its end where the current is zero but the voltage is maximum, it is said to be voltage-fed. A form of hertz antenna is the dipole, which is a half-wave antenna consisting of two quarter-wave arms, fed in the center. The VOR antenna found on almost all aircraft is a special form of dipole. It is horizontally oriented, and the arms form a "V", with its apex in the line of flight, usually facing forward.

2. Marconi

It is not always convenient to use a half-wave antenna on an aircraft, so a quarter-wave antenna is often used. With this type of antenna, all of the requirements for transmission can be met with a quarter-wave conductor by using a reflector to serve for the other quarter wave. Aircraft communication antennas are usually of the quarter-wave, vertically polarized type, and because of the wide range of frequencies they must handle, they are usually a compromise with regard to length. The skin of the aircraft serves as the reflector, or ground plane, to supply the other quarter-wave portion of the antenna. Sometimes VHF stations on the ground have their antenna mounted on top of a pole, and since there is no structure to serve as a reflector, four or more wires, each as long as the antenna, are installed to radiate out to form a ground plane. This type of antenna is illustrated in figure 11-20.

When a quarter-wave antenna is mounted on a fabric-covered airplane, it is wise to put a sheet of aluminum foil large enough to serve as the ground plane inside the airplane, under the fabric.

3. Loop

A vertically polarized antenna is omnidirectional; that is, it receives a signal equally well from any direction; and a horizontally polarized antenna will receive its signal best from a station on either side, perpendicular to the antenna. Its strength will be the least for a signal from a transmitter in line with its length.

The directional characteristics of an antenna may be enhanced by winding it in the form of a loop. A signal received from a transmitter in line with the plane of the loop will be received on side A (figure 11-21) before it is received on side B and the voltage induced in side A will be out of phase with that in side B. The antenna current will then be the difference between the current in the two sides.

When a signal is received from a transmitter directly broadside to the loop, the voltage induced into the two sides will be equal to each other, but opposite in polarity, and will cancel each other. This characteristic makes the loop antenna useful for directional finding.

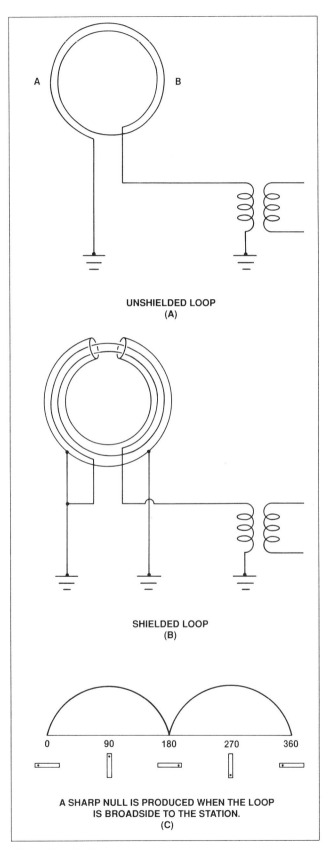

Figure 11-21. Loop antenna.

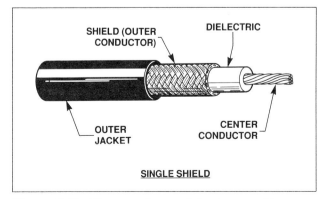

Figure 11-22. Diagram of a coaxial antenna cable.

VI. TRANSMISSION LINES

We can consider the antenna to be the electrical load and the radio transmitter to be the current generator. In order to get the maximum amount of power to the antenna, the load and the generator must be electrically matched; that is, the impedance of the load should equal the impedance of the source. Aircraft antennas are considered to have a nominal impedance of 50 ohms, and this should be matched by a transmission line having as near the same impedance as possible. Almost all aircraft antennas are connected to the transmitter with coaxial cable, which has a nominal impedance of about 53 ohms.

The coaxial cable most commonly used for aircraft radio equipment consists of a center conductor, a polyethylene dielectric, and a copper braid woven over the dielectric to form the outer conductor. Over the braid is a waterproof vinyl covering. Coax, as this type of cable is called, has the advantage that it produces no external field, and it is not susceptible to external fields. The construction of coax may be seen in figure 11-22.

Chapter XII

Communication Systems

During the first half of the life of aviation, electronics played an extremely small role. It was in World War I that the airplane itself was first recognized as being able to fill an actual need. Observation airplanes flew over the enemy lines and relayed their information back to the appropriate ground facilities by carrier pigeon. This method of communications was crude, but so were the airplanes of the time. The carrier pigeon was soon replaced with by a radio. Now the observer could tap out his message in Morse code and transmit the information more quickly and reliably than before.

When the war ended, so did the need for developing air-to-ground communications. This aspect of aviation development stagnated until the airlines started trying to fly a schedule that was not dictated by unknown weather conditions. The first civilian airborne radios were crude, but they did help the pilot find out what kind of weather lay ahead.

Communication was first accomplished using Morse code, then with voice transmissions in the low- and medium-frequency bands. These early radios used transmitters and receivers that were very tricky to tune and hard to hold on the frequency to which they were assigned. Next came crystal-controlled transmitters operating in the VHF frequency band. These units were equipped with a crystal plugged into the transmitter for each frequency, and a manually controlled tuner on the receiver.

This equipment was adequate when there were few airplanes in the air and just a few people to talk to on the ground. But as aviation grew, this early equipment was replaced with transmitters and receivers that were both crystal controlled. Newer equipment was introduced using frequency synthesizers to produce almost any frequency needed from a single highly accurate crystal. Today many communications radios are controlled by microprocessors that can store several frequency selections and tune both the transmitter and receiver at the touch of a single button.

I. SHORT RANGE COMMUNICATIONS

The vast majority of aeronautical communications for civilian aircraft is done in the very high frequency (VHF) band between 108.0 and 135.95 MHz. VHF communications is, because of the nature of its waves, restricted to line-of-sight distances. The line-of-sight nature of VHF communications limits its use to relatively short distances.

VHF communications transceivers consist of an AM transmitter and a single or double conversion superheterodyne receiver. Today's equipment provides 720 communication channels at 25 kHz spacing. The channel selector will tune both the transmitter and receiver. On many popular units the large outer knob will change the Megahertz portion of the frequency display, and the smaller, inner knob, will select the kilohertz portion of the frequency. One frequency, and one antenna, is used for both transmitting and receiving. This is known as single channel simplex operation.

A. VHF Transmitters

In figure 12-1, we have a block diagram of a VHF transmitter. The two oscillators are crystal-controlled and produce a stable frequency for the carrier. The high frequency oscillator has its frequency selected by the megahertz selector and the low frequency oscillator by the kilohertz selector.

The output of both oscillators is fed into a mixer and then filtered to get the desired frequency. In the VHF range, the carrier frequencies are higher than it is possible for a crystal to vibrate, so the output of the mixer is fed into a circuit which doubles the frequency as it comes from the mixer, and then it goes into another doubler. The frequency that reaches the power amplifier is therefore, four times that which came from the oscillator.

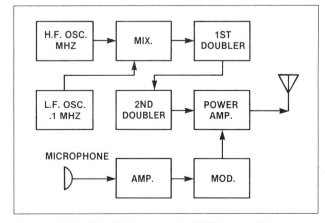

Figure 12-1. Block diagram of a VHF transmitter.

The sound waves that enter the microphone are converted into electrical signals and amplified, then taken into a modulator which works with the power amplifier to modulate, or superimpose, the audio signal on the carrier.

B. VHF Receivers

The simple superheterodyne receiver has all of the basic principles that makes for good communications, except that crowding of the available channels requires more precise control of the frequencies and closer spacing of adjacent channels. In the receiver illustrated in figure 12-2, we use double conversion, meaning that we have two local oscillators, one controlled by the megahertz selector and the other controlled by the kilohertz selector.

The first intermediate frequency amplifier, amplifies and passes a band of frequencies rather than one discreet frequency, as the simple superheterodyne receiver did. The output of the second oscillator heterodynes, or beats, with this first intermediate frequency to produce a second IF. This second IF is taken through the rest of the circuit as was done in the simple receiver.

II. LONG RANGE COMMUNICATIONS

Small general aviation aircraft are usually equipped only with VHF communication equipment. Larger business aircraft and transport category aircraft may also include high frequency (HF) communication equipment. The use of HF (2-30 MHz) communication equipment greatly extends the range at which the aircraft can maintain communication with ground stations. High frequency radio uses include long range contact with air traffic control agencies, time and frequency standard broadcasts, Omega navigation station status reports, weather and storm warning reports, radiotelephone service for personal messages and ARINC operational control services for messages relating to flying operations.

High frequency radio transmissions are able to cover great distances, but are still subject to a decrease in intensity because of the spreading of the waves. This occurs both as the signal travels upward toward the ionosphere, and as the reflected signal returns to earth. By using a transmission process known as single sideband (SSB) we can increase the effective range of HF signals.

If you recall the chapter describing radio transmitter operation, amplitude modulation (AM) is a process where the selected frequency (the carrier) and two sidebands are generated and transmitted. The upper sideband is the sum of the carrier frequency and the voice frequency (intelligence). The lower sideband is the difference between the carrier frequency and the voice frequency. The two sidebands are mirror images of each other. Each of these sidebands contain all of the intelligence to be transmitted.

It takes about two-thirds of the transmitter's power just to transmit the carrier, yet the carrier wave does not contain any of the intelligence (message) to be communicated. By electronically eliminating the carrier wave and one of the sidebands, a SSB transmitter manages to pack all of its power into the remaining sideband. The result is a system that has all of the "talk power" of AM transmitters having many more times the output.

HF communications is much more effective over long distances than the usual aircraft VHF frequencies. The HF channels are, as we have discussed above, subject to considerable atmospheric interference. The satellite communications system

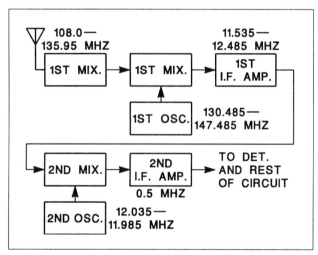

Figure 12-2. Block diagram of a double conversion superheterodyne receiver.

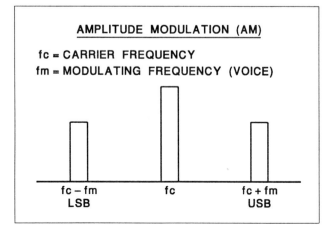

Figure 12-3. AM signal components.

SATCOM uses frequencies which are relatively static free, but are subject to the line-of-sight restrictions applying to VHF communications. This restriction is largely overcome by beaming the transmission to an orbiting satellite. The satellite amplifies the signal and retransmits it to the ground-based receiver. This system extends the range of a HF signal many miles.

III. SELCAL

SELCAL is short for Selective Calling and is a means by which the flight crew is relieved from continuously monitoring the aircraft radio receivers. The SELCAL decoder automatically listens for an assigned combination of tones indicating the particular aircraft on which the equipment is installed.

When a ground operator wishes to contact an aircraft equipped with SELCAL, they select the four tone code which has been assigned to that aircraft and transmits that code. The code is picked up by all receivers tuned to the frequency, but only the single unit which has that code assigned will respond. When the assigned tones are received, the SELCAL decoder activates an external alarm circuit to advise the flight crew of an incoming message.

IV. AUDIO INTEGRATING SYSTEMS

An audio system on board an aircraft may be as simple as a switching panel which allows you to select either of the two VHF transmitters, and direct the output to either the headphones or a speaker. On the other end of the scale, a transport category aircraft system may include:

1. *Flight Interphone*—for communications between flight deck crewmembers with each other or with ground stations.

2. *Service Interphone*—Provided intercommunication between cockpit, cabin crew and ground personnel.
3. *Passenger Address System*—Allows flight deck crewmembers and flight attendants to make announcement to the passengers.
4. *Passenger Entertainment System*—Allows the showing of movies and audio entertainment channels at passenger seats.
5. *Call System*—Used as a means for various crewmembers to gain the attention of other crewmembers, and to indicate that interphone communication is desired.
6. *Cockpit Voice Recorder*—Records cockpit audio on a 30-minute continuous loop tape.

V. EMERGENCY SYSTEMS

FAR Part 91.52 requires most aircraft in the general aviation fleet to be equipped with an emergency locator transmitter (ELT). This is a self-contained transmitter that starts transmitting automatically in the event of a crash. It transmits a down-sweeping tone on both 121.5 and 243.0 MHz, which are two of the universally used emergency frequencies for aviation. Both the FAA and all of the military flight facilities monitor these frequencies.

The ELT is usually mounted in the rear of the aircraft where it is least likely to be destroyed in the case of a crash, and the antenna is usually a thin flexible wire that is not likely to be broken off. Since the aircraft electrical power is often terminated by a crash, the ELT contains its own battery. The ELT battery must be replaced when the transmitter has a total of one hour of operation, or when the batteries reach fifty percent of the shelf life approved by the ELT manufacturer. The date of required replacement must be clearly marked on the outside of the ELT case and it also must be recorded in the aircraft maintenance records.

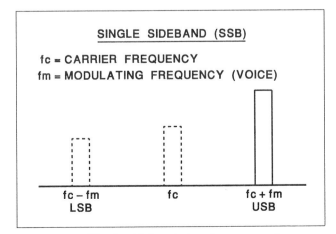

Figure 12-4. Single sideband signal components.

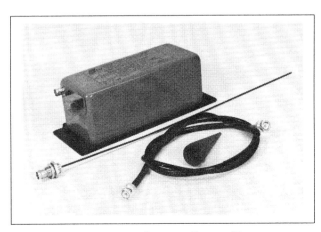

Figure 12-5. Emergency Locator Transmitter.

Chapter XIII

Navigation Systems

Instrument flight and radio navigation got its first real start in 1928 when the Guggenheim Foundation established the Full Flight Laboratory, headed by Jimmy Doolittle. Flight instruments as well as radio directional aids were developed that allowed Doolittle to prove, in 1929, that flight without reference to ground was indeed possible.

The first radio navigational instrument was a zero-center galvanometer that acted as a left-right indicator to show the pilot which side of a line to the transmitting antenna the nose of the airplane was pointing. By holding a heading that would keep the galvanometer needle centered, the pilot could fly over the ground station antenna.

Both navigation and communication equipment developed slowly through the 1930s, but lack of funds held back any serious development, and we entered World War II with very primitive equipment. The radio compass was the most highly developed radio navigational aid at the time. A direction-sensitive loop antenna received the signal from the ground and was rotated by a small motor and held in the position where the signal was at its weakest. The motor that rotated the loop also drove the pointer of the radio compass indicator, and this showed the pilot the relative bearing between the nose of the airplane and the radio transmitter antenna.

The four-course, low-frequency radio range was the chief navigation facility for the airways that crossed the United States in the 1930s. About every 50 to 150 miles along what was then called the colored airways, were low frequency transmitters with antenna systems that transmitted a four-course signal. In the north and south quadrants of the antenna field, the transmitter sent out the Morse code letter N (– ·), and in the east and west quadrants, the letter A (· –) was transmitted. The antenna was positioned so that these signals overlapped right down the center of the airway, and when the pilots flew where they overlapped, they heard both signals equally well, and the result was a continuous tone.

This equipment worked, but it required a great deal of skill on the part of the pilot, and the usable flight paths were limited to the courses to and from the station. Operating in the low frequency range, atmospheric static offered serious interference.

I. SHORT RANGE (VHF) NAVIGATION

After World War II, flying came into its own as a serious means of transportation, and the need for a reliable radio navigation system became apparent. In 1957, the Airways Modernization Board was created to lead in the development of a common system of radio navigation facilities. This system required the inherent accuracy of the system to be in the ground facilities, and it had to be usable by all aircraft. A user could buy equipment with a high degree of accuracy and reliability at a high cost, but the owner of a small airplane could buy inexpensive and simple equipment and use the same ground facilities. This system is what we know today as VOR, or Very high frequency, Omnidirectional Radio range. VOR has now been used for more than three decades, and its offshoots and elaborations fill our instrument panels with the equipment we will discuss in this chapter.

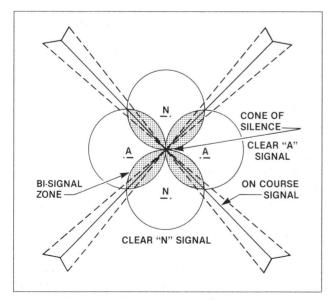

Figure 13-1. *One of the first practical types of radio navigation was the low frequency radio range. There were only four possible approaches to the radio facility.*

A. Omnirange Navigation

The very high frequency omnirange navigation system, was developed to provide an accurate, easy to operate navigation system in the United States. A VOR ground station transmits two signals in the very high frequency (VHF) range. One of these signals is amplitude-modulated at 30 Hz, and the other is frequency-modulated with a 10-kHz subcarrier and is fed into an antenna complex by a device known as a goniometer which rotates at 1,800 RPM. And as it does, it produces a signal which appears as 30-Hz AC when it is received.

The rotating signal and the fixed reference signal are in phase as the goniometer rotates through magnetic north, but as it continues to rotate, the signals get progressively out of phase until at magnetic south they are 180° apart. Anywhere along

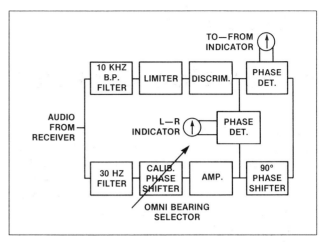

Figure 13-3. Block diagram of the VOR portion of an aircraft receiver.

a line magnetic east of the station, they are 90° out of phase. Magnetic south of the station, they are 180° out of phase, and at magnetic west, their phase difference will be 270°. They will be back in phase again at magnetic north.

Both of these signals are received on a horizontally polarized V-dipole antenna and are taken into the super-heterodyne receiver. After going through all of the required stages of mixing, converting into an intermediate frequency, amplifying, detecting and demodulating, the audio portion of the signal is fed into the VOR circuitry. A low-pass filter allows the amplitude modulated 30-Hz reference signal to enter one portion of the circuit, and the frequency modulated rotating signal passes through a 10kHz bandpass filter to enter another part of the circuit. This FM signal passes through a limiter and a discriminator and becomes 30-Hz AC.

The 30-Hz AM signal is fed into a calibrated phase shifter which is controlled by the pilot and is known to him as the omni bearing selector, the OBS. The dial of the OBS is calibrated in 360° increments and looks like the dial of a magnetic compass. If the airplane should be located magnetic north of the station with the phase shifter set on zero (or 360, this is the same), the two signals will be in phase. And when the aircraft is somewhere on a line from the station 120° clockwise from magnetic north, the two signals will be out of phase until the OBS is turned until it reads 120. Now the signals will be in phase and the course deviation indicator (CDI) needle will be in the center.

After the signal leaves the phase shifter, it is amplified to give it the same amplitude as the 30-Hz signal from the discriminator, and both of the 30-Hz signals are fed into a phase detector which drives

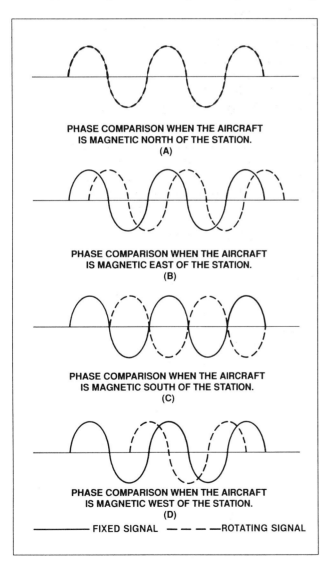

Figure 13-2. Phase comparison of signals in a VOR receiver.

the needle of the course deviation indicator, or left-right indicator as it is sometimes called.

There is another phase shifter and phase detector in parallel with the one that drives the course deviation indicator, and this one actuates the To-From indicator. The reference signal which has already had its phase shifted so it will agree with the rotating signal is again shifted, this time by exactly 90°, and it is again compared with the rotating signal. This shift tells whether the reference signal is leading or lagging the rotating signal, and its output causes the To-From indicator to indicate either to or from.

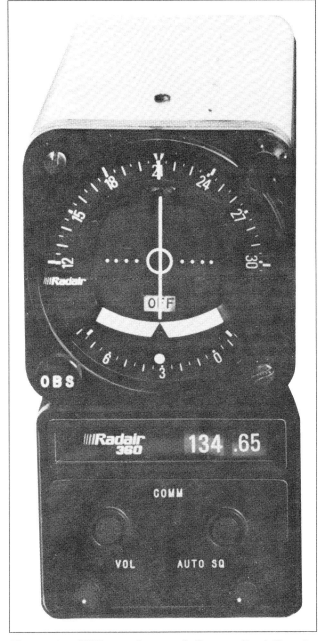

Figure 13-4. VOR receiver and Course Deviation Indicator (CDI).

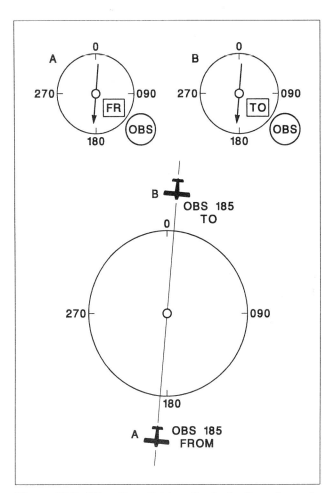

Figure 13-5. When the pilot has the indication shown in A, the aircraft will be south of the station. If the indication is that shown at B, the aircraft will be in position B.

Now for a real quick rundown on the way VOR is used: when the pilot tunes in the VOR station, the signal is received and fed into the VOR circuitry. The course deviation indicator will, in all probability, deflect off center, but by rotating the omni bearing selector, the pilot can center the needle. When it centers, and let's assume that it does with the OBS reading 185°, the To-From indicator will read either To or From, and for example, let's assume that it reads From. This means that the aircraft is on a line that corresponds to the 185° radial which is a line leading out of the VOR station 185° clockwise from magnetic north. Furthermore, it means that if the pilot will turn to a magnetic heading of 185°, he will go away from the station. If the To-From indicator had read To, it would have meant that the aircraft was on the 005° radial (the opposite side of the compass from the 185° radial). In that case, if the aircraft was turned to a magnetic heading of 185°, it would fly to the station.

B. DME/TACAN

The Tactical Air Navigation (TACAN) system was developed in the early 1950s to meet military requirements. It was superior to the VOR system in that in addition to bearing, it provided distance information. The weight and power requirements of the airborne equipment, and the fact that VOR stations were already in place made adoption of the TACAN system prohibitive for the civil fleet.

A compromise was reached in the late 1950s where VOR and TACAN ground facilities would be co-located forming a VORTAC navigation station usable by both military and civilian aircraft. The TACAN channels were paired with the VOR channels so that the TACAN could be tuned when selecting the VOR frequency. About this time, the first airborne DME (Distance Measuring Equipment) units were installed on airliners. In the early 1960s smaller and less costly DME equipment gave the general aviation fleet access to this additional navigation tool.

The principle of DME is quite simple, but only by the use of very complex electronic circuitry can the desired results be achieved. Figure 13-6 is a block diagram of the airborne and the ground equipment.

A signal, consisting of about 150 random spaced pairs of pulses per second, is generated and used to modulate the transmitter. These pulses are also fed into the search and track portion of the set where they are stored in a memory system until they are needed. The transmitter puts the modulation on a carrier of between 962 and 1024 MHz, or between 1151 and 1213 MHz.

The modulated carrier is fed to the preselector which acts as a switch connecting the antenna alternately with the transmitter and then with the receiver. When the preselector is connected to the transmitter, the modulated signal is transmitted from the aircraft. The ground station picks up the signal, and its preselector directs it into the receiver wide band amplifier and into a delay circuit where it is held for a specific period of time and released into the modulator. After modulation, the signal which uses a carrier of between 1025 and 1150 MHz is returned to the aircraft when the ground station preselector connects the transmitter to the antenna.

The airborne antenna picks up the signal, and the preselector directs it into the receiver. It is first amplified and then sent into the search and track circuits where the received signal is compared with the original that has been stored by the modulator. If the received signal does not coincide, the search circuit will continue to hunt for the proper pattern of pulses, and the indicator, during the search, will move to the end of its range and then back toward zero, and will repeat this until it finds the sequence it transmitted. When this particular pattern is found, the search and track circuit will lock onto it and track it. When the receiver locks onto the signal, the number of pulses sent out

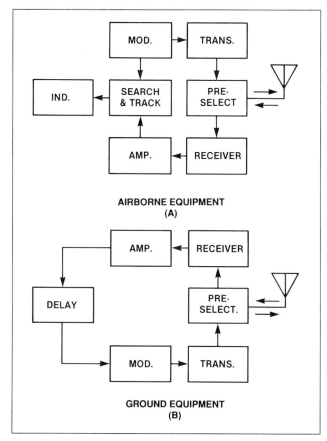

Figure 13-6. Block diagram of distance measuring equipment.

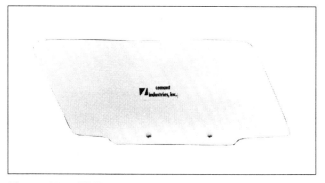

Figure 13-7. UHF blade antenna used for DME and for ATC radar transponder.

will drop from about 150 pairs per second to around 30 pairs per second, and the indicator will show the distance in nautical miles from the station.

The complex sequence of random codes generated by the airborne DME equipment is necessary since more than one aircraft may use the same ground station at any given time. The random code of the airborne equipment changes with each pulse transmitted and the receiver must identify that same code before using that information for display. There is virtually no possibility that two aircraft would continue to generate the same sequence of random codes while using the same ground DME station. This system therefore provides for accurate distance information regardless of the number of aircraft using the ground facilities.

The correlation between time and nautical miles is based on the assumption that a pulse of electrical energy travels through the air at a rate close to 161,900 nautical miles per second (186,000 statute miles per second). The time required for a pulse to leave the aircraft, reach the ground station, delay for a specific period of time, and return to the aircraft can be read directly in nautical miles to the station. This is naturally a slant-line distance and does not take into consideration the altitude of the aircraft above the station. Some DME display units incorporate information such as time to the station, and ground speed. This is accomplished through the use of a microprocessor which performs the necessary calculations.

DME is a pulse system, meaning that it transmits and receives pulses of UHF electrical energy, and it requires a short, vertically polarized antenna of either the blade or stub type.

C. Area Navigation

VOR allows us to fly from one radio facility to another using any of 360 radials, and this is a tremendous improvement over the older four-course radio range. But as the airways have become crowded, and aircraft need more radio fixes than are provided by the existing VORs, a new system of radio navigation called Area Navigation (RNAV) was developed.

The microprocessor which is the heart of all modern digital computers has made Area Navigation possible. The navigational computer unit (NCU) takes information from a VORTAC station and allows the pilot to program the computer so it can electronically move the radio facility to any location wanted, and use it as a waypoint or as a destination. For example, if the pilot wishes to fly directly from point A to point B (figure 13-8) but the nearest VORTAC facility is at point C, they can enter into the computer the radial from C on which B is located and its distance from C in nautical miles. Now, the course deviation indicator will accept this information as the location to which the pilot wishes to fly and it will direct them to it in much the same way as they would be directed by VOR.

The necessary information may be entered into the computer in the control-display unit, the CDU. Many CDUs use a keyboard-type of input, while others have knobs or switches. The information is displayed on the CDU in digital form and includes at least the frequency of the controlling VORTAC and the bearing and distance from the VORTAC to the waypoint that is selected. It is possible on most systems to program several waypoints into the system memory and use them as they are needed. Some RNAV CDUs also display the groundspeed in knots and time to the station in minutes.

The course deviation indicator (CDI) is similar to that used by the VOR, and it may be either a separate instrument or the needle on the horizon-

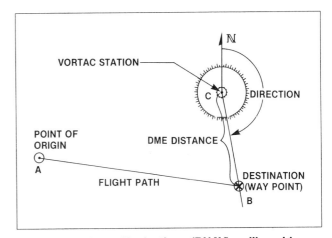

Figure 13-8. Area Navigation (RNAV) will guide an aircraft to a destination that does not have radio facilities. The VORTAC station may be electronically moved to a waypoint to which the RNAV will direct the pilot.

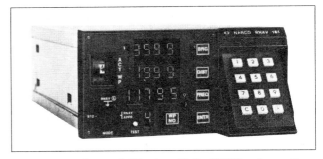

Figure 13-9. Control Display Unit (CDU) of an Area Navigation System.

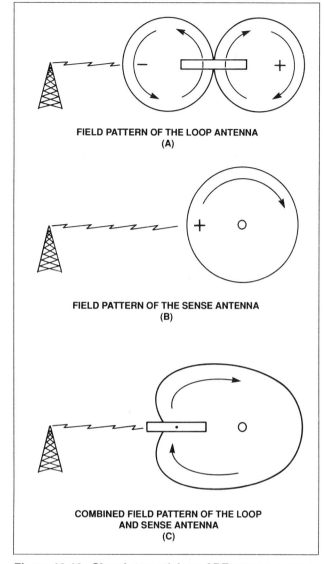

Figure 13-10. Signal strength in an ADF antenna system.

Figure 13-11. ADF receiver and indicator.

tal situation indicator. The basic difference between the RNAV display and that used for VOR is that with VOR, a needle deflection of one dot off center is an off-course condition of two and one-half degrees. The distance is very short near the station, but it increases with distance from the station. In the RNAV mode, a one-dot deflection of the needle is an off-course condition of one nautical mile, and this distance is constant regardless of the distance from the waypoint. In the approach mode, the sensitivity is amplified so that one dot represents only one-fourth of a nautical mile off course. This corresponds to the increased sensitivity of the VOR/LOC system when it is operating in the localizer mode.

II. LONG RANGE NAVIGATION SYSTEMS

A. Automatic Direction Finder

Because the VOR operates in the very high frequency range it is limited to line-of-sight operation. This results in reception problems making the VOR unusable at low altitudes, when high terrain such as mountains are between the aircraft and the ground station. For these conditions, and in some remote areas where VOR stations are not close enough to be usable, the Automatic Direction Finder (ADF) may be used as a handy backup.

ADF receivers usually operate in the frequency range from about 200 kHz up to around 1,600 kHz. This includes all of the low frequency non-directional radio beacons and compass locators, as well as the entire commercial broadcast band. In these frequency ranges the pilot is not limited to line-of-sight reception.

Loop antennas used with ADF systems are highly directional. It is possible to tell by the position of the loop the bearing of the station from the aircraft. ADF equipment uses a rotating loop and a stationary sense antenna to drive an indicator which shows the direction of the station relative to the nose of the aircraft.

The loop antenna has a signal strength such as we see in figure 13-10. When the current is maximum in the loop, the induced voltages in its two sides have opposite polarity. A second antenna, called a sense antenna, is used with the ADF system. The sense antenna is omnidirectional: that is, its signal strength is the same all of the way around. The function of this antenna is to add its omnidirectional signal to the directional signal of the loop and produce a field that is roughly heart shaped.

This is what is technically known as a cardoid pattern.

An ADF receiver normally has a function switch to select the mode of operation. If it is desired to receiver weather information or other voice communications, the mode selector is placed in the ANT position. In this mode, only the sense antenna is in use. This allows the signal to be received with equal strength regardless of its position relative to the aircraft. In the ADF position, both the loop and the sense antenna are in the circuit, and the indicator shows the relationship between the station being received and the nose of the aircraft. An ADF receiver and indicator are seen in figure 13-11.

We can trace the operation of an ADF using the block diagram in figure 13-12. The signal is received by both the loop and the sense antennas. The signal from the loop antenna is amplified and combined with the signal from the sense antenna. The resultant signal is mixed with a signal from the local oscillator to give us an intermediate frequency whose strength varies with the position of the loop antenna.

This IF voltage is further amplified, detected, and demodulated. The resulting audio frequency is again amplified and then fed into the speaker. A signal voltage is taken off just ahead of the AF amplifier, and is amplified enough to drive the loop motor which rotates the antenna to the null position. This is the point of lowest signal strength. The azimuth indicator is electrically driven from the loop drive motor, and indicates the relative position between the null position of the loop and the nose of the aircraft.

To use the ADF, the pilot selects the station to use and identifies it by its call signal. The loop drive circuit automatically rotates the loop until the signal is the weakest. This is the point of maximum sensitivity, and the azimuth indicator will show the pilot the direction of the station from the nose of the aircraft. To fly to the station, the pilot simply turns the aircraft until the indicator points to zero. Now, the station is directly ahead of the aircraft. If the pilot keeps the indicator reading zero, the aircraft will eventually arrive at the station.

Early ADF antennas were large open loops that collected ice and offered considerable wind resistance. As aircraft became faster the need for streamlining became greater. The loops were then made smaller and housed in streamlined plastic "footballs". The air-core loops were later replaced by loops using iron cores which were much more compact and could be mounted in smaller housings. More recently developed loops are fixed and are the ones used in most of the current production airplanes.

Two fixed coils, 90° to each other, are installed in either a flush or a streamlined housing. The signal from the transmitter is received by these coils and is carried into the receiver where two more coils produce identical fields. The fields in these coils produce a resultant field in a goniometer, or resolver, inside the receiver. This resultant field produces the directional signal.

The sense antenna used with ADF systems throughout the years has been a long wire. This wire usually extends from the top of the cabin to the vertical fin. Sense antennas of this type have been one of the weakest elements in the ADF system. They can be a source of precipitation static and are continually in danger of being carried away by a load of ice. The most recent development incorporates the sense antenna in the same unit as

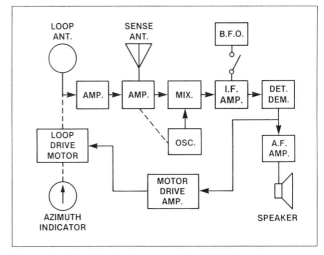

Figure 13-12. Block diagram of an ADF circuit.

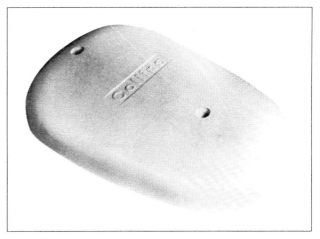

Figure 13-13. ADF loop antenna.

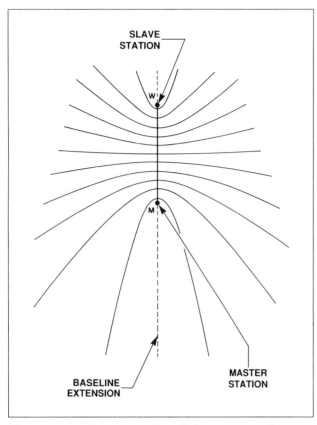

Figure 13-14. Representation of the master and secondary LORAN station time delay.

the fixed loop. Using this smaller antenna requires a great deal of amplification to compensate for its lower efficiency. Fortunately, obtaining this amplification is well within the realm of modern state-of-the-art electronics.

B. LORAN C

Long Range Navigation (LORAN) systems have gained popularity with pilots over the past several years for many reasons. LORAN systems offer accurate navigation, the ability to program your own waypoints and long distance reception to ground stations, all at a modest price compared to VOR or RNAV systems. LORAN receivers operate on a frequency of 100 kHz or within the low-frequency range. Radio transmissions at 100 kHz are ground waves and therefore follow the curvature of the earth and can be received greater distances than the VHF signals. This long distance characteristic of the loran systems provides accurate coverage of the United States and Canada with only nine ground stations.

The LORAN operation begins with choosing a master and two or more secondary ground stations. The master station transmits a series of pulsed signals. These signals are accurately sequenced with a specific time delay between the master and each secondary. This time delay is measured by the airborne receiver and the systems computer determines the aircraft location. At least two secondary and one master station must be used for the LORAN system to accurately determine the aircraft location. As shown in figure 13-14, the master and one slave can determine the aircraft location only along a hyperbolic line. The exact location along the first line must be determined by the intersection of a second hyperbolic line created by the master and another secondary. In some instances, more than one secondary is used to improve the systems accuracy.

The airborne equipment translates the time-delay signals into a latitude and longitude fix by using a map of the earth and the ground stations stored in the computers memory. This map can also store waypoints, airports and airways which may be of value to the pilot. The latitude and longitude fix can serve as a position fix to the pilot, however this information is somewhat difficult to navigate by. The loran system therefore provides information such as bearing and distance to a destination or way point, ground speed, and estimated time of arrival. Some systems provide a course deviation indicator (CDI) showing left-right heading information similar to that found on a VOR.

OMEGA STATIONS			
LETTER	NO.	LOCATION	LATITUDE	LONGITUDE
A	1	Aldra, Norway	66°25'N	13°08'E
B	2	Monrovia, Liberia	6°18'N	10°40'W
C	3	Haiku, HI, USA	21°24'N	157°50'W
D	4	La Mourne, ND, USA	46°22'N	98°20'W
E	5	La Reunion	20°58'S	55°17'W
F	6	Golfo Nuevo, Argentina	43°03'S	65°11'W
G	7	Australia	38°29'S	146°56'E
H	8	Tsushima, Japan	34°37'N	129°27'E
Each station transmits three basic frequencies: 10.2 kHz, 11.33 kHz, and 13.6 kHz. To prevent signal interference between stations, transmissions are timed such that only one station is transmitting a particular frequency at a time.				

Figure 13-15. Omega Stations.

C. OMEGA/VLF Navigation

The Omega Navigation System (ONS) is a long range navigation system used by aircraft and ships alike. While the concept was patented in 1923, it was not until 1968 that it was established that ONS was feasible and the setup of a worldwide network began.

Three time-multiplexed signals are transmitted omnidirectionally from each of eight stations strategically located around the world. In order to improve the reliability during times of less than ideal signal conditions, the airborne Omega receiver can also utilize the signal from any of seven VLF communication stations.

Each of the eight OMEGA stations transmits on four common frequencies and one frequency unique to that station. These signals are transmitted in a predetermined sequence and duration. Figure 13-17 shows the frequency, duration, and pattern for each of the eight stations.

All eight of the transmitters are phase-locked to a nearly absolute time standard. This standard is obtained by the use of atomic clocks at each of the station locations. This high degree of accuracy

VLF COMMUNICATION STATIONS				
NO. AND LOCATION	LATITUDE	LONGITUDE	FREQUENCY (kHz)	PWR (KW)
1—Maine	44°39'N	67°17'W	17.8	1026
2—Japan	34°58'N	137°01'E	17.4	48
3—Washington	48°12'N	121°55'W	18.6	124
4—Hawaii	21°26'N	158°09'W	23.4	588
5—Maryland	38°60'N	76°27'W	21.4	588
6—Australia	21°49'S	114°10'E	22.3	989
7—Great Britian	52°22'N	01°11'W	16.0	40
Each station transmits a specific frequency.				

Figure 13-16. VLF Communication Stations.

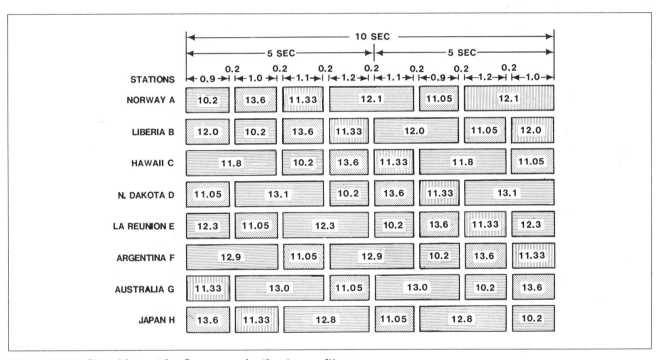

Figure 13-17. Signal format for Omega navigation transmitters.

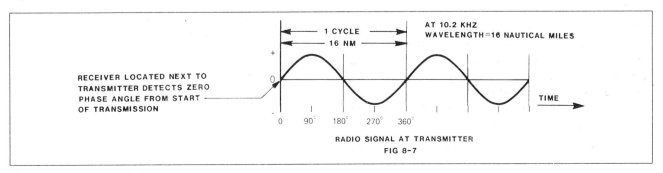

Figure 13-18. OMEGA signal at 10.2 kHz.

237

results in a maximum timing error between stations of 1 microsecond. This allows a maximum error in position fixing of 300m.

Figure 13-18 illustrates an Omega signal being transmitted on 10.2 kHz. The wavelength of one cycle at this frequency is 16 nautical miles. This means that the full cycle repeats itself each 16 miles. By measuring the phase angle of the received signal and computing information provided by the timing circuits, the distance from the station can be calculated.

Figure 13-19 shows the waveform of a signal coming from an OMEGA station. Each repetition of the wave form is called a lane. Circles have been drawn to illustrate the width of each lane from the transmitter site. To obtain usable navigational information, we must identify which lane we are in, and where in that lane we are in terms of degrees from true north.

If two stations are received the chart of their lanes would appear much like we see in figure 13-20. The receiver has determined that it is located in lane 4 of station A, and lane 6 of station B. This combination gives us two possible positions. These possibilities are shown in the illustration. This is known as RANGE RANGE or Rho Rho navigation.

By adding the signal from a third station we can eliminate one of the possibilities left with the Rho Rho setup. We can see in figure 13-21 that the receiver is still in lane 4 of station A, lane 6 of station B, and now we can see that it is in lane 3.4 of station C. This gives us three lines of position (LOPs) from each station. This type of navigation is referred to as RANGE RANGE RANGE (or Rho Rho Rho). Any combination of Omega and VLF stations may be used.

An even higher level of Omega Navigation may be obtained. It is known as hyperbolic. A description of this method would be too complex to address here. Hyperbolic navigation requires the use of three or more Omega stations. VLF station signals will not work here.

One last mode of navigation is possible with the Omega system. Should there be a loss of reliable signals from the ground stations, the receiver will take the last known position and plot a course using the magnetic heading and true airspeed information obtained from the aircraft's systems. This is known as dead reckoning mode.

The airborne equipment required for the Omega system consists of a Receiver Processor Unit (RPU), and Control/Display Unit (CDU), and the Antenna/Coupler Unit (ACU). The RPU is the heart of the systems. Omega broadcast signals from the ACU are processed together with other input sensors to give present position and guidance parameters as required. The CDU provides the interface between the flight crew and the Omega system.

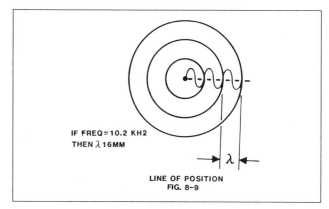

Figure 13-19. Each repetition of an Omega signal represents a lane.

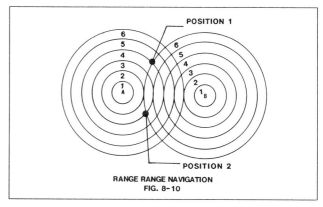

Figure 13-20. RANGE RANGE navigation.

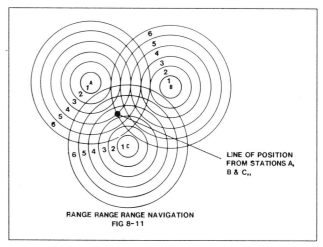

Figure 13-21. RANGE RANGE RANGE Navigation.

D. Inertial Navigation Systems

The main advantage of an inertial navigation system (INS) is that it requires no external radio signal to determine the aircraft position. The INS operates on the principals first introduced by Newton over 300 years ago. Newton's first, second and third laws of motion state: "A body continues in a state of rest, or uniform motion in a straight line, unless acted upon by an external force"; "The acceleration of a body is directly proportional to the sum of the forces acting on the body"; "For every action, there is an equal and opposite reaction".

From these laws, it can be seen that if an aircraft is parked at the terminal, it will remain at the terminal unless acted upon by another force, the engine thrust. It can also be seen that the acceleration of the aircraft in any direction is a function of all the forces which act on the aircraft. For example, the thrust, drag, lift, and various yaw, pitch and roll forces will accelerate the aircraft in different directions.

Newton's third law tells us that if we can measure the reaction force of an acceleration, we can determine the actual acceleration. The INS equipment measures the reaction forces of the various accelerations which act upon the aircraft. Through the INS computer, calculations are performed which determine the aircraft's velocity produced by the accelerations. The velocity of the aircraft can then be mathematically manipulated to determine displacement. In order for the INS system to operate, the initial starting location of the aircraft, latitude and longitude, must be programmed into the computer. From this initial reference, the INS equipment will constantly calculate and update the aircraft position simply by measuring the reaction force of the accelerations which act upon the aircraft.

In order to produce an accurate INS system, the acceleration forces of the aircraft must be carefully measured. Many of the reaction forces during flight can easily be felt by the passengers, such as the force which pushes people back into their seats during takeoff. The INS equipment uses an item called an accelerometer to measure even the slightest acceleration. The accelerometer (figure 13-22) can be thought of as a simple pendulum device which swings in the opposite direction of the acceleration.

Two accelerometers are needed for the system, one to measure North-South accelerations and one to measure East-West accelerations.

All accelerometers must incorporate some means of detecting the pendulum motion and some means of returning the pendulum to neutral position. Typically, a capacitive of inductive sensor can be used to detect the pendulum motion. The signal from this sensor is sent through an amplifier to the INS computer. The computer inputs this information and sends a signal back to a torque motor which restores the pendulum to its neutral position.

In order for the accelerometer to operate correctly in all flight attitudes, it must be kept level with the earths surface at all times. To do this, the accelerometer is mounted on a gimbal platform and held stable with gyroscopes. The gyroscopes sense the attitude changes of the aircraft and send signals to the gimbal motors through the computer. The gimbal motors move the gimbal platforms in the necessary manor to keep the accelerometer stable. Either rotating mass-type or laser gyroscopes can be used for sensing the aircraft attitude.

The INS equipment is often linked to the aircraft auto-flight system and the attitude, heading and weather displays. The communication between the various system components is typically done through a digital data bus system and discrete analog connections. The majority of the information is transferred through ARINC data bus systems. On the most modern aircraft, the position information calculated by the INS equipment is typically displayed to the flight crew through an electronic horizontal situation indicator. The flight crew typically controls the INS from a cockpit display called a control-display unit (CDU).

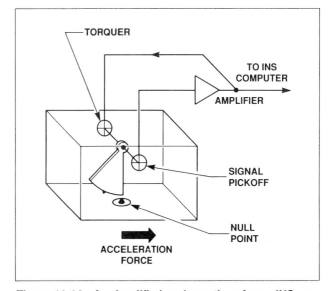

Figure 13-22. A simplified schematic of an INS accelerometer.

E. Global Positioning Satellite System

NAVSTAR Global Positioning System (GPS) is a military sponsored system providing worldwide navigational information, on land, on sea, and in the air. Development and testing of GPS took place between 1979 and 1985. The final phase—production and deployment—began in 1985 and is expected to be complete in 1993.

GPS consists of three major segments: Space, Control, and User. These segments are illustrated in figure 13-23. The space segment is the constellation of GPS satellites. When complete, the constellation will consist of 24 satellites (21 operational and three spares). The satellites operate in circular orbits 12,500 miles above the earth. They are arranged in six orbital planes, spaced about 55° apart. Their orbital period is slightly less than 12 hours. This arrangement is established so that at least four satellites will be visible from any point on the earth at any time.

The Control Segment includes a master control station in Colorado Springs, Colorado, and five monitor stations and three ground antennas located around the world. The ground stations continuously monitor all satellites in their view, and send the information received to the master control station. This allows the control station to transmit position updates to the satellites, as well as perform basic maintenance such as clock commands, power and attitude messages, new programming instructions, and regulate the gyroscopes used to stabilize the satellite.

The GPS User Segment consists of the antennas and receivers used by aircraft, ships and ground vehicles to determine their precise position and velocity. Accurate time-of-day data can also be received. Modern processor based electronics have made GPS receivers small enough to be held in one hand, and inexpensive enough to be available to nearly anyone who wonders, "Where in the world am I?".

Using GPS to identify the location of an aircraft would at first not appear to be much different than we discussed in the section on Omega navigation. We need to determine how far away from three or more stations we are. The difference with GPS is that the stations are constantly moving and because the signals travel through the atmosphere and the ionosphere, they don't travel at exactly the speed of light.

If we were able to receive only a single satellite, we could determine how far away from the satellite

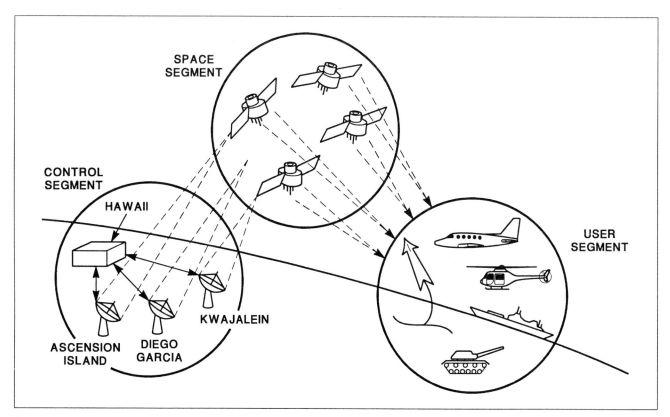

Figure 13-23. The GPS navigation system consists of three segments.

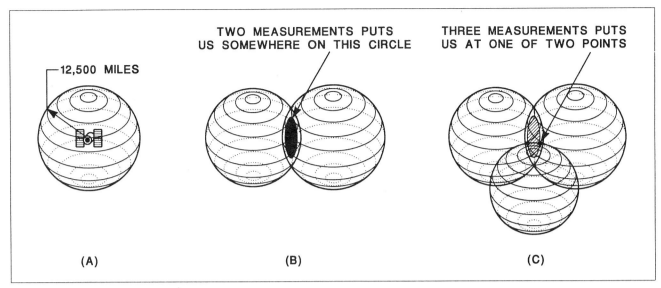

Figure 13-24. To accurately determine position using GPS four or more signals must be used.

we are by calculating the time it takes the signal transmitted by the satellite to reach us. Using the position information transmitted by the satellite the receiver could calculate the position of the satellite. This information would only be able to tell us that we are somewhere on a sphere with a radius of the distance calculated (figure 13-24(A)).

By receiving two signals and determining our distance from both satellites, we can see that our location falls within the circle described by the intersection of the two spheres (figure 13-24(B)).

Add a third satellite and the combination of the measurements will narrow down the possibilities to one of two choices (figure 13-24(C)). By determining that one position is unreasonable, we could be rather certain about our location. To make an exact determination, we must use a fourth signal.

Under ideal conditions the addition of a fourth signal would complete the problem, and our exact position would be known. If the fourth line of position does not pass exactly through the other three the receiver will detect the error. The receiver assumes that because the fourth line did not pass through the intersection of the other three that there is an error in its internal clock. The receiver will run a program that adjusts its internal clock until all four lines of position intersect at the same point. This is known as correcting clock bias.

Now that we known how our position is determined, how accurate is the information? Remember that the GPS is a military sponsored system, and they would really rather not see it work too well for non-military or hostile-military users. To accomplish this, the DoD designed the S/A (selective availability) mode of operation. The S/A mode intentionally degrades the signal accuracy. With the S/A mode off, GPS can obtain accuracies of 60-200 ft. When the S/A is activated, accuracies may be around 300 ft.

Differential GPS is a system where GPS information is processes along with information provide by a fixed ground station. This combination may be able to reliable provide accuracies of 2-5m. This type of operation is currently being used by the Coast Guard. The DGPS signals are transmitted piggyback on the Coast Guard's existing network of radio-beacon coastal stations. The possibilities exist to transfer this technology to signals broadcast on existing aviation NDB stations, thus bringing this type of GPS accuracy to airport terminal operations.

Airborne equipment for GPS navigation will be similar to that shown in figure 13-25. This illustration shows the Bendix/King KLN90 system components. This system consists of a panel mounted GPS sensor/navigation computer, a database cartridge, and an antenna. The illustrations shows other inputs which may be connected to the navigation computer.

The database cartridge is an electronic memory containing a vast amount of information on navaids, intersections, special use airspace, and other items of value to the pilot. The database may be updated by connecting it to a computer and using diskettes furnished by the equipment manufacturer, or by removing the obsolete cartridge and replacing it with the current one.

The antenna used with the GPS system is a "patch" type. This antenna is designed to always be mounted on top of the airplane (remember, you are receiving satellite stations).

III. MULTISENSOR NAVIGATION SYSTEMS

With the variety of short and long range navigational systems available, it would only seem logical that deciding which one to use, when, and being constantly on the watch for failures or inaccurate information would be quite a task. The multisensor navigation systems use microprocessor technology to manage this task.

The advantage of an integrated multisensor navigation management system is that the system can accommodate whatever sensors are required for any particular operation, and that the system can transition among the sensors automatically, while using a single control display unit.

Multisensor navigation systems may be connected to any combination of the following inputs:

VOR/DME	LORAN C
OMEGA/VLF	GPS
TACAN	Compass System
Inertial Navigation (INS)	Air Data Computer

Input/Output provisions may be made in the system to accommodate interfaces with such parameters as Altitude, Airspeed, Radio Magnetic Indicator (RMI), Compass, EFIS, Radar MAP overlay, and Flight Control Systems. This level of integration makes possible navigational functions such as long-range great circle navigation, approach capability, and vertical navigation.

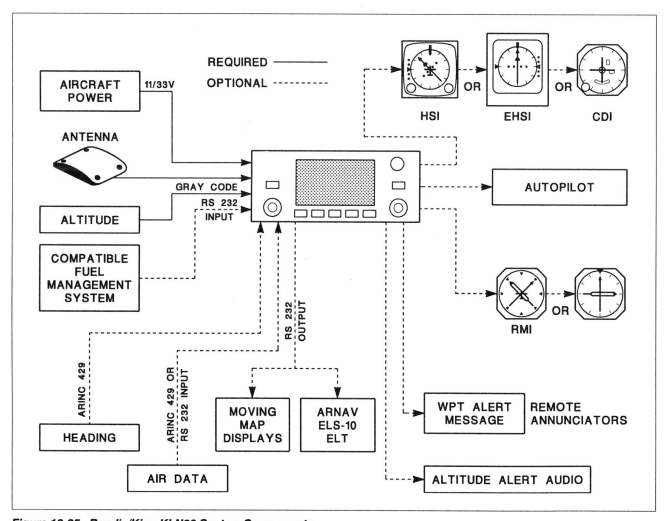

Figure 13-25. Bendix/King KLN90 System Components.

Chapter XIV

Terminal Navigation and Collision Avoidance Systems

I. INSTRUMENT LANDING SYSTEM

It is not enough to be able to fly from point A to point B if we are unable to land at the airport when we get there. To facilitate landing in conditions of low visibility, the instrument landing system (ILS) has been devised. This is, as its name implies, a complete system consisting of four basic components: the localizer, the marker beacons, the glide slope and the approach lights. The localizer directs the pilot down the center line of the runway, and the marker beacons locate the aircraft's position along the extended runway center line to tell how far away they are from touchdown. The glide slope is a slanting radio beam that allows the aircraft to descend at the proper angle to reach the minimum altitude the proper distance from the end of the runway. The complete ILS package includes the approach lights, but since they are not airborne components, we will not discuss them here.

A. Localizer

When the VOR/Localizer indicator is used for VOR, it shows the pilot whether or not they are on the course line selected. Each dot the needle moves over is an indication of being 2-½° off of the course. In other words, when the needle swings over full scale, the aircraft is ten degrees off course. This tolerance is acceptable when flying along the airways, but for a final approach, we must have a signal with a much closer tolerance. We get this with the localizer which provides the steering information for an instrument approach. The localizer uses the same instrument and receiver as is used for the VOR, but the localizer portion of the circuitry is quite different from that used for VOR, and the sensitivity of the indicator is four times as high. A full-scale deflection of the indicator needle in the localizer mode indicates a two and one-half degrees deviation from the runway center line.

The ground facilities for a localizer are altogether different from those used for VOR, but both systems operate in the same frequency range. Terminal VORs, the low-powered transmitters usually located on an airport, transmit on frequencies between 108.2 and 111.8 MHz, and all are in increments of even ¹⁄₁₀ MHz. The localizer frequencies are in the same range, but operate from 108.1 to 111.9 MHz, and all of them are frequencies in odd tenths

Figure 14-1. Indicator for both VOR and ILS.

243

of a megahertz increments. The Morse code identifiers used for localizer and VORs are usually the same, but the codes for the localizer are prefaced by the code letter "I", which is two dots in the Morse code.

Rather than being a phase comparison system like VOR, the localizer is a voltage comparison system. The localizer antennas are located at the far end of the runway they serve, and they radiate two carriers on the same frequency. One carrier is modulated with a 90-Hz signal and the other with a 150-Hz signal. The one modulated with the 150-Hz tone transmits its signal to the right side of the runway as viewed from its approach end, and the one with 90-Hz modulation produces a similar pattern on the left side of the runway.

Since there is only one carrier frequency, the receiver picks up the signal and processes it through the RF, IF, and AF amplifiers. In the AF stages, two filters separate the 90-Hz signal from the 150-Hz, and convert both of them into DC with opposing polarities. The two DC voltages are then fed into the course deviation indicator where they are compared. Rather than being calibrated in terms of Left and Right, the indicator has two colored segments at the end of the indicator needle. The segment on the left side of the dial is blue while that on the right side is yellow.

These colors indicate only the side of the runway on which the aircraft is located. Looking at figure 14-3, we see that if the aircraft is in position A, it is on the yellow side of the runway and will be receiving the 90-Hz signal stronger than the 150-Hz signal, and the needle will be driven to the right side of the dial. This may sound backward, but there is logic to it. If the aircraft is on the right side of the runway, in position B, the 150-Hz signal will be the stronger and when it is converted into DC, it drives the needle to the left side of the dial, into the blue segment. When the aircraft is right down the center line of the runway, it will receive both the 90- and the 150-Hz signals with equal intensity and the resulting DC voltages will cancel each other and the needle will stay in the center of the dial.

The reason for the blue and yellow segments identifying the sides of the runway rather than left or right is that the localizer does not have the capability of reverse sensing, the way VOR has. When using VOR, if the pilot is flying toward the station with the OBS set so the To-From indicator reads from, they would correct an off-course condition by turning away from the needle. But, by turning the OBS 180°, the To-From indicator will indicate To, and an off course correction is made by turning toward the needle, a much more natural response.

The localizer circuitry does not have this capability but it is set up so when the pilot is making a front-course approach, that is, when they are approaching the airport from over the outer marker, the instrument will have normal sensing. When the aircraft is off-course to the right, the needle will be in the blue segment, and a turn toward the needle will bring the aircraft back onto the localizer center line. When the pilot makes a back-course approach, that is, when approaching the runway from the end opposite that used for the normal approach, and drifts off course into the blue segment it will be necessary to turn away from the needle to get back onto the center line.

B. Glide Slope

The correct descent angle for the approach is established by following the glide slope. Like the localizer, the glide slope transmits two signals on the same frequency, one modulated with 90-Hz and

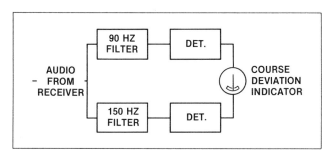

Figure 14-2. The course deviation indicator measures the relative strength of the two modulations to provide localizer information.

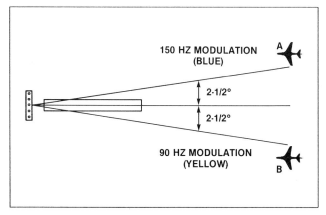

Figure 14-3. The left side of the runway, from the normal approach end, is in the 90-Hz modulated field, while the right side is in the 150-Hz modulated field.

one with 150-Hz. The carrier for the glide slope is in the UHF band between 329.3 and 335.0 MHz. Instead of requiring the pilot to tune the glide slope receiver, each ILS has the glide slope paired with its localizer so that when the pilot tunes the localizer frequency, the glide slope receiver is automatically tuned to its proper frequency.

The antenna pattern for the glide slope is positioned in such a way that the 150-Hz signal is transmitted below the 90-Hz signal, and they overlap by about 1-½°. This overlap occurs at a nominal 3° above the horizontal.

As the pilot makes an ILS approach, at the altitude specified for the approach, and while outside of the outer marker, the glide slope indicator needle will point upward, indicating that the glide slope is above. At the outer marker, the pilot should intercept the glide slope and the needle should come to center. The pilot then begins the descent and keeps the needle centered. If the aircraft should go above the prescribed altitude, the needle will point downward, and any time the aircraft sinks below the glide slope, the needle will point upward.

The circuitry for the glide slope is essentially the same as that used for the localizer. A twenty channel crystal-controlled receiver is tuned with the same control used to select the localizer frequency. Ninety- and 150-Hz filters select the modulation being received and convert this into DC. The DC from the two filters is then compared and the output drives the needle either up or down.

The null, or the point of equal signal, occurs not only on the correct glide slope, but because of the reflection of the signal from the earth, it will produce several nulls, or false on-course signals above the true slope. There should be no danger from these false glide slopes, because they are all above and are steeper than the true slope. On a properly executed ILS approach, the first time the glide slope needle centers will be the time the aircraft passes over the outer marker at the altitude specified on the approach plate.

The antenna for the glide slope is a UHF dipole, and since the length of the antenna varies inversely as the frequency, this antenna is much shorter than the VHF dipole used by the localizer and the VOR. Some glide slope antennas are mounted inside the cabin, some are above or below the fuselage or in the nose, and some are mounted on the inside of the windshield behind the sun visors. In some cases the glide slope antenna is combined into one unit with the VOR antenna.

C. Marker Beacons

Compass locators are, as the name implies, used to locate the markers, but for the actual approach, a more precise method of position location is needed, and this is supplied by the marker beacons.

A marker beacon uses a single-frequency transmitter with a 75-MHz carrier frequency, and a power of two to three watts. The signal from the marker beacon is transmitted by a highly directional antenna which directs its signal straight up in a fan-shaped pattern. The carrier transmitted by the outer marker is modulated with a series of 400-Hz dashes at the rate of two dashes per second. The middle marker is modulated with a series of 1,300-Hz alternating dots and dashes, and the airways, or Z-marker, is modulated with a series of 3,000-Hz dots. These 3,000-Hz markers may be placed at the point where the decision height should be reached during an ILS approach, or they may be used to indicate the point on a back-course approach where the descent should begin. Those markers used on a back-course are modulated with a series of two-dot groups.

The receiver for the marker beacon is a fixed frequency superheterodyne. The signal is received and is carried through the RF, IF and AF stages, but instead of going to the speaker, it is sent through a series of filters. If the signal being received is

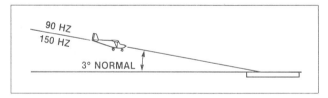

Figure 14-4. Antenna pattern for a glide slope projects the glide slope upward approximately 3° from the surface.

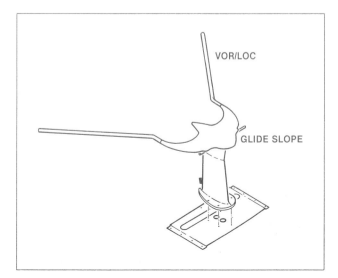

Figure 14-5. Glide slope antenna.

modulated with 400-Hz, it passes through the 400-Hz filter and causes a blue light on the instrument panel to light up. If the aircraft is over the middle marker, the 1,300-Hz signal passes through the 1,300-Hz filter and turns on the amber light. Any time the 3,000-Hz signal is received, it causes the white light on the panel to light up.

The fixed frequency of the marker beacon receiver allows the use of a simple tuned antenna, usually mounted on the belly of the aircraft.

II. MICROWAVE LANDING SYSTEM

While the Instrument Landing System (ILS) has been the world's standard precision approach system since 1948, it does have limitations. Expansion of the ILS system is restricted by the following factors:

1. The VHF/UHF frequencies used for ILS have technical limitations for this function. At these frequencies, antennas cannot be built large enough to be completely independent of ground effects. This results in problems such as costly site preparation, interference, multipath reflections and weather effects.

2. The ILS has a maximum of 40 channels available. This limits the number of approaches available in some areas.

3. The ILS provides only a single, narrow ± 3° to ±6° course and a single glide slope limited to 4° on the upper side.

A new landing system, named for the frequency range in which it operates, has been developed to overcome these challenges. The Microwave Landing System (MLS) operates at 5031 to 5090.7 Mhz. This is 50 times the ILS localizer frequency and 16 times the ILS glide slope frequency. The first MLS system Instrument Flight Rules (IFR) operations were authorized in Valdez, Alaska on February 28, 1984.

MLS offers a number of distinct advantages over the ILS system. They include:

1. Smaller antennas may be used, saving cost in site preparation.
2. MLS is less susceptible to interference of various types.
3. MLS is less sensitive to the effects of weather and terrain.
4. 200 channels are available, compared to the 40 maximum ILS channels.
5. Small antennas permit the use of scanning beam techniques which minimize multipath reflection problems.

The basic components of the MLS system are the approach azimuth facility, approach elevation facility, and DME.

The azimuth facility and the DME are collocated at the rollout end of the runway, about 1,000 ft. beyond the end of the runway and on an extended

Figure 14-6. Marker beacon lights in the instrument panel.

Figure 14-7. Marker beacon antennas.

center line. The elevation azimuth facility is located to one side of the runway between 400 and 1,000 ft. from the threshold.

The azimuth antenna will transmit a signal with a 2° beam that scans ± 60°. The elevation antenna uses a beam width of 1.5° and scans from the surface to + 20°. The relative area of coverage for both the MLS and ILS systems may be seen in figure 14-8.

An aircraft flying anywhere in the MLS coverage area has the necessary azimuth, elevation, and distance information to accurately derive its position. This characteristic of MLS makes it possible for aircraft to fly curved or descending approaches. The old straight-line ILS approaches are no longer necessary.

The airborne MLS equipment requires a receiver and antenna separate from current ILS equipment. MLS airborne equipment solves navigational equations by timing the intervals between the sweeps of the azimuth and elevation signals. Information from the MLS may be switched into standard instruments and the display will be identical to ILS. Additionally the MLS information may be integrated into advanced display and autopilot systems.

III. RADAR TRANSPONDERS

With increased air traffic and higher aircraft speeds, traffic separation and positive identification of aircraft by controllers is a necessity.

A. 4096 Code Transponders

Air traffic control radar on the ground sends out pulses of extremely high-frequency, high-power electrical energy that radiate outward into space and do not bounce back from the ionized layer of the atmosphere that surrounds the earth. But, when these pulses strike a metal object such as an aircraft, they do bounce back. The energy that returns to the radar is extremely weak, in the nature of microwatts. After the proper amplification in the receiver, this signal is used to leave a trace on the phosphorescent radar scope being watched by the air traffic controller.

The controller must sort out the targets on the scope, and aircraft not equipped with a transponder must be identified by having the pilots make the appropriate radar-identifying turns, and once the aircraft is identified, a small plastic "shrimp boat" is placed over the dot as it progresses across the scope. The ATC transponder eliminates the need for the radar identification turns by instantaneously identifying a particular aircraft on the controllers scope. A mode "C" transponder coupled with an encoding altimeter will also display the altitude of the aircraft on the radar scope.

A modern air traffic control radar system uses a special interrogator antenna attached to the primary radar antenna and rotates with it. The interrogator sends out a pulse of energy on a frequency of 1030 MHz. The aircraft receives this interrogation and transmits a coded response on a frequency of 1090 MHz. This response is received on the ground by the primary radar, and instead of the aircraft appearing on the scope as a small dot, it appears as a bright signal or double slash because of the energy transmitted by the aircraft transponder.

The pilot selects the four-digit code requested by the controller, and all of the aircraft flying in

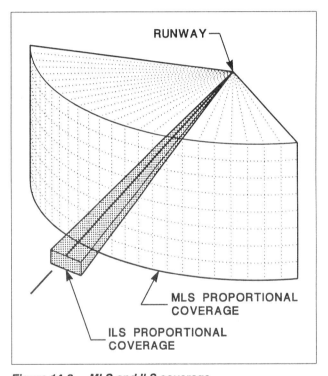

Figure 14-8. MLS and ILS coverage.

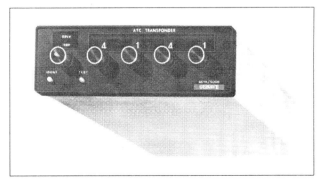

Figure 14-9. ATC transponder.

a particular level of the airspace or operating on a certain type of clearance can be quickly sorted out by the controller. Much safer flight can thus be maintained along our crowded airways and in the busy terminal areas.

In order to positively identify an aircraft, the controller may ask the pilot to "squawk ident". The pilot responds by momentarily pressing the ident button on the control head and, for a period of about twenty seconds, a special identification pulse is sent out each time the transponder is interrogated. An ident signal causes the return on the controller's scope to be filled in between the two slashes and immediately and positively identifies the aircraft.

If a pilot encounters an emergency situation and wants to notify the controller of the problem, they select code 7700. When the ground station receives this reply to its interrogation, both slashes become wider and brighter and the controller can take the necessary action to aid the pilot in distress.

The operation of the transponder is similar to that of the DME previously described. Follow figure 14-10 and see that the interrogation signal on 1030 MHz is received by the antenna, passed through the preselector, then is amplified by the video amplifier, and is detected. It then passes into the decoder which is set to respond only to the interrogation pulse selected by the pilot's control. The output of the decoder is fed into a keyer which produces a pulse, and this in turn causes the encoder to produce a series of pulses appropriate to the code selected. The transponder is capable of replying to the ground interrogator in any of 4096 codes. These transponder pulses modulate the 1090-MHz carrier, which is amplified in the RF amplifier. The preselector then directs this energy into the antenna and prevents any of this signal from entering the receiver portion of the circuit.

When the transponder is operating in mode A, it provides identification information only, but when the pilot selects mode C, the interrogation is answered by a code produced in the encoding altimeter and responds with information that produces a readout on the controller's scope showing the altitude of the aircraft in 100 ft. increments. On many corporate or commercial-type aircraft, the altitude information is sent to the ATC transponder from the central air data computer.

The only indication the pilot has of the transponder operating is the winking light on the face of the control head. This light blinks each time the transponder responds to an interrogation from the ground radar. Installation of the transponder is similar to that of DME. Some small transponders fit into the instrument panel, and others have only the control head on the panel, and the actual unit itself is remotely located in the avionics equipment rack. The antenna is a short blade or stub and is located on the belly of the aircraft as far as practical from any other antenna, and in a position that will not be shielded by the landing gear when it is extended. Some transponder installations are interconnected with the DME system by a suppressor bus. This unit prevents the DME and transponder from transmitting simultaneously.

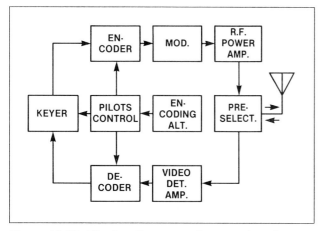

Figure 14-10. Block diagram of a radar beacon transponder.

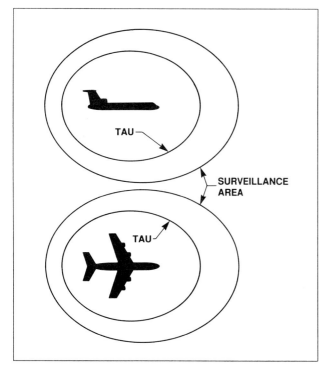

Figure 14-11. TCAS surveillance area.

B. Mode S Transponders

The latest development in transponder technology is the Mode Select transponder. Each Mode S equipped aircraft is assigned a unique address code. Ground radar installations are able to selectively interrogate specific aircraft using its unique address code. This technology offers two immediate applications:

1. It provides Air Traffic Control with improved aircraft surveillance and reporting accuracy.
2. Mode S reduces interference in identity and altitude reporting.

C. Traffic Alert and Collision Avoidance System

Traffic Alert and Collision Avoidance System (TCAS) is a system designed to alert the pilot that another aircraft is in the vicinity and that there is a threat for a midair collision. There are three types of TCAS systems in use or currently under development. The TCAS I system provides a warning to the flight crew of a potential collision with another aircraft within the surveillance area. TCAS I provides no collision avoidance information. It simply tells the pilot to take extra precautions since there is another aircraft close to his flight path. TCAS II provides the warning associated with TCAS I, and it also offers the flight crew vertical maneuvers to avoid a crash. TCAS III, currently under development, will offer a warning, and both vertical and horizontal maneuvers for collision avoidance.

The TCAS system is designed to provide airborne separation and/or collision avoidance. To do this, a surveillance area is defined around the aircraft as shown in figure 14-11. The TCAS equipped aircraft will identify other aircraft from their transponder signal. The TCAS transmitter power and receiver sensitivity will determine the surveillance envelope for your aircraft.

There is also an area known as TAU which defines the boundaries of the minimum time the flight crew can respond to and avoid a collision. Through a very complex set of calculations, the TCAS system communicates with other transponder-equipped aircraft within the surveillance area and determines if a collision is likely. If conditions are right for a potential collision, the TCAS equipment will alert the flight crew and offer avoidance maneuvers on TCAS II & III systems.

A typical TCAS system employs both top and bottom mounted antennas. An ATC/TCAS II control computer, a TCAS II receivertransmitter, and mode S transponder located in the electrical equipment bay. The TCAS control panel and the display unit are located so they are easily accessible and visible to the flight crew.

D. Datalink Communications

The technology employed in Mode S transponders makes possible the rapid transmission of compressed bursts of communications. This may be used to enable on-board processors to coordinate collision avoidance information between TCAS-equipped aircraft.

Operating much like a computer modem, the Mode S datalink makes possible the exchange of other information. This may consist of weather forecasts, safety advisories and terminal data.

Chapter XV

Weather Warning Systems

I. WEATHER RADAR

Airborne weather radar uses a pulse-echo system to detect potentially hazardous weather which may threaten flight safety. Through the use of weather radar, a pilot can visually identify water vapor, or clouds, and determine their severity over 100 miles away. This allows ample time to reroute a flight to provide passenger comfort and safety.

A. System Operation

The name radar is made up of the two functions of the equipment: Radio Detection and Ranging. The basic operating principle of weather radar is the generation of pulses of extremely high-frequency energy in the range between 9,000 and 10,000 MHz. These pulses are transmitted with a peak power of usually between 5 and 10 kilowatts. Because of their extremely high frequency, these pulses are carried to the antenna through wave guides rather than through wires. From the antenna, they are radiated into space in a carefully shaped beam formed by the reflector in which the antenna is mounted.

The pulses of energy travel until they hit either a solid object or liquid water. They then bounce back and are picked up by the antenna from which they were transmitted. The return pulses have an extremely low level of power and are amplified before they are directed into the indicator. The indicator for weather radar is a cathode ray tube (CRT), similar to those used in television sets and oscilloscopes. Monochrome indicators of the older analog-type radar paint an image of the return signal CRT. Electrons cause the material on the surface of the tube to phosphoresce, or glow each time the antenna sweeps. You can see the sweep rotating or oscillating over the face of the tube. By the time the glow starts to disappear on one sweep, another sweep comes by to re-excite it. If moisture is detected by the radar receiver, the electrons emitted toward the surface of the CRT will create an image of the storm, displaying both its distance and location from the aircraft.

The newer digital displays use a horizontal sweep of the electron beam to display the radars image. The horizontal sweep method uses deflection plates to direct the electrons toward the phosphorous sur-

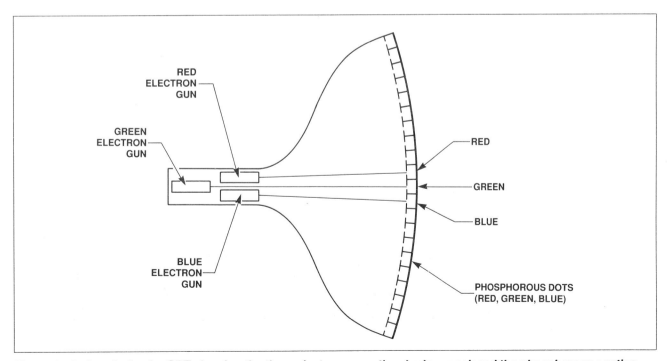

Figure 15-1. A typical color CRT showing the three electron guns, the shadow mask and the phosphorous coating.

face of the CRT. The electrons strike the phosphorous material starting in the upper left side of the tube, they move horizontally across the screen until they reach the upper right corner. The electron beam then moves back to the left side of the screen to begin another horizontal sweep across the CRT. This sweep sequence continues well over 1,000 times a second until the entire CRT screen has been covered. The process repeats and the CRT screen appears as though it has one continuous image painted on the screen.

Digital radar has the big advantage over analog radar in that its display can offer shades of images on a monochrome set, and different colors can be used to indicate different degrees of severity on color radar. A green return indicates the minimum rainfall (1-4mm per hour), yellow areas have a rainfall of between 4 and 12mm per hour, and red areas have heavy rainfall, greater than 12mm per hour.

B. Color Radar

A modern color CRT display works on the same principle as described above for the horizontal sweep monochrome display. However, there are three separate electron guns in the color display: blue, red and green. The electron guns themselves do not actually have a color, they are simply aimed at a specific area of the CRT phosphorous material. The red electron gun is aimed at the red phosphorous dots, blue at the blue dots, and green at the green phosphorous dots. As shown in figure 15-1, each horizontal line of the phosphorous contains hundreds of blue, red and green dots, a shadow mask helps to direct the electron beams onto there specified color dot. A full range of colors can be achieved by illuminating the various color dots in the correct location and with the correct intensity.

Basic weather radar has the limitation that its signal is reflected only by the liquid water inside the storm, and it cannot detect water vapor, lightn-

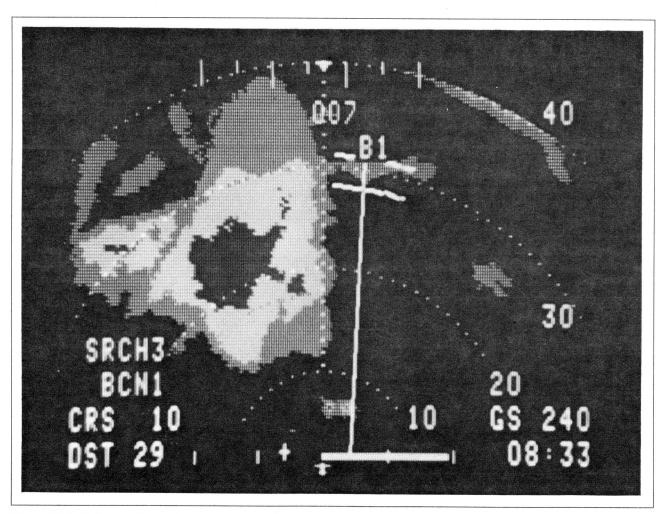

Figure 15-2. Color weather radar.

ing or wind shear, which are of extreme importance when picking your way through a frontal system. This limitation causes the effectiveness of weather radar to be pretty much determined by the experience and knowledge of the pilot using it.

C. Doppler Radar

Doppler radar has recently been introduced to help the pilot detect turbulence within a storm system. Since turbulence is often present even when moisture of the storm is very low, the Doppler radar measures the actual shifting of the moisture to detect the amount of turbulence. Doppler radar measures the frequency shift, or Doppler shift, of the reflected signal to determine the amount of turbulence percent. The color Doppler radar displays the turbulence, or wind shear, as the color magenta.

D. Radar Components and Operation

The complete weather radar installation consists of basically four subassemblies: the indicator, the receiver/transmitter (R-T), the antenna in its radome, and the wave guides that connect the R-T to the antenna. Each of these assemblies must be carefully inspected on each routine maintenance inspection for condition and security of mounting. When maintenance must be performed on the radar, it is preferable to do as much as possible in the aircraft, as the entire system operates as a unit, and it is quite possible that any one of the major subassemblies could function properly on the test bench by itself, and yet fail to operate as it should when connected into the system.

Since the energy transmitted by a radar is both a very high frequency and power level, the final waveform is sent to the antenna through a hollow tube called a wave guide. The wave guide is made of aluminum to a specific rectangular dimension and shaped to connect the antenna to the receiver transmitter unit. The wave guide connects to the antenna and directs the incoming and outgoing signals to and from the antenna. Two different antennas are currently used for radar transmission, the concave antenna and the flat plate, or phased array, antenna. Both antennas direct the radar signal into a narrow directional beam which can be pointed in a specific direction or scanned across the horizon. The flat plate collector is more efficient since it produces less "noise" to the sides of the transmitted signal.

The antenna of a radar system is housed inside a radome that must be not only structurally strong enough to withstand all of the flight loads imposed on it, but it must also be electrically transparent. That is, it must not reflect nor attenuate any of the pulses of electrical energy being transmitted nor received by the antenna. The repair and painting of radomes is therefore very critical and one should always consult the correct manufacturer's service information prior to maintenance.

Weather radar is a pulse electronic system and, as such, it requires special knowledge and equipment to service it. When it is operated on the ground, it must be turned on to the Standby condition to warm up the circuits. The radar must not be turned ON when there are people or buildings within 100 yd. of the antenna sweep. And when on the ground, the antenna must be tilted to its full UP position. It is extremely important that radar never be operated when the aircraft is being fueled or defueled. The pulses of energy transmitted from the radar antenna are strong enough that they can seriously injure a person struck by them, and they can be reflected from a nearby building with enough energy to ruin the receiver circuitry.

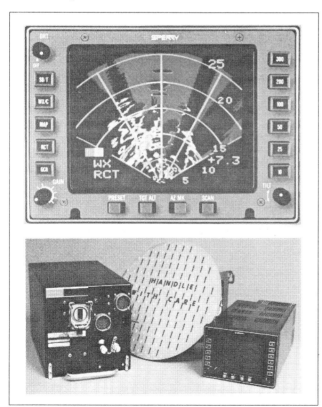

Figure 15-3.
(A) A typical weather radar display/control unit.
(B) The three units of a typical color radar system, from left to right: the receiver/transmitter, the antenna, and the display/control unit.

E. Radar Frequency Ranges

There are two frequency ranges used by aviation aircraft radar: the X-band with an operating frequency of 9,333 gigahertz, and the C-band with a frequency of 5.44 gigahertz. At these frequencies, the energy travels on the surface of the conductor and by enclosing it in hollow wave guides, the maximum amount of energy can be conducted. It is important that these wave guides not be dented nor kinked, and all connections must be tight. A wave guide must never be used as a handhold.

The X-band radar is more common in general aviation aircraft and the C-band systems are often found on larger commercial aircraft. With X-band frequencies, most of the transmitted energy is reflected by even small amounts of precipitation, therefore, very little energy can travel through one storm to detect another which may be behind the first. Because of this high amount of reflected energy, the X-band frequencies do provide a greater distance and provide a higher resolution display than C-band radars. The C-band frequencies will provide more penetration of a storm in order to "see" activity beyond the intervening precipitation. Therefore, X-band radar is better for weather avoidance maneuvers and the C-band radar is better for penetration into areas of known precipitation.

II. STORM SCOPE SYSTEMS

The Storm Scope is an alternative weather mapping system which works on the principles similar to a basic ADF receiver to detect a storms activity. Every storm has a certain level of moisture, and as that moisture moves within the storm, or in the form of rain, there is a buildup of static electricity. The static electricity will discharge after it builds to a certain level. The discharge of static electricity will produce a radio wave which is easily detected by an AM radio receiver.

A storm scope is a special radio receiver which can detect the direction from which a signal was transmitted. The intensity of a storm can be determined by "counting" the number of static discharges within the storm. The greater the number of discharges per minute, the more intense the storm. The distance to the storm can be derived by measuring the strength of the static signal received. Through the use of microprocessor technology, the storm scope receiver translates the input information to a display unit which can easily be read by the pilot.

The storm scope system typically contains three units—the antenna, the receiver (or computer/processor) and the display. The antenna is typically mounted to the bottom of the aircraft and the computer/processor is found in the electrical equipment. The display unit is a 4" unit consisting of either a CRT or liquid crystal display.

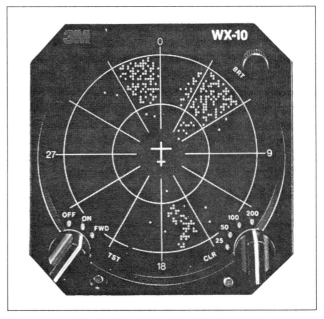

Figure 15-4. Typical storm scope display.

Chapter XVI

Electronic Instrumentation

In recent years, a major change has occurred in the design and construction of aircraft instruments. The older style instruments containing rotating mass gyros, delicate bearings, hairsprings and other extremely sensitive components have been replaced with electronic solid-state devices. The new "strap down" systems contain no moving parts and boast of greater reliability and higher time before failure.

I. COMPUTERIZED FUEL SYSTEMS

A recent development in fuel flow instruments is the processor based system that uses a small turbine wheel in the fuel line to the fuel control unit. As fuel flows through this line, it spins the turbine and a digital circuit reads the number of revolutions in a specified time period. This information is converted to a fuel flow rate. This flow rate may be electronically compensated for any idiosyncrasies of the specific system.

In addition to the fuel flow rate, these units may also calculate the amount of fuel remaining on board, and the amount of time the flight may be continued at the current fuel flow rate. By integrating this system with the DME or other navigation information, the range of the aircraft can be displayed. This permits the flightcrew to determine if they have enough fuel to complete the flight.

II. RADIO ALTIMETER

The radio altimeter has been devised to make instrument approaches safer. These instruments are sometimes called radar altimeters. Regardless of what they are called, they operate on the principle of the equipment directing an electrical signal to the ground and receiving it after it has bounced off of the surface. By measuring the lapsed time between transmission and reception, the equipment can give the pilot a readout of the height of the aircraft above the ground.

There are three types of radio altimeter systems in use today. One system uses a low-power continuous wave (CW) that is frequency modulated with a continuously varying signal. The CW is transmitted toward the ground and is received when it bounces back. By comparing the frequency of the received signal with that being transmitted, it is possible for the equipment to determine exactly how long a time elapsed between transmission and reception. By knowing this time, it is a simple matter for the electronic circuitry to convert the lapsed time into feet above the terrain.

A second method of measuring height electronically is by the use of a pulse of high frequency electrical energy much like that used in radar. This pulse is directed to the ground and it bounces back to the receiver. The time required for the pulse to travel to the ground and back is divided by two, and then by the speed the electrical wave travels, to find the height of the aircraft.

The third system uses a modified pulse method in which a continuous stream of high-frequency pulses is transmitted. The equipment then measures their travel time to the ground and back.

All of these systems have some provisions for damping the indication so the indicator needle will not jump as the aircraft flies over buildings and trees. For increased utilization, most radio altimeters are capable of being coupled with the flight directors to give the pilots an indication of their height on the last portion of an approach to landing.

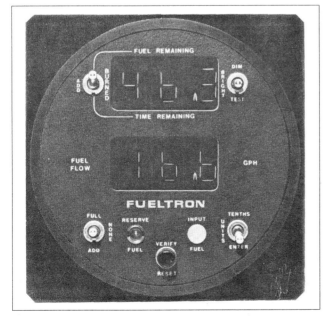

Figure 16-1. *This fuel flow indicator shows not only the rate of fuel flow, but also the amount of fuel and the time remaining at the present flow rate.*

Almost all of the altimeter systems have the provision for turning on an indicator light and sounding an aural warning when the aircraft reaches the decision height the pilot has programmed into it. This warns that the aircraft has reached the altitude at which the pilot must either have the runway in sight or must execute a missed approach.

III. ELECTRONIC FLIGHT INSTRUMENT SYSTEMS

Electronic Flight Instruments Systems (EFIS) have several advantages over conventional instruments. First and foremost is the advantage of increased reliability. EFIS reduces instrument panel clutter by combining several instruments into one unit. Pilot work load can be reduced by removing unnecessary information form the EFIS displays. Weather radar information can be overlapped with the navigational mapping.

The four displays shown in figure 16-4 are each an interchangeable CRT driven by their associated symbol generator. The center symbol generator is used as a redundancy check between the left and right systems. If one or more of the units fail, the other(s) will control the electronic displays. The symbol generator receives signals from the various instrument and navigational sensors located throughout the aircraft. The weather radar receiver/transmitter receives signals from the aircraft's radar antenna and sends information to both the pilots and copilots symbol generators. The display controller allows the pilot to select the appropriate system configuration for the current flight situation.

The use of digital based microprocessor electronics allows us to replace several mechanical instruments with the modern electronic flight instruments displaying the information on one or more cathode ray tubes (CRT). The signals sent to the various components of the systems are typically linked through a digital data bus. A data bus is made up of a twisted pair of insulated wires surrounded by an outer shielding. The electrical signals sent through the data bus consist of short pulses of voltage on or voltage off; binary ones and zeroes. These pulses are extremely short in duration. A typical system is capable of transmitting signal pulses which last only 10 microseconds.

One common bus system is known as ARINC 429. ARINC stands for Aeronautical Radio Incorporated, and the 429 is a code for the digital data standard. This system uses a 32-bit word for all information transmitted over the data bus. As seen in figure 16-5, the 32-bit word is made up of one

Figure 16-2. Radar (radio) altimeter.

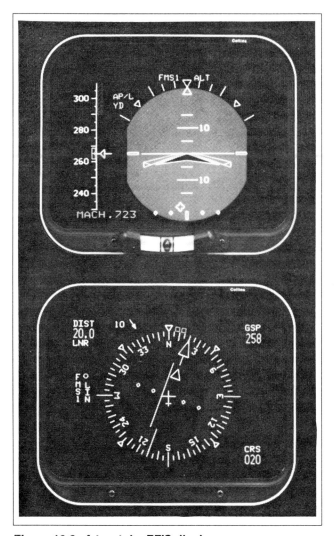

Figure 16-3. A two tube EFIS display.

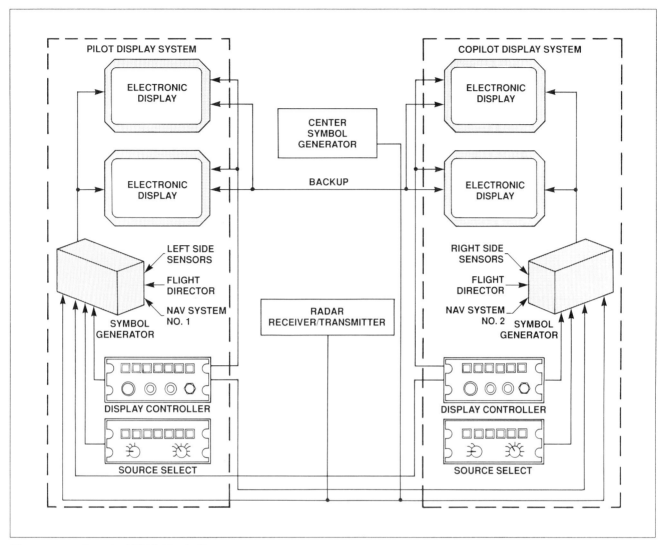

Figure 16-4. A typical EFIS block diagram.

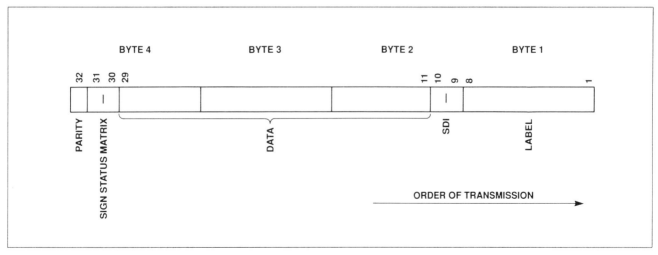

Figure 16-5. An ARINC 429 standard data word.

parity bit, a sign status matrix, the data the source destination indicator, and the label or identification of the sending unit.

A. Electronic Attitude Director Indicator

The electronic attitude director indicator (EADI) like its mechanical predecessor, displays much of the basic flight data needed to maintain a smooth and comfortable flight. From basic pitch and roll information to approach decision height, the information is displayed in a color format which is sunlight readable. There are both full-time and part-time displays which are available on most EADIs. The full-time displays give the pilot information needed for flight control. The part time displays offer information typically needed for runway approach or basic navigation.

The full-time displays include the aircraft symbol which is used as a reference for pitch and roll information. To determine the aircraft's attitude, the pilot must compare the aircraft symbol to the attitude sphere. The amount of pitch and roll (10, 20, etc.) is indicated on the attitude sphere. The attitude source indicator is also displayed to inform the flight crew which symbol generator is currently driving this EADI.

The part-time displays include the rising runway which appears at 200 ft. above ground level is used during an approach. The glide slope and localizer guide the pilot to the touchdown point. The marker beacon and radio altimeter displays are shown in the lower left corner of the display. The decision height is displayed in the lower right corner.

It should be noted that this EADI is only one of several versions currently available. The information displayed on any particular system may vary; however, the basic configuration of the instrument will remain very similar to that discussed here.

B. Electronic Horizontal Situation Indicator

The modern electronic horizontal situation indicator (EHSI) is modeled after the older electromechanical version and displays much of the same information. As with the EADI, the EHSI is capable of displaying both full-time and part-time information dependent on the current mode of operation. The primary function of an EHSI is to display navigational information.

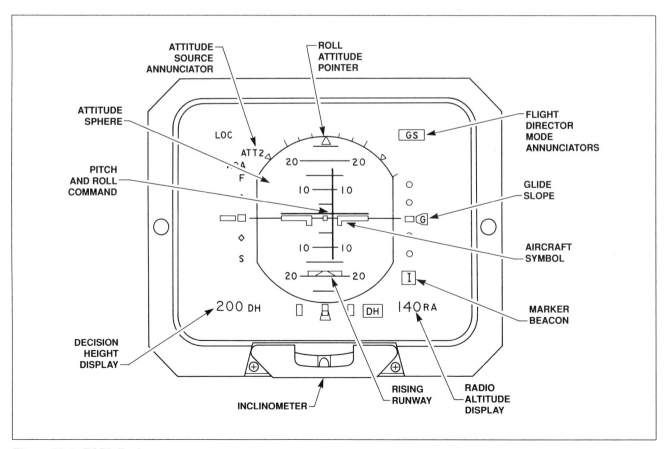

Figure 16-6. EADI display.

The electronic horizontal situation indicator can be set for one of four modes of operation. These modes are Plan, Map, VOR and ILS. The various modes are selected by the flight crew at the EFIS display controller. In the Plan mode, the CRT displays enroute flight information entered into the flight management system in order that the flight crew can get a visual reference of their recorded flight plan.

In the Map mode, the EHSI will display the currently operating flight plan showing waypoints, VORs, airports, etc. A real-time magnetic compass rose and other pertinent navigational information is also displayed on a moving map display. The map changes as the aircraft changes its position during flight. This mode can also display the weather radar if selected. The Map display of the EHSI is most commonly used during enroute portions of the flight.

The VOR and ILS display modes of the EHSI system display the information from the current navigational facility (VOR or ILS). The wind speed, VOR or ILS frequency received, and the current aircraft heading are also displayed.

C. Electronic Systems Monitoring Displays

In an effort to further reduce instrument panel clutter and pilot workload, many of the traditional engine and system instruments have been replaced with electronic monitors. These monitors rely on computers to receive data inputs from the various systems of the aircraft. The computer analyzes the information and transmits a digital signal to one or more CRT displays which inform the flight crew of the various systems conditions. The monitoring systems are also used to alert the flight crew of any system malfunctions and in some cases provide corrective actions.

1. Electronic Centralized Aircraft Monitor

The electronic centralized aircraft monitor system (ECAM) is comprised of two CRT display units, a left and right symbol generator, an ECAM control panel, discrete warning light display unit, two flight warning computers, and a data analog converter.

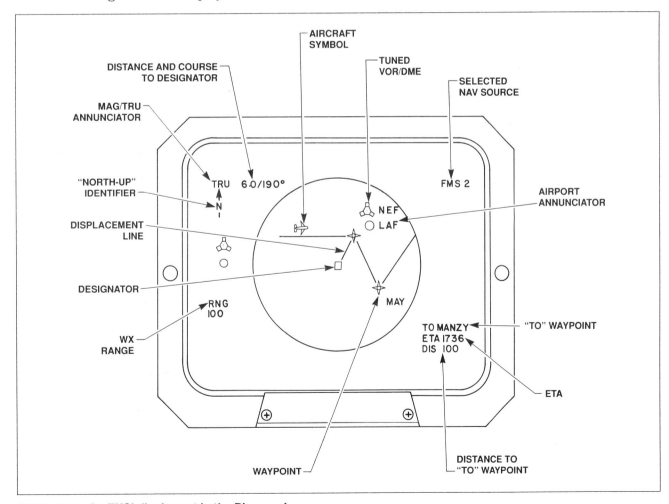

Figure 16-7. An EHSI display set in the Plan mode.

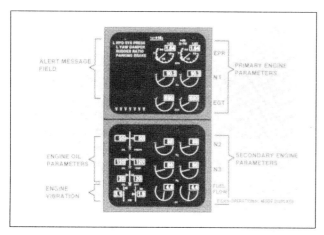

Figure 16-8. An electronic systems monitoring display from a Boeing 757 aircraft.

The various components communicate through data bus systems as shown in figure 16-9. The left CRT contains information on systems status, warnings and any associated corrective actions. The right CRT displays information in a pictorial format such as control surface positions. All information displayed on either CRT is shown in a digital format.

There are four basic modes of operation for the ECAM system. The manual mode will display pictorial diagrams of various aircraft systems. The other three modes operate automatically and are referred to as the flight phase, advisory and failure modes. The flight mode displays information related to the current phase of flight, such as pre-flight, take-off, climb, enroute, descent and landing. The flight mode information is displayed on the right-hand CRT. The advisory mode information is displayed on the

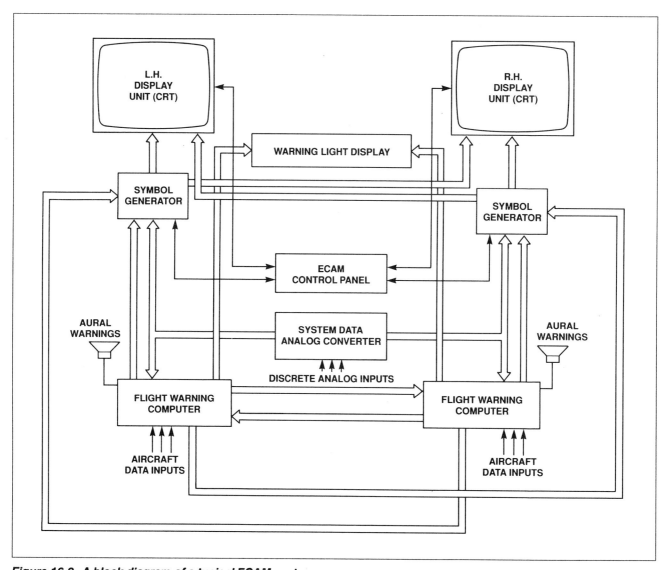

Figure 16-9. A block diagram of a typical ECAM system.

left CRT and contains information of concern to the flight crew, yet not critical.

The failure mode of operation automatically takes precedence over all other modes. The failure mode displays any information which may be considered critical to the flight safety. The left-hand CRT displays appropriate information and corrective action while the right unit displays the status of the failed system. If a failure occurs, the flight crew is also alerted to the problem through a aural and visual warning.

2. Engine Indicator and Crew Alerting System

The Engine Indicator and Crew Alerting System (EICAS) is used to monitor the engine and various other systems of the aircraft similar to the ECAM; however, the EICAS uses both a digital and analog format to display the information. The engine indicator and crew alerting system consists of two CRT displays, a left and right system computer, a display selector panel, discrete caution and warning displays, and a standby engine indicator.

The CRTs of the EICAS system are arranged one on top of the other and a discrete annunciator is located in front of both the pilot and copilot. The standby engine display is a liquid crystal display unit located in the center area of the instrument panel. The liquid crystal display shows the engine EGT, EPR, and N_1 speed.

As seen in figure 16-10, the upper CRT of the system displays primary engine in both analog and digital formats. Warnings, cautions and advisories are also displayed on the upper CRT. The lower CRT is normally blank during flight unless specific information is selected by the flight crew. The lower CRT displays information such as systems status, aircraft configuration, fluid quantities, various temperatures and maintenance information.

In the event of one CRT failure, all needed information is displayed on the operable CRT. In this case, the analog display of information is removed and the system is automatically converted to the compact mode. In the event the second CRT fails, the flight crew can still find critical engine information on the liquid crystal display.

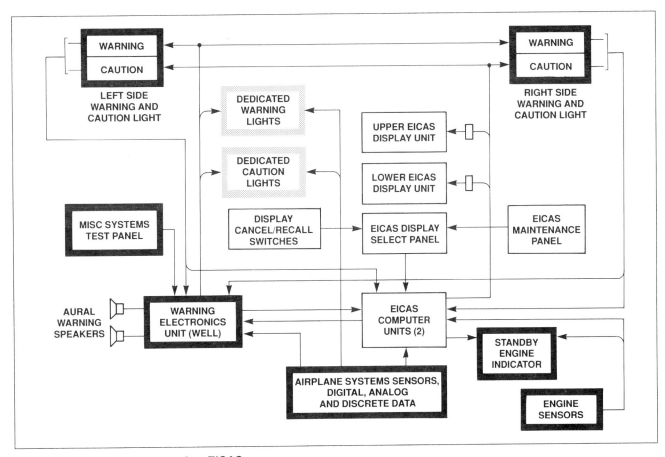

Figure 16-10. Block diagram of an EICAS.

Chapter XVII

Electronic Flight Systems

I. AUTOPILOT BASICS

A. Autopilot Functions

The early automatic pilots were large and cumbersome and had limited performance, but the modern generation of autopilots have smaller and more efficient gyros and digital electronic circuits. Their bulk is greatly reduced and their utility is increased countless times. In fact, an ultra-modern autopilot system is often referred to as an auto-flight system, since the system controls virtually all phases of flight including landing gear operation, throttle settings, navigational headings and flight control operations. A modern auto-flight system can virtually complete an entire flight from takeoff to landing if the proper programming is entered into the ships flight management system.

Modern autopilots vary in size and complexity from the most simple wing leveler, that uses a single canted rate gyro to control a pneumatic servo attached to the aileron control cable, to the exotic auto-flight control systems that have inputs from not only a sophisticated air data computer, but from all of the electronic navigation systems as well.

In order to better understand the operation of the autopilot, we will break its function down into four logical categories: error sensing, correction, follow-up and command.

1. Error Sensing

Some method is required to determine when the flight condition of the aircraft differs from that commanded by the pilot. Almost all modern aircraft use a gyro of some type for this purpose, and there are two ways the error signal can be generated.

a. Attitude Gyros

All of the earliest autopilots and many modern ones use a directional gyro and an artificial horizon to provide a stable reference from which the error signal is measured.

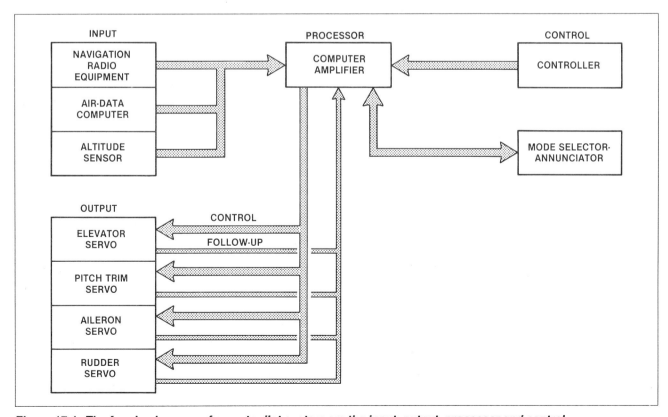

Figure 17-1. *The four basic areas of an autopilot system are the input, output, processor and control.*

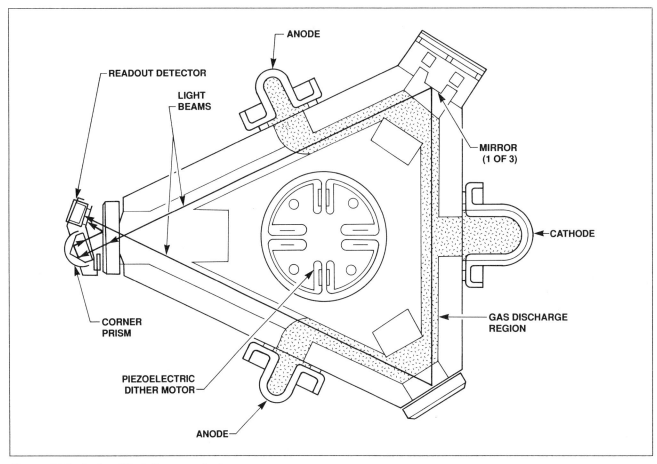

Figure 17-2. A simplified diagram of a laser gyro.

Pickoffs which take the signal from the gyro may be in the form of a pneumatic valve or some electronic device such as a variable resistor, a variable capacitor or variable inductor.

For directional control, the heading "bug" on the directional gyro is set to the heading the pilot wants to fly and when the autopilot is turned on to the Heading mode, the aircraft will turn to this heading, going there by the shortest direction, and, once established, will generate an error signal any time the heading varies from that set on the directional gyro.

Roll and pitch are sensed by the gyro horizon, and when these modes are engaged, an error signal will be generated any time the nose of the aircraft pitches up or down or any time a wing drops from level flight.

b. Rate Gyros

An attitude gyro senses an error proportional to the amount the aircraft is off the desired attitude, but rate gyros sense the rate at which the aircraft has departed from the desired flight condition. Most rate-sensitive autopilots use a modified turn coordinator which senses rotation about both the roll and yaw axes to produce the error signal. Electronic or pneumatic pickoffs measure the precession of the canted, or tilted, gyro, which is proportional to the rate of yaw and roll. This signal is fed into a computer that controls the servos to apply the appropriate correction to the flight controls.

A new concept in gyroscopes technology, known as the strap-down laser gyro, has recently been introduced on many corporate and commercial aircraft. This system gets its name "strap-down" from the fact that there are no moving parts in the gyro mechanism; the entire system is electronic. The laser gyro produces two low-power, helium-neon laser beams and directs them into a triangular path. The triangular path is formed by placing mirrors in the appropriate corners of the unit.

In one of the corners, there is the sensor unit made up of a prism and photo diodes. The prism directs the laser beams into a detector which can sense the frequency difference between the two beams. The two beams are in phase if the unit is stationary; the two laser beams have different frequencies if the gyro, or aircraft, has moved along the axis on which the gyro is mounted. The photo diodes convert this frequency shift into a digital signal which can be used by the auto-flight system. As with any gyro system, there would be three laser gyros for each complete system, one for each axis, longitudinal, lateral and vertical.

The laser gyro integrated with an accelerometer can be used for both rate sensing and attitude sensing. If the aircraft is equipped with strap-down laser gyros, it is very likely that there would be no rotating mass-type gyros used for other instruments. Laser gyros are typically part of a complete electronic instrument and auto-flight system as will be discussed later in this chapter.

c. Pitch Error Sensing

It is easy to sense pitch deviation with attitude gyros, but the smaller systems that use a rate gyro to sense roll and yaw deviation use inertial and dynamic forces to measure pitch deviation.

When the nose of the aircraft pitches up or down, there is an inertial force that is felt by an accelerometer, and after the initial change in attitude, the resulting change in airspeed and vertical speed is sensed by bellows to produce an error signal that is amplified and used to drive the servos in the elevator system.

d. Altitude Deviation Sensing

Sensing deviation from straight and level flight makes the autopilot useful, but by sensing deviation from a given pressure level makes it even more useful.

(1) Altitude Hold

When pilots select the altitude hold mode on the autopilot, they trap a sample of air at the pressure level at which they are flying, and when the aircraft deviates from this pressure level, an error signal is generated by a differential pressure sensor. This signal is sent to the elevator servo to produce a pitch change in the correct direction to return the aircraft to the desired altitude.

(2) Altitude Select

The altitude hold mode will hold the aircraft at the pressure level it was flying when the mode was engaged, but the altitude select mode allows the pilot to select the altitude at which to fly. If the aircraft is not at the selected altitude when the mode is engaged, an error signal is generated which directs the elevator servo to change the pitch attitude of the aircraft in the correct direction to seek the desired altitude. When this altitude is reached, the error signal will disappear.

2. Correction

Once it has been established that an error in the aircraft's attitude or altitude exists, the autopilot sets about to correct the situation. The signal from the error sensor is usually too weak to be used directly, so it must be amplified. Pneumatic pickoffs use either a purely mechanical advantage form of amplification, or the small pressure changes may be directed to a large area servo piston to produce the amount of physical force needed for the control system.

Autopilots using hydraulic servos use an air signal from the pickoff to operate a sensitive hydraulic selector valve to direct hydraulic fluid under pressure to one side of the servo piston or the other.

Electrical signals are the easiest to amplify, and the output of the amplifier can be used to drive the servo motors in the control system. The modern auto-flight systems utilize a digital electronic system. These signals are sent to a central flight computer which controls the servos.

Some of the smaller aircraft can be flown by the autopilot by using only one servo in the aileron system to control the aircraft in both roll and yaw, but other aircraft require a servo in the rudder system as well. A three-axis autopilot is one that has servos in the aileron, rudder and elevator controls.

Servos vary in complexity with the size of the aircraft in which they are installed, and with the aerodynamic forces they must control. The simplest servo is pneumatic and uses a diaphragm moved by either suction or a positive air pressure from the gyro pickoff. This diaphragm is attached to the control cable by a clamp and it pulls on the cable at the command of the autopilot.

Electric servos for light aircraft may use either a reversible DC motor driving a capstan through a reduction gear, or it may use a single direction DC motor which drives two gears turning in opposite directions.

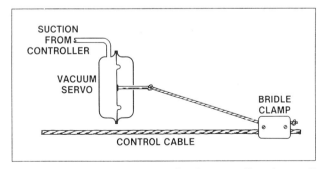

Figure 17-3. Pneumatic servo for the autopilot of a small aircraft.

Electrically operated clutches controlled by the autopilot connect the capstan to one of the other gears to drive the capstan in the direction required to properly execute the correction signal. Large aircraft usually have AC motor-driven servos with the control cables wrapped around the capstans. In any aircraft using an autopilot system, there must be some means of disconnecting the auto pilot servo in the event of a system failure. Some type of slip clutch is always installed in the system if the pilot could not manually override the autopilot with a reasonable amount of effort.

3. Follow-up

To cause the control surface to move in the correct direction is not enough. For example, if an airplane under control of the autopilot drops its left wing, the gyro will sense an error and signal the servo to lower the left aileron. If there were no follow-up system, the left aileron would move down and the left wing would rise until it becomes level; then the gyro would sense the level condition and send a signal to the servo to bring the aileron back to its streamlined position. But the inertia of the aircraft would keep the left wing rising until it caused the gyro to sense a left wing-up condition, and it would then send a signal to the servo to raise the left aileron and bring the wing back down. This oscillation will continue and cause the aircraft to rock its wings violently.

To prevent this undesirable condition, all types of autopilots have some form of follow-up system that stops the control surface movement when the desired deflection is reached, and then it brings the surface back to its streamlined position as the wing comes back to level flight attitude with no over- or undershoot.

a. Displacement Follow-up

This type of follow-up system stops the control surface movement once sufficient displacement has been reached. For example, let's assume that the left wing drops. The gyro senses an error and sends a signal to the aileron servo to move the left aileron down. When the aileron has moved down an amount proportional to the amount the wing has dropped, the follow-up system will produce a signal equal in intensity, but opposite in direction to the error signal, and it will cancel the error signal. The left wing is still down and the aileron is deflected, but since the signal from the follow-up system has canceled that from the controller, there will be no more control movement. Now, aerodynamic action raises the wing, and an error signal will be generated which is opposite the original one and the aileron will be brought back to its streamlined position as the wing comes level. There will be no overshoot.

b. Rate Follow-up

The displacement follow-up system considers the amount the airplane has deviated from its desired condition to determine the amount the control surface should move, but the rate system works on the basis of how fast the airplane deviated from its original condition. This assumes that the faster the deviation, the farther the airplane would eventually deviate.

Let's consider the same situation we had with the displacement system. If the left wing drops rapid-

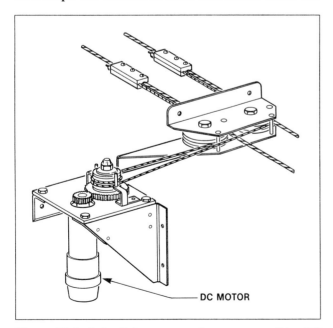

Figure 17-4. *Autopilot servo using a reversible DC electric motor.*

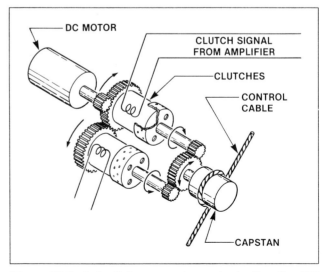

Figure 17-5. *Autopilot servo using a single-direction DC motor and two clutches to drive the capstan.*

ly because of a gust of wind, the rate gyro will sense this rapid departure from straight and level flight and send a signal to the servo to lower the left aileron. The left aileron moves down an amount that is proportional to the rate of deviation, and as it moves down, the wing stops dropping and starts back up. The recovery will be slower than the original roll, and it will generate a signal opposite to the original. The aileron will move back toward the neutral position, and by the time the wing is level, the aileron will be streamlined.

4. Command

In order for the autopilot to fly the aircraft as the human pilot wants, a command system is incorporated.

When the autopilot is in the Heading mode and the pilot changes the position of the heading "bug" on the directional gyro, an artificial error signal is injected into the system and the aircraft will turn until its new heading agrees with that commanded by the directional gyro.

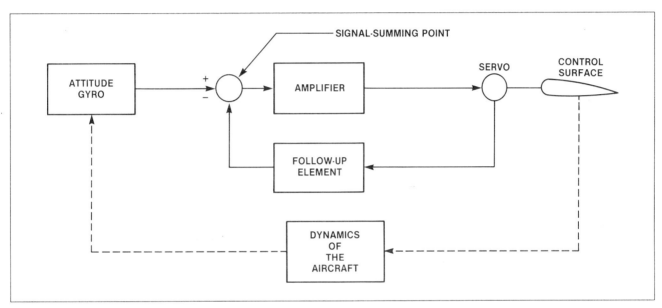

Figure 17-6. Displacement follow-up system for an autopilot.

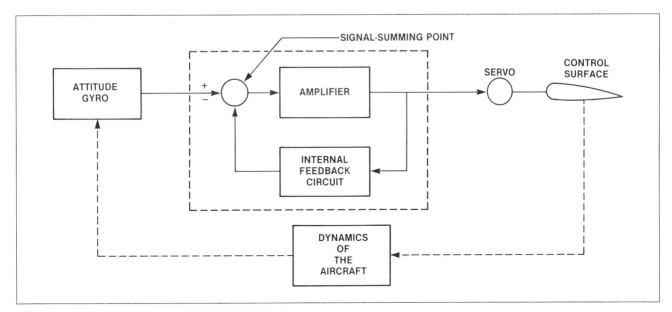

Figure 17-7. Rate follow-up system for an autopilot.

Most complete autopilots are coupled with the radio navigational systems so that they can receive the signals and fly the airplane accordingly. For example, when the automatic pilot is placed in the VOR mode and a usable signal is received from an omni station, the autopilot will sense an error signal and turn the airplane to a heading to intercept the desired radial, computing the maximum intercept angle. As the aircraft nears the radial, the error signal decreases and disappears when the radial is reached. The airplane will then track the radial, and anytime it gets off, an error signal will be generated to bring the airplane back on. In the LOC mode, the autopilot senses an error signal anytime the aircraft is off of the localizer center line, and it will command a correction to bring the aircraft back onto it.

II. FLIGHT MANAGEMENT SYSTEMS

The most modern aircraft incorporate a Flight Management System (FMS) as part of the auto-flight system. The FMS has an alpha-numerical keyboard and CRT display located in the center console of the flight deck. This unit is used by the pilot to input various flight data. The main memory unit

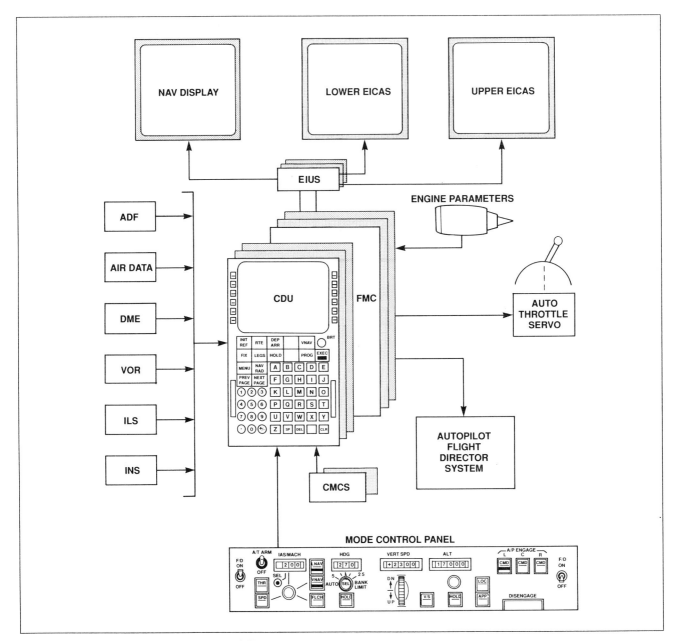

Figure 17-8. A block diagram of a typical flight management system.

of the FMS is the Flight Management Computer (FMC). The FMC memory can store information, such as navigational aids, airports, waypoints, runways and taxiways, standard departure and arrival routes. Virtually all the information needed to operate an aircraft in a specific geographical area can be stored by the FMC.

For a typical flight, the pilot would select the entire flight plan from the navigational database stored in the FMC. The database is updated on a regular basis using a portable database loader. The loader is basically a computer disc drive unit which is used to input the updated navigational data into the FMC. If the desired flight plan is not currently stored in the navigational memory, the pilot can enter individual waypoints, altitudes, etc. to create a new flight plan.

In figure 17-8, we have a block diagram of a complete autopilot system as is used by an executive jet-type aircraft. The sensors take the data from the navigation radio equipment, the air data computer and the attitude gyros and feed all of this information into the computer amplifier unit. The controller and mode selector also send their command signals into the computer amplifier where the actual condition is compared with that commanded by the pilot. Any error signals that result are fed to the servos so they will deflect the proper control surface to restore the aircraft to the condition called for by the pilot. The feedback system between the servos and the computer-amplifier prevents overshooting or undershooting of the controls which would produce undesirable oscillation.

III. AUTOPILOT SYSTEM MAINTENANCE

In general, the maintenance of an auto-flight system can constitute a large portion of the down-time of the entire aircraft. The complexity of the system makes troubleshooting operations difficult and component repair often requires special equipment. The A&P technician must, however, have a good understanding of the system in order to make logical determinations of the needed repairs. For the most part, a technician dealing with a particular autopilot system would have extensive factory training on the that system.